# Arithmetic Groups and Their Generalizations

## What, Why, and How

AMS/IP

# Studies in Advanced Mathematics

Volume 43

# Arithmetic Groups and Their Generalizations

## What, Why, and How

Lizhen Ji

American Mathematical Society • International Press

Shing-tung Yau, General Editor

2000 *Mathematics Subject Classification.* Primary 11F06, 22E40.

For additional information and updates on this book, visit
**www.ams.org/bookpages/amsip-43**

**Library of Congress Cataloging-in-Publication Data**

Ji, Lizhen, 1964–
Arithmetic groups and their generalizations : what, why, and how / Lizhen Ji.
p. cm. — (AMS/IP studies in advanced mathematics ; v. 43)
Includes bibliographical references and index.
ISBN 978-0-8218-4675-9 (alk. paper)
1. Arithmetic groups. 2. Linear algebraic groups. I. Title.

QA171.J5 2008
512.7—dc22 2008009816

AMS softcover ISBN 978-0-8218-4866-1



∞ The paper used in this book is acid-free and falls within the guidelines
established to ensure permanence and durability.
Visit the AMS home page at `http://www.ams.org/`
Visit the International Press home page at URL: `http://www.intlpress.com/`

10 9 8 7 6 5 4 3 2 1 14 13 12 11 10 09

To my wife Lan Wang
for her constant support and understanding

"What is... Symmetry?"
by Lena Min Ji

*(showing both global and local symmetries)*

# Contents

# Preface

In one guise or another, many mathematicians are familiar with certain arithmetic groups, such as $\mathbb{Z}$ and $\mathrm{SL}(n,\mathbb{Z})$. But their relatives, for example, $S$-arithmetic groups such as $\mathrm{SL}\left(n,\mathbb{Z}\left[\frac{1}{p_1},\dots,\frac{1}{p_m}\right]\right)$, where $p_1,\dots,p_m$ are prime numbers, and their analogues over function fields such as $\mathrm{SL}(n,\mathbb{F}_p[t])$, where $\mathbb{F}_p$ is a finite field and $t$ is a variable, may not be so well-known. The purpose of this expository book is to explain through some brief and informal comments what these groups are, why they are important to study, and how they can be understood and applied to many fields such as analysis, geometry, topology, number theory, representation theory and algebraic geometry.

We try to emphasize the point of view that it is the group action on good spaces that makes the group interesting and understandable, and that it is also the group action which makes the spaces invloved more interesting and useful. In fact, problems naturally arise and are solved by such pairs of groups and spaces. For example, though symmetric spaces are important and interesting in themselves, their quotients by arithmetic groups, which are locally symmetric spaces of finite areas and sometimes called arithmetic locally symmetric spaces, are much richer in structures and have more applications. One obvious reason is that symmetric spaces of noncompact type have trivial topology, but cohomology groups of locally symmetric spaces are usually nontrivial and carry a lot of valuable information both for the arithmetic groups and other related objects, in particular when the locally symmetric spaces can be interpreted as moduli spaces in algebraic geometry and number theory. This point of view can be seen clearly in comparison between the Poincaré upper half plane $\mathbb{H}$ and its arithmetic quotients such as modular curves and Shimura curves.

Though Riemannian symmetric spaces and locally symmetric spaces are probably the most natural spaces associated with arithmetic subgroups of Lie groups, we also try to emphasize that it is natural and important to consider other related spaces such as pseudo-Riemannian symmetric and locally symmetric spaces, buildings (both spherical and Euclidean type) and Teichmüller spaces, which can feed back to symmetric and locally symmetric spaces, besides their own interests. In fact, for $S$-arithmetic groups such as $SL\left(n,\mathbb{Z}\left[\frac{1}{p_1},\dots,\frac{1}{p_m}\right]\right)$, both symmetric spaces and buildings are needed simultaneously in order to understand them.

We also discuss some related groups such as tree lattices, building lattices, CAT(0)-groups, mapping class groups, and outer automorphism groups of free groups, and their related spaces such as Teichmüller spaces and moduli spaces of Riemann surfaces, outer spaces, by emphasizing their similarities to arithmetic groups and symmetric spaces, and also to $S$-arithmetic groups and Bruhat-Tits buildings. By putting all these related groups and spaces together at one place, results from one class can motivate and suggest results for another class.

Hopefully the reader will be convinced of and appreciate more the importance and ubiquity of arithmetic groups in mathematics, and consequently, learn some non-obvious connections between different subjects through them, by the snapshots on motivations, informal descriptions of basic objects and applications, and the many references included in this book, which might also serve as a partial guide to the huge literature on arithmetic groups and related topics.

# Acknowledgements

The original motivation to write this book, or rather article, was to provide an informal guide to the students attending an instructional conference titled *Geometry, analysis and topology of discrete groups and locally symmetric spaces* held in Beijing from July 17 to August 4, 2006, at the Morningside Center of Mathematics. The introduction to the conference describes its purpose:

*Locally symmetric spaces and discrete subgroups of Lie groups have played a fundamental role in many branches of modern mathematics. Various aspects of these important objects are often studied by different groups of people using different methods. It would be beneficial and fruitful to bring together experts in all these areas to exchange their results, techniques, to develop possible collaborations, and to show the power and beauty of locally symmetric spaces and discrete subgroups of Lie groups.*

A shorter and incomplete version of this article was circulated at the conference. Encouragement from several people, including L. Carbone, R. Spatzier, T. Kobayashi, E. Leuzinger, and P. Gunnells, helped me not only to finish this article, but to turn it into a book. Although the work is still not as complete nor as detailed as I had planned it to be, or as it could be[1], I hope that it might still be of some interest to students and non-experts who would like to learn something informal about arithmetic groups and related topics (as one learns from a colleague over tea time or at colloquium dinners), and who wonder about where to look up relevant references.[2] [3]

---

[1] Ideally, this book should give precise definitions of important concepts introduced, state significant theorems, explain their applications, and describe in some detail of interactions between various topics introduced below.

[2] I have always been wondering how and where arithmetic groups are used in some unexpected areas of mathematics, and how some other mathematical objects and methods are motivated by arithmetic subgroups and their associated symmetric and locally symmetric spaces. The above three weeks long conference on geometry, analysis and topology of discrete subgroups and locally symmetric spaces with talks by experts from different subjects around the world was very instructive to me. Writing this book has also been a very instructive and rewarding, though demanding and tiresome, experience to me.

[3] It should also be pointed out and emphasized right from the beginning that there are many subjects connected with arithmetic subgroups and related discrete subgroups, and a huge literature on many important topics in these subjects. Though efforts have been made and experts have been consulted to make the coverage of this book and its references comprehensive, I would like to apologize in advance to people whose important

I would like to thank R. Spatzier for very helpful references, discussions and comments on an earlier version (in particular about the Zimmer program), V. Jones for references and suggestions on rigidity in the theory of von Neumann algebras associated with countable groups (in particular lattices of semisimple Lie groups), K. Brown for suggesting to include outer automorphism groups of free groups and alerting me to their close relations with mapping class groups and arithmetic groups, S. Zelditch for some references on quantum chaos, R. Miatello for some references on automorphic forms, J. Schwermer for some references on cohomology of arithmetic groups, K. Vogtmann for some references on the outer automorphism groups of hyperbolic groups, G. Prasad for some references and conversations on automorphism groups of buildings, T. Januszkiewicz for conversations on CAT(0)-groups, L. Saper for comments and references on precise reduction theory and cohomology groups, S.T. Yau for comments and references on rigidity of Hermitian locally symmetric spaces, F. Luo for comments on the automorphism group of the curve complex of surfaces, A. Bloch for helpful conversations and references, J. Milne and S. Gelbart for very helpful comments, Peter Li and M. Olbrich for very helpful comments and references, S. Donkin, S. Friedberg and P. Igodt for helpful references, A. Deitmar for carefully reading an earlier version and many helpful comments, U. Bunke and Shihai Yang for many comments on the first preliminary version circulated at the conference and some references, and M. Goresky for some constructive suggestions. I would also like to thank A. Ranicki for suggesting to turn this article into book form, which greatly encouraged me to speed up the writing and revising process.

Especially, I would like to thank T. Kobayashi for many helpful comments and suggestions, in particular on unitary representations and discontinuous groups for pseudo-Riemannian cases, and for his substantial contributions to §4.8, §11.9, §15.5 and §15.6, E. Leuzinger for contributing to the subsections §3.3.13, §13.9 and §17.9, L. Carbone for contributing to the subsections §5.10, §12.14, §18.1 and §18.2, and S. Zelditch for contributing to the subsection §12.6.

I would also like to thank the Morningside Center of Mathematics in Beijing, in particular Lo Yang and Xiaoning Li, for making the conference mentioned above run smoothly, and thank the speakers of the conference for their interests and help in various ways. Otherwise this project would certainly not be finished since it is really beyond my ability and knowledge to give a comprehensive survey and guide on the vast and broad subjects of arithmetic groups and related topics, and the huge, still rapidly expanding, literature on them.

This long and non-standard article (or rather, short and unusual book) is closely related to the joint book with A. Borel [BoreJ]. The working

---

contributions and papers are not mentioned (or not discussed adequately) in this little book due to the lack of knowledge and of ability of the author to describe them.

experience with him from 1997 to 2003 has had a lot of influence on both the writing and contents of this book. Besides doing mathematics proper, he shared with me many of his other insights. I would also like to thank M. Goresky and R. MacPherson for conversations over the years about arithmetic groups, their associated cohomology groups and compactifications, which instilled in me a desire to learn more about general arithmetic groups, locally symmetric spaces and related results.

Finally, I would like to thank S.T. Yau for his intensive student seminar during my graduate school years, where many papers in broad areas of geometry and analysis were lectured on and discussed. Those years opened up my appetite for learning more mathematics and understanding better relations between different parts of mathematics, and allowed me to be immersed in a very stimulating environment.[4] They also fully prepared me for attending multi-seminar talks and multi-conferences nonstop.

During the preparation of this book, I have been partially supported by NSF grant DMS 0604878. I also enjoyed the hospitality of the Morningside Center of Mathematics in Beijing and the Center of Mathematical Sciences in Hangzhou during a part of the summer of 2006.

Lizhen Ji
Ann Arbor, November 2007.

[4]Here I quote several pieces of simple and profound advice from S.T. Yau to his students:

(1) If you need something from a new subject or area, just learn it!
(2) Nothing is as simple as you may think, but not as hard as you might fear either.
(3) Study and learn mathematics for the joy and excitement of doing it.

CHAPTER 1

# Introduction

Arithmetic groups are basic objects in many subjects of mathematics. They first arose in number theory in connection with the reduction theory of quadratic forms. They have also played a fundamental role in automorphic forms and automorphic representations. They are also special among discrete groups of Lie groups and provide basic examples in combinatorial group theory. The fact that arithmetic groups are discrete subgroups of Lie groups leads to their natural actions on symmetric spaces and buildings associated with the Lie groups, which in turn lead to important arithmetic locally symmetric spaces.

In fact, their actions on symmetric spaces and buildings allow one to understand their structures in more depth, such as cohomological properties and asymptotic (or large scale) geometry. One striking example is that actions of arithmetic groups on symmetric spaces allow us to show finiteness properties such as finite generation, finite presentation and other cohomological finiteness, for example $FP_\infty$, finite virtual cohomological dimension; on the other hand, their actions on the associated spherical Tits buildings are needed to show that they are duality groups, a generalization of the Poincaré duality groups, and consequently to determine the precise value of their virtual cohomological dimension. Cohomology groups of arithmetic subgroups of semisimple linear algebraic groups are important in understanding algebraic $K$-groups of integers.

Arithmetic quotients of symmetric spaces, in particular of bounded symmetric domains, are special manifolds and many moduli spaces are given in terms of them. They tend to have rigid properties due to rich algebraic structures induced from abelian subgroups of the arithmetic subgroups.

One important example to keep in mind is the arithmetic subgroup $SL(n, \mathbb{Z})$ of the Lie group $SL(n, \mathbb{R})$. The symmetric space associated to the semisimple Lie group $SL(n, \mathbb{R})$ is $\mathcal{SP}_n = SL(n, \mathbb{R})/SO(n)$, the space of positive definite symmetric matrices (or quadratic forms) of determinant 1 endowed with an invariant Riemannian metric given by $ds^2 = Tr(Y^{-1}dY)^2$, where $Y \in \mathcal{SP}_n$. The quotient $SL(n, \mathbb{Z})\backslash\mathcal{SP}_n$ has finite volume and hence $SL(n, \mathbb{Z})$ is not a small subgroup of $SL(n, \mathbb{R})$ in certain sense.

The quotient $SL(n, \mathbb{Z})\backslash SL(n, \mathbb{R})$ can be identified with the moduli space of lattices in $\mathbb{R}^n$ of co-volume 1, i.e., the unimodular lattices, and the quotient $SL(n, \mathbb{Z})\backslash\mathcal{SP}_n = SL(n, \mathbb{Z})\backslash SL(n, \mathbb{R})/SO(n)$ can be identified with the moduli space of unimodular lattices in $\mathbb{R}^n$ up to rotations of $\mathbb{R}^n$.

A probably more intrinsic interpretation is to identify $\mathcal{SP}_n = SL(n, \mathbb{R})/SO(n)$ with the moduli space of marked flat metrics on the torus $(S^1)^n$ of total volume 1, and $SL(n, \mathbb{Z})\backslash\mathcal{SP}_n$ is the moduli space of flat metrics on $(S^1)^n$. This moduli interpretation of $SL(n, \mathbb{Z})\backslash\mathcal{SP}_n$ is important for some considerations of its geometry, for example, to identify its compact subsets in terms of the geometric invariants of the flat metrics.

Besides these natural interpretations of the quotient $SL(n, \mathbb{Z})\backslash\mathcal{SP}_n$, the action of $SL(n, \mathbb{Z})$ on the contractible space $\mathcal{SP}_n$ is important to understand many properties of $SL(n, \mathbb{Z})$.

The fruitful interaction between arithmetic groups and symmetric spaces can be generalized to other groups such as $S$-arithmetic groups, mapping class groups of surfaces, outer automorphism groups of free groups, and other geometric or transformation groups.

An important example of $S$-arithmetic subgroup is $SL(n, \mathbb{Z}[\frac{1}{p}])$, where $p$ is a prime number. Clearly, $\mathbb{Z}[\frac{1}{p}]$ is not a discrete subgroup of $\mathbb{R}$, and consequently $SL(n, \mathbb{Z}[\frac{1}{p}])$ is not a discrete subgroup of $SL(n, \mathbb{R})$ and does not act properly on the symmetric space $\mathcal{SP}_n$. In particular, the quotient $SL(n, \mathbb{Z}[\frac{1}{p}])\backslash\mathcal{SP}_n$ is non-Hausdorff.

To overcome this problem, let $\mathbb{Q}_p$ be the field of $p$-adic numbers, and $SL(n, \mathbb{Q}_p)$ be the associated $p$-adic Lie group. The group $SL(n, \mathbb{Z}[\frac{1}{p}])$ can be naturally embedded into $SL(n, \mathbb{Q}_p)$ but the image is not a discrete subgroup either. On the other hand, under the diagonal embedding, $SL(n, \mathbb{Z}[\frac{1}{p}])$ becomes a discrete subgroup of the product $SL(n, \mathbb{R}) \times SL(n, \mathbb{Q}_p)$. Instead of acting on a symmetric space, the group $SL(n, \mathbb{Q}_p)$ acts properly and isometrically on a simplicial metric space $X_p$, the Bruhat-Tits building associated with it. Hence $SL(n, \mathbb{Z}[\frac{1}{p}])$ acts isometrically and properly on the product $\mathcal{SP}_n \times X_p$ with a quotient of finite volume (when suitably interpreted.) This action can also be used to study $SL(n, \mathbb{Z}[\frac{1}{p}])$ as in the case of the $SL(n, \mathbb{Z})$-action on $\mathcal{SP}_n$ above, to show its finiteness properties and that it is a virtual duality group and hence to determine its virtual cohomological dimension.

Let $S_{g,p}$ be an orientable surface of genus $g$ with $p$ punctures. Let $\text{Diff}(S_{g,p})$ be the group of all diffeomorphisms of $S_{g,p}$, $\text{Diff}^+(S_{g,p})$ be the subgroup of all orientation preserving diffeomorphisms, and $\text{Diff}^0(S_{g,p})$ the identity component of $\text{Diff}(S_{g,p})$ (or $\text{Diff}^+(S_{g,p})$) which is a normal subgroup. Then $Mod_{g,p}^+ = \text{Diff}^+(S_{g,p})/\text{Diff}^0(S_{g,p}) = \pi_0(\text{Diff}^+(S_{g,p}))$, which is the component group of $\text{Diff}^+(S_{g,p})$, is the mapping class group associated with $S_{g,p}$, and $Mod_{g,p} = \text{Diff}(S_{g,p})/\text{Diff}^0(S_{g,p})$ is the extended mapping class group. They are also denoted by $\Gamma_{g,p}^+$ and $\Gamma_{g,p}$. Assume $3g - 3 + p > 0$. Let $T_{g,p}$ be the Teichmüller space of marked hyperbolic metrics, or equivalently the space of marked complex structures, on $S_{g,p}$. (Note that when the Euler characteristic of $S_{g,p}$ is negative, for each complex structure on $S_{g,p}$, there is a unique hyperbolic metric on $S_{g,p}$ conformal to it, by the famous uniformization theorem for Riemann surfaces.) It is known that $T_{g,p}$ is a complex manifold of dimension $3g-3+p$ and is diffeomorphic to $\mathbb{R}^{6g-6+2p}$. It also admits several natural metrics such as the Teichmüller metric and Weil-Petersson metric. Then $\Gamma_{g,p}^+$ and $\Gamma_{g,p}$ act on $T_{g,p}$ properly and isometrically by changing the markings. The quotient $\Gamma_{g,p}^+\backslash T_{g,p}$ is the moduli space of hyperbolic metrics on $S_{g,p}$, or the moduli space of algebraic curves of genus $g$ with $p$ punctures. The group $\Gamma_{g,p}^+$ and its action on $T_{g,p}$ can be studied by methods similar to those for arithmetic groups. In fact, then $g=0$ and $p=0$, $\Gamma_{g,n}^+ = SL(2,\mathbb{Z})$, and the associated Teichmüller space of marked complex structures can be identified with the upper half plane $\mathbb{H}^2 = SL(2,\mathbb{R})/SO(2)$. So $SL(n,\mathbb{Z})$ and $\Gamma_{g,p}^+$ are two different generalizations of $SL(2,\mathbb{Z})$.

There is another natural generalization of $SL(2,\mathbb{Z})$ or rather $GL(2,\mathbb{Z})$. In fact, let $F_n$ be the free group on $n$ generators, and $Out(F_n) = Aut(F_n)/Inn(F_n)$ be the group of outer automorphisms of $F_n$. When $n=2$, $Out(F_n) = GL(2,\mathbb{Z})$. $Out(F_n)$ also contains a subgroup $SOut(F_n)$ of index 2 such that $SOut(F_2) = SL(2,\mathbb{Z})$.

We note also that $GL(n,\mathbb{Z})$ is the automorphism group of the free abelian group $\mathbb{Z}^n$, and the extended mapping class group $\Gamma_{g,p}$ of the surface $S_{g,p}$ is the outer automorphism group of the surface group $\pi_1(S_{g,p})$. This gives another relations between them.

A natural space where $Out(F_n)$ acts properly is the outer space of marked normalized metric graphs with the fundamental group equal to $F_n$.

The above interpretation of $\mathcal{SP}_n$ as the marked flat metrics of total volume 1 on the torus $(S^1)^n$ gives a yet another common thread of these three class of groups: *arithmetic groups, mapping class groups of surfaces, outer automorphism groups of free groups.*

One of the basic points we want to emphasize here is that in order to study a group, it is crucial to find a good space where it acts properly with some suitable control on the quotient. Arithmetic groups and symmetric spaces provide a good model for the above groups related to arithmetic groups. Indeed, both results and methods used to prove them can be generalized to these groups.

Another point we want to emphasize is that there are many other naturally related groups which are not arithmetic groups. Even among discrete subgroups of Lie groups, there are smaller subgroups, for example, non-cofinite volume subgroups. These groups are important in lower dimensional topology and include Fuchsian groups of the second kind and co-convex non-cocompact Kleinian subgroups. It is also worthwhile to point out that finitely generated subgroups of $GL(n,\mathbb{Q})$ form a natural class of groups, but they are not arithmetic subgroups in general. Another example is provided by realizing that the symmetric space $\mathcal{SP}_n$ is a simply connected and nonpositively curved Riemannian manifold, i.e., a so-called Hadamard manifold, and the Bruhat-Tits building $X_p$ associated with the $p$-adic Lie group $SL(n,\mathbb{Q}_p)$ is also a contractible metric space of nonpositive curvature when suitably interpreted, i.e., a so-called CAT(0) geodesic metric space. In fact, there is a rich supply of discrete groups acting isometrically and properly on CAT(0)-spaces such as trees and their products.

All these connections make it worthwhile to study arithmetic groups, their actions on symmetric spaces, and methods used to understand the arithmetic groups, the quotients of symmetric spaces by arithmetic groups, which are arithemtic locally symmetric spaces, and other aspects of discrete groups and symmetric spaces, for example, large scale geometry and compactifications. We hope to give some motivations of problems considered in this expository book and brief descriptions of many results cited.

Since there are many references and some related papers study different aspects of similar problems, it is often difficult to pick out the main references to convey the whole pictures of some topics. Consequently, we often list many of them for some topics simultaneously. On the other hand, we try to provide and emphasize some expository or survey papers whenever available, hoping that the reader could find more references from such papers.

The index at the end of this book is extensive. Besides the usual role of an index as a guide to find locations of important topics and concepts, we have tried to list many relations between different subjects and topics there. Relations between different topics are indicated in the index by pairs of the primary topic names and the secondary names, and they can be searched and located by both names.

Hopefully the table of contents at the beginning will make clear the topics under discussion, relations between various parts and the global structure of this book. The basic plan is that we start with special examples of arithmetic groups and their diverse applications. Then we discuss general arithmetic subgroups and locally symmetric spaces, and many different aspects of using and understanding them. Finally, we describe generalizations of arithmetic groups and symmetric spaces.

CHAPTER 2

# General comments on references

There are many books and papers cited in this book. In general, we have tried to cite some books and survey articles to cover the general themes at the beginning or the end of every section or subsection, and more specific papers near each topic. Some references appear several times at different places and in different topics when they are appropriate or desired. The detailed index near the end also lists and shows relations between different topics and where they can be found.

For the convenience of the reader, we have listed here some books of more general interests on symmetry, geometry and discrete groups. Many more specific references will be provided below in various sections.

*Expository books:*

Symmetry is an important aspect of nature and mathematics, and is described by group theory. According to the Erlangen program of Klein, an essential aspect of geometry is to study actions of groups and their invariants.

Some elementary books on topics related to group actions and symmetry are [Wey] [AsG1] [Arm] [NiS] [Shaf] [Ron4]. See also [Wein1] on applications of arithmetic groups to algorithms and spaces of Riemannian metrics on manifolds. For a history of Lie groups and linear algebraic groups, see [Bore11] and [Haw].

*Basic books on discrete groups:*

There are many books and survey articles on discrete groups acting on spaces of constant sectional curvature, for example, [Bea] [Rat] [BeneP] [Ohs] [Th1] on the geometry and topology of three dimensional hyperbolic manifolds, [MaclR] on arithmetic subgroups acting on three dimensional hyperbolic space, [ElsGM] on all aspects of such arithmetic groups acting on the three dimensional hyperbolic space and associated locally symmetric spaces, and [ViS] giving a survey on construction and structures of both arithmetic and non-arithmetic groups acting on spaces of constant sectional curvature. For the action of discrete isometry subgroups on the space of constant sectional curvature for indefinite metric, see [Wol1] [Kob1] [KobY].

There are fewer books on discrete subgroups of higher rank Lie groups. A comprehensive introduction to discrete subgroups of Lie groups is [Ra5]. For the reduction theory of arithmetic groups, the classical book is [Bore4]. Besides a detailed summary of reduction theory of arithmetic subgroups, the book [PlR1] also contains many number theoretic aspects of arithmetic subgroups. A shorter and more elementary introduction to reduction theory together with some arithmetic applications is given in [Hum2]. A preliminary version of a book on geometry and dynamics of arithmetic groups is [Morr1]. Certainly, the survey article on arithmetic groups [Serr1] is highly recommended, which gives an overview of what arithmetic groups are through various examples and cohomological properties of arithmetic groups. For rigidities of lattice subgroups of semisimple Lie groups, see the books [Most1] [Marg1] [Zi1]. For some related rigidity conjectures about von Neumann algebras associated with lattices of semisimple Lie groups, see [Jon] [Pop1].

Several people have contributed profoundly to many aspects of the modern theory of arithmetic groups. They include C. Siegel, A. Borel, A. Selberg, Piatetski-Shapiro, R. Langlands, G. Mostow, G. Margulis. For the former four, one can read their collected or selected works [Si5] [Bore21] (also [Pr6]) [Sele3] [PiaS3] to gain global and historic points of view of their deep contributions. See the website

*http://sunsite.ubc.ca/DigitalMathArchive/Langlands/intro.html*

for the collected works of Langlands. For the last two, their books [Most1] [Marg1] give good summaries of part of their pioneering works on rigidity and global geometry of arithmetic groups. (See also the article [Ji9] for a summary of the work of Margulis and some related results.) G. Shimura made fundamental contributions to arithmetic algebraic theory of Hermitian locally symmetric spaces, the theory of so-called Shimura varieties (see his collected works [Shi2] and also the long expository article [Miln3].) It is also interesting to read books and special issues dedicated to some of them, for example, [AuBG] [How] [HejST] [CoJPS] to see some impacts of their ideas and results.

CHAPTER 3

# Examples of basic arithmetic groups

Though $\mathbb{Z}$ is a familiar group, we use it to motivate many questions later. It has played many roles such as the ring of integers, the free group on one generator, a discrete cofinite subgroup of the Lie group $\mathbb{R}$, and the fundamental group of the aspehrical manifold $\mathbb{Z}\backslash\mathbb{R}$. Its connections with different fields might be more than one expects on the first thought. One basic point in this section is to show that the embedding of $\mathbb{Z}$ in the Lie group $\mathbb{R}$ and its proper action on the homogeneous space $\mathbb{R}$ provide rich structures for the discrete group $\mathbb{Z}$ and means to understand them.

## 3.1. $\mathbb{Z}$ as a discrete subgroup of $\mathbb{R}$

$\mathbb{Z}$ is an infinite cyclic group. Its first enhanced (or its most important) structure comes from embedding into $\mathbb{R}$, a connected real Lie group. Besides the standard embedding

$$\mathbb{Z} \hookrightarrow \mathbb{R}, \quad 1 \mapsto 1,$$

for each positive number $\ell > 0$, there is an associated embeding

$$\mathbb{Z} \hookrightarrow \mathbb{R}, \quad 1 \mapsto \ell.$$

Three immediate reasons for considering the embedding of $\mathbb{Z}$ into $\mathbb{R}$ are the following:

(1) $\mathbb{R}$ is a connected Riemannian manifold and its Riemannian distance induces a distance function on $\mathbb{Z}$. The quotient space $\mathbb{Z}\backslash\mathbb{R}$ is also a Riemannian manifold.
(2) The quotient space $\mathbb{Z}\backslash\mathbb{R}$ has the properties:

$$\pi_1(\mathbb{Z}\backslash\mathbb{R}) = \mathbb{Z}, \quad \text{and } \pi_i(\mathbb{Z}\backslash\mathbb{R}) = \{1\}, \quad \text{for } \ i \geq 2,$$

i.e., $\mathbb{Z}\backslash\mathbb{R}$ is the classifying space for $\mathbb{Z}$ (or $K(\mathbb{Z},1)$-space). So $\mathbb{Z}\backslash\mathbb{R}$ is a special topological space.
(3) The space $\mathbb{Z}\backslash\mathbb{R}$ is a homogeneous space of $\mathbb{R}$. The $\mathbb{R}$-action on it allows one to study representation theory of $\mathbb{R}$ on periodic functions on $\mathbb{R}$ (each embedding of $\mathbb{Z}$ gives a periodic structure with period equal to $\ell$.)

The group $\mathbb{Z}$ *together with an embedding into the Lie group* $\mathbb{R}$ is the most basic example of arithmetic groups. Since arithmetic deals with integers, the terminology *arithmetic group* is natural.

**3.1.1. Poisson summation formula.** Probably the most important consequence of the embedding $\mathbb{Z} \hookrightarrow \mathbb{R}$ above is the *Poisson Formula* (also called *Poisson summation formula, Poisson relation*), which relates items (1) and (3) in the previous sub-subsection.

If $f$ is a rapidly decreasing function and $\hat{f}$ its Fourier transformation defined by

$$\hat{f}(y) = \int_{-\infty}^{\infty} e^{-2\pi\sqrt{-1}xy} f(x)dx,$$

then the *Poisson formula* states that

$$\sum_{n\in\mathbb{Z}} f(n) = \sum_{n\in\mathbb{Z}} \hat{f}(2\pi n).$$

The left hand side is a summation over the length spectrum of $\mathbb{Z}\backslash\mathbb{R}$, i.e., the collection of lengths of closed geodesics; on the other hand, the right hand side concerns the eigenvalues of the Laplace operator $-\frac{d^2}{dx^2}$ of the Riemannian manifold $\mathbb{Z}\backslash\mathbb{R}$.[5] In fact, the set of oriented closed geodesics of $\mathbb{Z}\backslash\mathbb{R}$ is parametrized by $\mathbb{Z}$, and their lengths are equal to $|n|$, $n \in \mathbb{Z}$, where two geodesics of the opposite orientations have the same lengths. The eigenvalues of the Laplace operator of $\mathbb{Z}\backslash\mathbb{R}$ are $4\pi^2 n^2$, $n \in \mathbb{Z}$.

Therefore, this Poisson summation formula relates *geometric data* to *spectral data* of the arithmetic quotient $\mathbb{Z}\backslash\mathbb{R}$ of the Lie group $\mathbb{R}$. There is also a more general Poisson formula for each embedding $\mathbb{Z} \hookrightarrow \mathbb{R}$ where 1 is mapped to $\ell > 0$.

**3.1.2. Riemann zeta function.** One application of the Poisson formula above concerns the meromorphic continuation of the Riemann zeta function $\zeta(s)$.

In fact, one can define the *theta function*

$$\theta(t) = \sum_{n\in\mathbb{Z}} e^{-n^2 t}, \quad t \in \mathbb{C} \quad \text{with} \quad \mathrm{Re}(t) > 0.$$

Then the Poisson formula gives a transformation law for $\theta(t)$ under the transformation $t \to \frac{1}{t}$:

$$\theta(t) = \frac{1}{\sqrt{t}}\theta\left(\frac{1}{t}\right).$$

Since the Riemann zeta function $\zeta(s)$,

$$\zeta(s) = \sum_{n=1}^{\infty} \frac{1}{n^s}, \quad s \in \mathbb{C} \quad \text{with} \quad \mathrm{Re}(s) > 1,$$

[5]In the following, by an eigenvalue of a Riemannian manifold, we mean an eigenvalue of the Laplace operator of the Riemannian manifold. Since the Laplace operator is a canonical differential operator associated the Riemannian manifold, hopefully this will not cause any confusion.

can be expressed in terms of the Mellin transformation of the function $\theta(t)$ defined by

$$\int_0^\infty \theta(t) t^s \frac{dt}{t},$$

(Note that the Mellon transform of $e^{-ct}$, $c > 0$, is equal to $c^{-s}\Gamma(s)$, where $\Gamma(s)$ is the standard $\Gamma$-function), both the meromorphic continuation of $\zeta(s)$ from the half-plane $\mathrm{Re}(s) \geq 1$ to the whole complex $s$-plane, and the functional equation of $\zeta(s)$ relating $\zeta(s)$ to $\zeta(1-s)$ can be proved easily from the Poisson summation formula. See [Kobl] and [Mum2] for details.

For every $t$ with $\mathrm{Re}(t) > 0$, the function $\varphi_t(x) = \exp(-tx^2)$ and its derivatives are rapidly decreasing. Note that this is a family of functions on $\mathbb{R}$ parametrized by $t$. We can also take other suitable family of rapidly decreasing functions $\varphi_t(x)$ on $\mathbb{R}$, and the function

$$\theta_\varphi(t) = \sum_{n \in \mathbb{Z}} \varphi_t(n)$$

can also considered as a theta function.

## 3.2. $\mathbb{Z}^n$ and lattices in $\mathbb{R}^n$

A natural generalization of the group $\mathbb{Z}$ is $\mathbb{Z}^n$, $n \geq 1$, which is the free abelian group on $n$-generators. It is also the product of $n$-copies of $\mathbb{Z}$.

As in the case of $\mathbb{Z}$, it is fruitful to embed $\mathbb{Z}^n$ into the connected Lie group $\mathbb{R}^n$. Together with the embedding into $\mathbb{R}^n$, $\mathbb{Z}^n$ becomes an arithmetic subgroup. In fact, each such embedding $\mathbb{Z}^n \hookrightarrow \mathbb{R}^n$ gives an integral structure on $\mathbb{R}^n$.

For the standard embedding $\mathbb{Z}^n \hookrightarrow \mathbb{R}^n$, the lengths of closed geodesics of $\mathbb{Z}^n\backslash\mathbb{R}^n$ are given by $|v|$, where $v \in \mathbb{Z}^n$, since the lifts in $\mathbb{R}^n$ of closed geodesics of $\mathbb{Z}^n\backslash\mathbb{R}^n$ are straight line segments whose end points are identified under translation by $\mathbb{Z}^n$, and the eigenvalues of the flat Riemannian manifold $\mathbb{Z}^n\backslash\mathbb{R}^n$ are $4\pi^2|v|^2$ by using the Fourier expansion of periodic functions, where $v \in \mathbb{Z}^n$.

Therefore, counting the lengths of geodesics and the eigenvalues of the Riemannian manifold $\mathbb{Z}^n\backslash\mathbb{R}^n$ is equivalent to counting integral points $v \in \mathbb{Z}^n$ in balls with center at 0.

Briefly, for any $R > 0$, let $N(R)$ be the number of points $v \in \mathbb{Z}^n$ with $|v| \leq R$. It is easy to see that the leading term of $N(R)$ as $R \to +\infty$ is equal to the volume of the ball in $\mathbb{R}^n$ of radius $R$.

But it is a difficult problem to find sharp bounds on the reminder of $N(R)$. Even the case of $n = 2$ is not completely solved. It is related to the famous *Gauss circle problem* counting integral points inside circles. See [GruL] and also [Berg, §1.8, §9.5] for more details. (A proof of the optimal error term bound for the circle problem was recently announced in [CapS].)

The study of eigenvalues and eigenfunctions on the torus $\mathbb{Z}^n\backslash\mathbb{R}^n$ endowed with the flat metric is important for general Riemannian manifolds too. The basic reason is that a Riemannian manifold can be cut into small pieces, most of them are similar to cubes in $\mathbb{R}^n$; and eigenfunction problems on the cube $[-1,1]^n$ in $\mathbb{R}^n$ with either the Dirichlet or Neumann boundary conditions are related to eigenfunctions on $\mathbb{Z}^n\backslash\mathbb{R}^n$. For example, bounds on the counting function $N(R)$ for $\mathbb{Z}^n\backslash\mathbb{R}^n$ above together with the minimax principle and monotonicity properties of eigenvalues, in particular, the Dirichlet-Neumann bracketing, can be used to prove the *Weyl law* on the asymptotics of the counting function of eigenvalues of a compact Riemannian manifold (see §§3.4.1, 3.4.5, 12.4 below and [Cha] for details and references). Another application of the idea of localization to cubes is to prove the regularity of solutions of elliptic operators on compact manifolds (see [War, Chap. 6]).

**3.2.1. Lattices in $\mathbb{R}^n$.** Besides the standard embedding $\mathbb{Z}^n \hookrightarrow \mathbb{R}^n$, there are many other embeddings. In fact, for any $n$ linearly independent vectors $v_1, \dots, v_n$ of $\mathbb{R}^n$, there is an embedding of $\mathbb{Z}^n$ into $\mathbb{R}^n$ by mapping the standard generators $e_1, \dots, e_n$ to $v_1, \dots, v_n$. Denote the image by $\Lambda$.

Then $\Lambda$ is a discrete subgroup of $\mathbb{R}^n$ such that the quotient $\Lambda\backslash\mathbb{R}^n$ has finite volume. In other words, $\Lambda$ is a *lattice* in $\mathbb{R}^n$, which is defined by these two properties. Every lattice in $\mathbb{R}^n$ can be obtained in this way as an image of $\mathbb{Z}^n$.

Since $\mathbb{R}^n$ has the canonical inner product, each embedding gives an inner product on $\mathbb{Z}^n$. Conversely, an inner product on $\mathbb{Z}^n$ gives an embedding of $\mathbb{Z}^n$ into $\mathbb{R}^n$, which is unique up to rotations in $\mathbb{R}^n$. In some papers, this is the definition of a lattice of a vector space.

The standard embedding $\mathbb{Z}^n \hookrightarrow \mathbb{R}^n$ gives the standard *integral* structure and *rational* structure $\mathbb{Z}^n \otimes_{\mathbb{Z}} \mathbb{Q}$ on $\mathbb{R}^n$. Different lattices $\Lambda \subset \mathbb{R}^n$ give different integral structure of $\mathbb{R}^n$, but could induce the same rational structure $\Lambda \otimes_{\mathbb{Z}} \mathbb{Q}$. For example, for every nonzero integer $n$, $\frac{1}{n}\Lambda$ is also a lattice of $\mathbb{R}^n$ and different from $\Lambda$, when $|n| \neq 1$ but defines the same $\mathbb{Q}$-structure as the lattice $\Lambda$.

The inner product on $\mathbb{R}^n$ is important in these problems. Otherwise, as abstract groups, all pairs $(\Lambda, \mathbb{R}^n)$ are isomorphic. In fact, the isomorphisms can be given by linear transformations (but not orthogonal linear transformations in general).

In the above discussion, we have fixed the inner product (or the quadratic form) on $\mathbb{R}^n$ and considered arbitrary lattices $\Lambda$. On the other hand, we could have fixed the standard lattice $\mathbb{Z}^n \subset \mathbb{R}^n$ and arbitrary inner products (or quadratic forms).

These are the basic objects of the subject called *geometry of numbers.* One fundamental problem in this subject is to find integral (or lattice) points inside convex and symmetric domains of $\mathbb{R}^n$ under a low bound on the volume of the domains. Some basic and important books on geometry of numbers are [Cass1] [Gru] [GruL] [Si3]. Closely related books on quadratic forms are [Cass2] [Si4].

The convex domains in the geometry of number are often not polytopes. But counting integral points inside polytopes is important for many other applications both in pure and applied mathematics. See [DeL] for a survey.

The ring $\mathbb{Z}$ is the ring of integers of the field $\mathbb{Q}$ of rational numbers. For any number field $k$, i.e., a finite extension of $\mathbb{Q}$, there is the ring $\mathcal{O}_k$ of integers, which contains $\mathbb{Z}$ as a subring and is a finitely generated module over $\mathbb{Z}$. They define lattices under suitable embeddings of $k$ into $\mathbb{R}$ and $\mathbb{C}$. The group of units in $\mathcal{O}_k$ also induces lattices, i.e., the famous Dirichlet unit theorem. Similarly, orders of $\mathcal{O}_k$, i.e., subrings of finite index, give rise to lattices as well. Methods from geometry of numbers have been crucial in understanding number fields. For applications of methods from geometry of numbers to algebraic number theory and more information about lattices or orders in number fields, see [Sen] [BoSh] [Cohn].

As in the case of $\mathbb{Z} \hookrightarrow \mathbb{R}$, there is also a theta function $\theta_\Lambda(t)$ for every lattice $\Lambda$ in $\mathbb{R}^n$. This function carries a lot of information about $\Lambda$ and is useful for the problem of counting lattice points. The basic idea is that the theta function $\theta_\Lambda(t)$ is a modular form and its Fourier coefficients contain counting functions. Once the theta function is expressed in terms of other modular functions whose Fourier coefficients can be computed relatively easily, we can obtain information about the counting functions. It is also useful for the problem of sphere packing discussed below. See [Mum2] and [ConS] for more about theta functions and [Sar1] for the applications for counting mentioned here.

Besides number theory, lattices of $\mathbb{R}^n$ also occur naturally in algebraic geometry, for example, in describing the moduli spaces of polarized $K3$ surfaces (see [BarHPV] [CarMP] for example), and in understanding structures of finite groups of symplectic automorphisms of $K3$ surfaces (see [Kon] and the survey article [Dol1], where reflection groups connected with lattices and their many applications in algebraic geometry are also described. See also [Dol2] for related topics). One basic reason is that for a compact smooth manifold $M$, the torsion-free part $H^i(M, \mathbb{Z})$ is a lattice in $H^i(M, \mathbb{R}) \cong H^i_{dR}(M, \mathbb{R})$, which can also be identified with the space of harmonic $i$-forms on $M$ if $M$ is a Riemannian manifold (note that this latter identification gives a norm on the cohomology group.) If the manifold $M$ is a Kähler manifold, then the Hodge decomposition of $H^i(M, \mathbb{C})$ is taken into account and lattices in various components are used in studying the manifold.

**3.2.2. Sphere packing.** One problem in this subject concerns the density of sphere packing on lattices. Briefly, given a lattice $\Lambda$ in $\mathbb{R}^n$, we can place balls of the same radius with the center at the lattice points. The radius is maximized under the condition that there is no overlap between the interior of the balls. (This radius is equal to half of the minimal norm of the nonzero vectors in the lattice.)

Given such a sphere packing, there is a well-defined *density* of the packing, which is the proportion of the space covered by these balls. A natural question asks which lattices have maximal density. It can be seen easily that the standard lattice $\mathbb{Z}^n$ does not have the maximal density since $\mathbb{Z}^n$ can be built up from layers of $\mathbb{Z}^{n-1}$, and when a new layer is added on, the holes between different balls in the old layers are not filled in to increase the density. Instead, the balls are placed on the top of the existing layers. Clearly, moving the new ones into the holes will increase the density of the packing. In dimension 3, the best lattice packing is obtained this way, which has also the optimal density among all sphere packings in $\mathbb{R}^3$.

A basic point in this subject is that lattices constructed from or related to the root lattices of Lie algebras enjoy good properties. For example, it can be shown relatively easily that the root lattices for $SL(n, \mathbb{R})$ give maximal density lattice packing in $\mathbb{R}^{n-1}$ for $n \leq 3, 4$. But this is not true for some other higher dimensions. Since the space of lattices in $\mathbb{R}^n$ can be identified with $GL(n, \mathbb{Z})\backslash GL(n, \mathbb{R})$, this problem is also closely related to the reduction theory of arithmetic subgroups such as $SL(n, \mathbb{Z})$ discussed below.

A general, comprehensive introduction to this important field is the book [ConS], which contains over 100 pages of references. See also survey papers [Slo] [ConGS].

**3.2.3. Poisson summation formula for lattices in $\mathbb{R}^n$.** For each lattice $\Lambda \subset \mathbb{R}^n$, there is also a Poisson formula. The formula is still the same as in the case of $\mathbb{Z}$, i.e., one side is a summation over the lattice $\Lambda$ representing the lengths of geodesics in the flat Riemannian manifold $\Lambda\backslash\mathbb{R}^n$, and the other side is over the dual lattice $\Lambda^* = \{v \in \mathbb{R}^n \mid \langle v, \lambda \rangle \in \mathbb{Z}, \text{ for all } \lambda \in \Lambda\}$, representing the eigenvalues of the Riemannian manifold $\Lambda\backslash\mathbb{R}^n$. See [Berg, §9.5] [Cha] for examples.

In this case, both the geometry and spectral theory of the Riemannian manifold $\Lambda \subset \mathbb{R}^n$ is more complicated. For example, there could be high multiplicity of both lengths of geodesics and eigenvalues.

This Poisson formula is basic in harmonic analysis of periodic functions on $\mathbb{R}^n$. It is also important in studying the problem of sphere packing in higher dimension (see [ElC] [CoKu]).

In number theory, the Poisson formula for both $\mathbb{Z} \subset \mathbb{R}$ and lattices in the general dimension is often formulated in terms of the embedding of $\mathbb{Q}$ into the ring $\mathbb{A}$ of adeles. Briefly, the ring of adeles $\mathbb{A}$ is defined as follows. It is well-known that $\mathbb{R}$ is the completion of $\mathbb{Q}$ with respect to the standard absolute value $\|\cdot\|_\infty$. For every finite prime number $p$, there is associated an ultra-norm $\|\cdot\|_p$ on $\mathbb{Q}$; and the completion of $\mathbb{Q}$ in this norm gives the field of $p$-adic numbers $\mathbb{Q}_p$. The closure of $\mathbb{Z}$ in $\mathbb{Q}_p$ is the ring $\mathbb{Z}_p$ of $p$-adic integers. Then $\mathbb{A}$ is the restricted product $\mathbb{R} \times \prod_p' \mathbb{Q}_p$ with respect to $\mathbb{Z}_p$, where the restricted product means that for every adele $a = (a_\infty, a_2, a_3, \ldots, a_p, \ldots)$, all but finitely many of $a_p$ belongs to $\mathbb{Z}_p$. An important property is that $\mathbb{A}$ is a locally compact group and contains $\mathbb{Q}$ as a discrete subgroup. One motivation for introducing adeles is to treat all evaluations or places (finite or not) of $\mathbb{Q}$ on the equal footing. For many applications, it is fruitful to compare the embedding $\mathbb{Q} \hookrightarrow \mathbb{A}$ with the standard embedding $\mathbb{Z} \hookrightarrow \mathbb{R}$. The harmonic analysis on $\mathbb{R}/\mathbb{Z}$ is the study of periodic functions, and harmonic analysis on the locally compact and abelian groups $\mathbb{A}$ and $\mathbb{A}/\mathbb{Q}$ have been very important in number theory, in particular the zeta function $\zeta(s)$ and more generally the zeta function of a number field. Besides the zeta function of number fields, it has also been used to study the Tate type zeta function. See [CassF] [KnL1] [RamV].

**3.2.4. Weil-Siegel formula.** The Poisson summation formula is also important in the Weil-Siegel formula, which identifies the integral of a certain theta series as the special value of an Eisenstein series. It has applications to the Langlands program, in particular to special values and poles of automorphic $L$-functions.

The formula is expressed in terms of adeles. We briefly mention the idea. Assume that $\mathbf{G}$ is a connected semisimple linear algebraic group defined over $\mathbb{Q}$. Let $\pi$ be a representation of $\mathbf{G}$ on a vector space $V$ defined over $\mathbb{Q}$. Let $\mathbb{A}$ be the ring of adeles over $\mathbb{Q}$. For a suitable function $f$ on $V(\mathbb{A})$ with rapid decay, define a function

$$\varphi_{f,\pi}(g) = \sum_{\gamma \in V(\mathbb{Q})} f(\pi(g^{-1})\gamma)$$

on $\mathbf{G}(\mathbb{A})$. Then $\varphi_{f,\pi}(g)$ is invariant under $\mathbf{G}(\mathbb{Q})$. Note that the embedding $\mathbb{Q} \subset \mathbb{A}$ as a discrete subgroup is similar to the embedding $\mathbb{Z} \subset \mathbb{R}$ above, and hence the above summation can be considered as an analogue of a summation over $\mathbb{Z}$, and $\varphi_{f,\pi}(g)$ can be considered as a theta function. The Poisson summation formula says that

$$\varphi_{f,\pi} = \varphi_{\hat{f},\tilde{\pi}},$$

where $\hat{f}$ is the Fourier transform of $f$ as above, and $\tilde{\pi}$ is the contragredient representation (see §3.1.1 and also [CassF] [RamV]). Then the point of the Weil-Siegel formula can approximately be stated that the integral of $\varphi_{f,\pi}(g)$ over the quotient $\mathbf{G}(\mathbb{Q})\backslash\mathbf{G}(\mathbb{A})$ can be identified with an Eisenstein series. See [KudR1-3] [HarrK] [Ya] for the correct statements, where a pair of dual groups is needed.

If the integral does not exist, then one needs to truncate the function. This leads to a truncated Poisson summation formula, as studied in [Lev1] and [Lev2].

**3.2.5. Voronoi formula.** A closely related formula is the Voronoi summation formula. Briefly, the Poisson summation formula can also be formulated as expressions for truncated sums of the type $\sum_{a \leq n} f(n)$ in terms of suitable transformation of $f$. The key idea of the Voronoi summation formula is to consider weighted sum $\sum_{a \leq n \leq b} a_n f(n)$, where $a_n$ could be the coefficients of modular forms or other number theoretic sequences, and to express it in terms of suitable transformations involving $f$ and $a_n$. For its history and precise expressions, see expository papers [MilW3] [MilW5] [Schm2] and also other related papers [MilW1] [MilW2][MilW4] [GoldL]. For other results on and generalizations of the Voronoi formula, see [BeiB].

**3.2.6. Generalizations of the Poisson summation formula.** As pointed out before, the Poisson summation formula relates the spectral and geometrical invariants of the compact flat Riemannian manifolds $\mathbb{Z}^n \backslash \mathbb{R}^n$. A vast generalization of the Poisson formula to general compact manifolds was given in [DuG]. The basic idea here is to relate the singularities of the Fourier transformation of the spectral measure of the Laplace operator of a compact Riemannian manifold to the lengths of closed geodesics. This is called the Poisson relation, which plays an important role for studying questions in *spectral geometry*, a branch of Riemannian geometry studying relations between eigenvalues of the Laplace operator (or more generally Laplace-Beltrami operator.) For example, better estimates on the error term of the counting function of the eigenvalues of compact Riemannian manifolds can be obtained using the Poisson relation and the method of wave equations. See [Ivr] [DuG] and also [Ber].

An important problem in spectral geometry is to construct and classify isospectral manifolds, i.e., non-isometric Riemannian manifolds which have the same spectra (or eigenvalues). The first such isospectral manifolds were constructed in [Milno2] using the theta functions of lattices in $\mathbb{R}^n$. The Poisson formula is crucial for some results in these direction. For example, for the Poisson formula for nilpotent groups and applications to classifications of isospectral nilmanifolds, see [Pes1-2] [Gord] and the references there.

A generalization of the Poisson relation to noncompact locally symmetric spaces of $\mathbb{Q}$-rank 1 is given in [JiZ1] [JiZ2]. The formula in this case relates the singularities of the Fourier transformation of the continuous spectral measure of a locally symmetric space to the normalized lengths, the so-called sojourn times, of scattering geodesics, i.e., geodesics which are distance minimizing eventually in both directions and have to run away to infinity. It is expected that a similar relation holds for general locally symmetric spaces of higher $\mathbb{Q}$-rank, and scattering flats are also needed besides the scattering geodesics and give rise to multi-dimensional scattering times. One possible way to do this is to use the factorization of the spectral measures (or rather scattering matrices) of the higher dimensional continuous spectrum in terms of the 1-dimensional ones and then apply the results in [JiZ1] [JiZ2]. See also [PetS] for Poisson relation and related problems on unbounded domains in $\mathbb{R}^n$.

The generalized Poisson relation here can also be interpreted as a wave equation generalization of the Kuznetsov formula in the theory of automorphic forms [Brug1] [Iw1] (for [BruM] for a summary and generalizations).

## 3.3. The modular group $SL(2,\mathbb{Z})$

As mentioned earlier, a natural generalization of the group $\mathbb{Z}$ is $\mathbb{Z}^n$, but $\mathbb{Z}^n$ is still an abelian arithmetic group. They share many similar properties, for example, the Poisson summation formula and both giving rise to flat Riemannian manifolds. To get really new results and different kinds of Riemannian manifolds, we need to consider non-abelian generalizations.

A natural non-abelian generalization is the *modular group* $SL(2,\mathbb{Z})$:

$$SL(2,\mathbb{Z}) = \left\{ \begin{pmatrix} a & b \\ c & d \end{pmatrix} \mid a, b, c, d \in \mathbb{Z}, ad - bc = 1 \right\}.$$

Since $\mathbb{Z}$ is naturally embedded into $\mathbb{R}$ as a discrete subgroup, $SL(2,\mathbb{Z})$ is also naturally embedded into the Lie group $SL(2,\mathbb{R})$ as a discrete subgroup. The group $SL(2,\mathbb{Z})$ together with the embedding $SL(2,\mathbb{Z})$ into $SL(2,\mathbb{R})$ is an arithmetic subgroup. It can not be overemphasized that when we think of $SL(2,\mathbb{Z})$, for example in connection with quadratic forms, we automatically view it as a subgroup of the Lie group $SL(2,\mathbb{R})$.

We can consider similar questions as in the case of $\mathbb{Z} \subset \mathbb{R}$. On the other hand, there are some differences.

(1) The quotient $SL(2,\mathbb{Z})\backslash SL(2,\mathbb{R})$ is not compact.
(2) The space $SL(2,\mathbb{R})$ is not simply connected, and $SL(2,\mathbb{Z})$ contains nontrivial torsion elements. Hence, $SL(2,\mathbb{Z})\backslash SL(2,\mathbb{R})$ is not a classifying space for $SL(2,\mathbb{Z})$.

To overcome the problem of non-simply connectedness, we replace $SL(2,\mathbb{R})$ by the quotient $SL(2,\mathbb{R})/SO(2)$ by a maximal compact subgroup $SO(2)$, which can be identified with the upper half plane

$$\mathbb{H} = \{x + iy \mid x \in \mathbb{R}, y > 0\} \cong SL(2,\mathbb{R})/SO(2),$$

and is hence contractible. (This is related to the general fact that the topology of a noncompact Lie group with finitely many connected components is contained in the topology of its maximal compact subgroups. See [Most4] and [Hoc, Theorem XV.3.1].) In fact, $SL(2,\mathbb{R})$ acts transitively on $\mathbb{H}$ via the linear fractional transformations and the stabilizer at $i$ is equal to $SO(2)$. Hence $\mathbb{H} = SL(2,\mathbb{R})/SO(2)$. We also note that the Poincaré metric $ds^2 = y^{-2}(dx^2 + dy^2)$ of $\mathbb{H}$ is a $SL(2,\mathbb{R})$-invariant Riemannian metric, which is the unique metric of constant negative curvature up to scalar multiplication.

### 3.3.1. Fundamental domain of $SL(2,\mathbb{Z})$ in the upper half plane.

To understand the action of $SL(2,\mathbb{Z})$ on the upper half plane $\mathbb{H}$, in particular the quotient space $SL(2,\mathbb{Z})\backslash\mathbb{H}$, it is helpful to get a good fundamental region (or domain).[6] Recall that a domain $\Omega$ in $\mathbb{H}$ is a *fundamental domain* for $SL(2,\mathbb{Z})$ if the following conditions are satisfied:

---

[6]Though a domain is usually an open subset, we use it in a more general sense. In a following, we often take it to be a closed subset for simplicity, i.e., to avoid ambiguity in the definition of fundamental domains.

(1) $\mathbb{H} = SL(2,\mathbb{Z})\Omega$,
(2) its interior $\mathrm{Int}(\Omega)$ is mapped injectively into the quotient $SL(2,\mathbb{Z})\backslash\mathbb{H}$,
(3) the restricted map $\Omega \to SL(2,\mathbb{Z})\backslash\mathbb{H}$ is finite-to-one.

It is often implicitly assumed that the set of boundary points of $\Omega$, i.e., $\overline{\Omega}-\Omega$, is of measure 0. Sometimes, we also add a condition that every bounded subset of $\mathbb{H}$ is covered by finitely many translates of $\Omega$ (see [FeN]). Hence, $SL(2,\mathbb{Z})\backslash\mathbb{H}$ is obtained from $\Omega$ by suitable identification along its boundary. By a good fundamental domain $\Omega$, we mean that the boundary structure of $\Omega$ is simple and the gluing pattern on the boundary is clear.

Such a fundamental domain for $SL(2,\mathbb{Z})$ is well-known and is given by the subset:

$$\Omega = \left\{x+iy \in \mathbb{H} \mid -\frac{1}{2} \leq x \leq \frac{1}{2}, x^2+y^2 \geq 1\right\}.$$

(In the quotient, the left side boundary of $\Omega$ is identified with the right side. The interior of $\Omega$ is mapped injective to an open and dense subsset of $SL(2,\mathbb{Z})\backslash\mathbb{H}$. To get a bijection with $SL(2,\mathbb{Z})\backslash\mathbb{H}$, we need to remove basically half of the boundary points.)

Since it contains the vertical strip given by $-\frac{1}{2} \leq x \leq \frac{1}{2}$, $y > 1$, and only the two sides of this region are identified in the quotient, it implies the following result.

COROLLARY 3.3.1. *The quotient $SL(2,\mathbb{Z})\backslash\mathbb{H}$ is noncompact, and has finite area with respect to the hyperbolic metric $ds^2 = y^{-2}(dx^2+dy^2)$.*

From the identification of the boundary of $\Omega$, it is easy to see that $SL(2,\mathbb{Z})\backslash\mathbb{H}$ is homeomorphic to the plane $\mathbb{C}$. In fact, there is a canonical map, called the j-invariant of elliptic curves, which identifies $SL(2,\mathbb{Z})\backslash\mathbb{H}$ with $\mathbb{C}$. See [Silv].

For detailed discussions of fundamental domains in the upper half plane, see [Mag].

**3.3.2. Reduced quadratic forms and reduction theory.** The above description of the fundamental domain $\Omega$ of $SL(2,\mathbb{Z})$ allows us to pick a special point in each $SL(2,\mathbb{Z})$-orbit. It is known that the space of positive quadratic forms in two variables of determinant 1 can be identified with $\mathbb{H}$. Since the quadratic forms corresponding to points in an $SL(2,\mathbb{Z})$-orbit are equivalent in the sense that they give the same quadratic form under a linear change of variables with integral coefficients (hence their values on $\mathbb{Z}^2$ represent the same set of integers), this allows us to pick out a so-called *reduced quadratic form* in each equivalence class of quadratic forms. This is called the *reduction theory of quadratic forms* and was studied by Gauss and others before him. See [ScO] [Cox] [Golm] (and also [Fri3]) for motivations and history of the reduction theory, which plays a fundamental role in the study of geometry and topology of arithmetic groups.

Because of this connection with quadratic forms, determination of a fundamental domain of an arithmetic subgroup acting on an associated symmetric space is called the reduction theory of the arithmetic subgroup. It is often difficult to get a precise fundamental domain, and suitable approximated sets are often sought for and used. They are often called *fundamental sets*. See §4.13 below.

**3.3.3. Generators of $SL(2,\mathbb{Z})$.** Under the action of $SL(2,\mathbb{Z})$, the translates of the fundamental domain $\Omega$ give an equivariant tessellation of $\mathbb{H}$. Elements which identify sides of the fundamental domain give a set of generators. Then another corollary of the description of the fundamental domain $\Omega$ of $SL(2,\mathbb{Z})$ is that $SL(2,\mathbb{Z})$ is generated by $S = \begin{pmatrix} 0 & 1 \\ 1 & 0 \end{pmatrix}$ and $T = \begin{pmatrix} 1 & 1 \\ 0 & 1 \end{pmatrix}$. We can also show that $PSL(2,\mathbb{Z}) = \mathbb{Z}/2\mathbb{Z} \star \mathbb{Z}/3\mathbb{Z}$, where $PSL(2,\mathbb{Z}) = SL(2,\mathbb{Z})/\{\pm 1\}$.

Another famous $SL(2,\mathbb{Z})$-equivariant tessellation of $\mathbb{H}$ is given by ideal triangles with vertices at infinity belonging to $\mathbb{Q} \cup \{\infty\}$. The point is that every three points in the boundary $\partial\mathbb{H} \cup \mathbb{R} \cup \{\infty\}$ determine a unique ideal triangle in $\mathbb{H}$, and the extended $SL(2,\mathbb{Z})$-action on $\partial\mathbb{H} \cup \mathbb{R} \cup \{\infty\}$ preserves the rational points $\mathbb{Q} \cup \{\infty\}$. The ideal triangle with vertices $0, 1, i\infty$ and its translates under $SL(2,\mathbb{Z})$ form a $SL(2,\mathbb{Z})$-tessellation of $\mathbb{H}$, which is called the *Farey tessellation* and is related to symbolic dynamics in §3.3.8 below. For more discussions about tessellations of $\mathbb{H}$, see [Mag].

**3.3.4. Volume formula for modular curves and locally symmetric spaces.** The area of $SL(2,\mathbb{Z})\backslash\mathbb{H}$ can be computed by the explicit description in §3.3.1 of the fundamental domain $\Omega$ of $SL(2,\mathbb{Z})$ in $\mathbb{H}$ and the Gauss-Bonnet formula.

On the other hand, for other natural locally symmetric spaces, for example, related to the groups $SL(n,\mathbb{Z})$, $n \geq 3$, the computation of their volumes is more complicated, though they can sometimes be computed explicitly. They are related to special values of the zeta function $\zeta(s)$. See [ScO, Chapter 10] and [GilG] for more details. See also [GrunH] [Borel6] for volume of hyperbolic 3-dimensional manifolds and see [Milno3] [AleVS] for volume of polyhedra in hyperbolic spaces.

For the most general volume formula of quotients of symmetric spaces and Lie groups by arithmetic and $S$-arithmetic subgroups, see [Pr1]. In principle, explicit formulas of volume of locally symmetric spaces associated with congruence subgroups can be derived from the adelic formula in [Pr1], though such derivations have not been carried out explicitly in [Pr1]. (Note that the formula in [Pr1] is given as infinite products over primes. To see relations with special values of zeta functions, we note that zeta functions and more general $L$-functions admit Euler products, i.e., special products over primes.)

**3.3.5. Moduli space of lattices.** Besides its action on $\mathbb{H}$, the Lie group $SL(2,\mathbb{R})$ also acts on the space of unimodular lattices in $\mathbb{R}^2$. Recall that a lattice $\Lambda \subset \mathbb{R}^n$ is called a *unimodular lattice* if the volume of $\Gamma\backslash\mathbb{R}^n$ is equal to 1. Since $SL(2,\mathbb{R})$ acts transitively on the space of unimodular lattices of $\mathbb{R}^2$, and the stabilizer of the standard lattice $\mathbb{Z}^2$ is equal to $SL(2,\mathbb{Z})$, it follows that the space of unimodular lattices in $\mathbb{R}^2$ can be identified with the homogeneous space $SL(2,\mathbb{Z})\backslash SL(2,\mathbb{R})$. This identification gives a natural moduli interpretation of this homogeneous space. The orthogonal group $SO(n)$ acts on the set of unimodular lattices and it is also natural to identify lattices which are rotations of each others. Then $SL(2,\mathbb{Z})\backslash SL(2,\mathbb{R})/SO(2) \cong SL(2,\mathbb{Z})\backslash\mathbb{H}$ is the moduli space of equivalence class of unimodular lattices in $\mathbb{R}^2$.

We note also that each unimodular lattice in $\mathbb{R}^2$ gives a flat Riemannian metric on the torus $S^1 \times S^1 = \mathbb{Z}^2\backslash\mathbb{R}^2$ of total area 1, and two equivalent ones give rise to isometric flat Riemannian manifold. Hence, $SL(2,\mathbb{Z})\backslash\mathbb{H}$ is also the moduli space of all such flat metrics on $S^1 \times S^1$. (The moduli space of marked flat metrics of volume 1 can be identified with $\mathbb{H} = SL(2,\mathbb{R})/SO(2)$, where a marking refers to a choice of a basis of $H^1(\mathbb{Z}^2\backslash\mathbb{R}^2,\mathbb{Z})$. This is related to Teichmüller spaces below.)

This relation between the unimodular lattices and $\mathbb{H}$ and the description of the fundamental domain $\Omega$ of $SL(2,\mathbb{Z})$ in $\mathbb{H}$ in §3.3.1 allow us to show that the densest lattice sphere packing in $\mathbb{R}^2$ is given by the hexagon lattice packing. This is the connection between dense sphere packing and reduction theory mentioned before. See [Sie3] and [Ji6] for explanation of this and related topics.

**3.3.6. Compactifications.** Since the quotient $SL(2,\mathbb{Z})\backslash\mathbb{H}$ is noncompact, a natural problem is to compactify it. One simple reason for finding compatifications is that it is easier to study and use compact spaces. For example, the structure of at infinity and compactifications of $SL(2,\mathbb{Z})\backslash\mathbb{H}$ can be used for understanding behaviors of functions, in particular automorphic forms, at infinity.

Another reason is that for any torsion-free subgroup $\Gamma$ of $SL(2,\mathbb{Z})$ of finite index, suitable compactifications of $\Gamma\backslash\mathbb{H}$ can be used to construct good models of $K(\Gamma,1)$-spaces.

See the introduction of [BoreJ] for motivations and history of compactifications of modular curves and locally symmetric spaces.

There are two general approaches to compactifications:

(1) Add some ideal points at infinity of $SL(2,\mathbb{Z})\backslash\mathbb{H}$.
(2) Remove some non-compact part from $SL(2,\mathbb{Z})\backslash\mathbb{H}$ to get a compact space whose interior is diffeomorphic to $SL(2,\mathbb{Z})\backslash\mathbb{H}$.

For the method (1), we can either add one point to $SL(2,\mathbb{Z})\backslash\mathbb{H}$, and the resulting space is homeomorphic to the compact space $\mathbb{C}\cup\{\infty\} = \mathbb{C}P^1$. In this case, if $SL(2,\mathbb{Z})\backslash\mathbb{H}$ is interpreted as the moduli space of elliptic curves, then the ideal point $\{\infty\}$ represents the rational curve $\mathbb{C}P^1$ minus two points, ie., the punctured plane $\mathbb{C}^\times$, or the rational curve $\mathbb{C}P^1$ with a self-intersection point. If $SL(2,\mathbb{Z})\backslash\mathbb{H}$ is identified with the moduli space of flat metrics on the torus $\mathbb{Z}\backslash\mathbb{R}$ normalized so that the injectivity radius is equal to 1, then the ideal point represents the noncompact complete flat Riemannian manifold $\mathbb{Z}\backslash\mathbb{R}^2$, which can be identified with $\mathbb{Z}\backslash\mathbb{C} \cong \mathbb{C}^\times$. We can also add a circle at infinity which parametrizes the set of geodesics that go through the cusp of $SL(2,\mathbb{Z})\backslash\mathbb{H}$ to the infinity. This can be thought of as a real blow-up of the previous compactification, where the ideal point is replaced by all directions from this point. Such moduli interpretations of the compactification is important for general cases, which are often difficult.

The method (2) is to cut off a cusp neighborhood and obtain a compact submanifold with boundary contained in $SL(2,\mathbb{Z})\backslash\mathbb{H}$, whose interior is homeomorphic to $SL(2,\mathbb{Z})\backslash\mathbb{H}$.[7] Equivalently, it can be constructed by removing horoballs near the rational boundary points of $\mathbb{H}$ in a $SL(2,\mathbb{Z})$-equivariant way so that what remains, denoted by $\mathbb{H}(\varepsilon)$, has a compact quotient $SL(2,\mathbb{Z})\backslash\mathbb{H}(\varepsilon)$ under $SL(2,\mathbb{Z})$, which is a compact submanifold with boundary. It is also important to note that $\mathbb{H}$ can be equivariantly deformation retracted to $\mathbb{H}(\varepsilon)$. The point here is that even though the space $SL(2,\mathbb{Z})\backslash\mathbb{H}(\varepsilon)$ is smaller, it contains all the topology of $SL(2,\mathbb{Z})\backslash\mathbb{H}$.

**3.3.7. Deformation retraction to co-compact subspaces.** The fundamental domain $\Omega$ above in §3.3.1 gives a $SL(2,\mathbb{Z})$-equivariant simplicial tesselation of $\mathbb{H}$. The spine of this simplicial complex is a tree invariant under $SL(2,\mathbb{Z})$, whose quotient under $SL(2,\mathbb{Z})$ consists of one edge and two end points. (Note that there is one vertex of the spine for every 2-dimensional simplex, and one edge for every pair of adjacent 2-dimensional simplexes.) This also allows one to conclude

$$PSL(2,\mathbb{Z}) = \mathbb{Z}/2\mathbb{Z} \star \mathbb{Z}/3\mathbb{Z}$$

as mentioned above. See [Serr1]. This is related to the beautiful combinatorial group theory of actions of discrete groups on trees in [Serr4].

It should be pointed out that the deformation retract to the spine here is the smallest possible dimension, i.e., the dimension of the spine is equal to the virtual cohomological dimension of $SL(2,\mathbb{Z})$, which is equal to 1. This is not the case with the retraction to $\mathbb{H}(\varepsilon)$ above. On the other hand, for sufficiently small $\varepsilon$, $\mathbb{H}(\varepsilon)$ is a manifold with boundary whose boundary components are contractible and parametrized by $\mathbb{Q}\cup\{i\infty\}$, and these facts can be used to show that $SL(2,\mathbb{Z})$ is a virtual duality group of dimension 1. But clearly, the spine is not a manifold with boundary.

**3.3.8. Geodesic flow, continued fractions and symbolic dynamics.** An integral part of a Riemannian manifold is the structure of geodesics, and the geodesic flow provides a natural dynamical system.

Geodesics or equivalently the geodesic flow on $SL(2,\mathbb{Z})\backslash\mathbb{H}$ can be coded by symbolic sequences which are given by the cutting patterns with the Farey

[7] The spaces constructed by method (2) are not exactly compactifications of $SL(2,\mathbb{Z})\backslash\mathbb{H}$. On the other hand, their interiors are diffeomorphic to $SL(2,\mathbb{Z})\backslash\mathbb{H}$ and hence can be considered as compactifications in a suitable sense.

tessellation of $\mathbb{H}$ by ideal triangles with rational points as vertices (recall that the ideal triangles of the Farey tessellation consist of the ideal triangle with vertices $0, 1, i\infty$ and translates under $SL(2, \mathbb{Z})$). Hence the geodesic flow can be studied by symbolic dynamics. These symbolic sequences are related to the continued fractions of the boundary points of $\mathbb{H}$. See [Seri1-3] [KatoU] for explanations and the history of this subject. See also [Lal] [BirS] [Pat3] [Fri3] for related results.

If a geodesic in $\mathbb{H}$ connect two irrational boundary points, then its image in $SL(2, \mathbb{Z})\backslash\mathbb{H}$ will not escape to infinity through the cusp. Rather it oscillates and goes deeper and deeper into the cusp. The depths of such excursions are related to the integers which appear in the continued fraction expansions of these irrational numbers. See [Sul1] and the generalization to higher rank spaces in [KleM1]. Such geodesics are related to limiting modular symbols. If geodesics in $\mathbb{H}$ connect rational boundary points, i.e., points belonging to $\mathbb{Q} \cup \{i\infty\} \subset \partial\mathbb{H}$, then they give modular symbols of $SL(2, \mathbb{Z})\backslash\mathbb{H}$. See §3.4.8 and the references there.

For analogous results over fields of positive characteristic, see [Pau3]. For extensions to non-constant negative curvatures, see [HePa].

**3.3.9. Congruence subgroups.** The above discussions show that $SL(2, \mathbb{Z})$ is an important arithmetic subgroup of $SL(2, \mathbb{R})$. But there are also other subgroups $\Gamma$ of $SL(2, \mathbb{Z})$ of finite index which are important for various purposes.

A particularly important class of subgroups consists of the principal congruence subgroups, which are defined as follows. For every level $N \geq 1$, define the principal congruence subgroup of level $N$ by

$$\Gamma(N) = \{g \in SL(2, \mathbb{Z}) \mid g \equiv Id \mod N\}.$$

It is known that for $N \geq 3$, $\Gamma(N)$ is torsion-free. (The same conclusion holds for the congruence subgroups of $SL(n, \mathbb{Z})$, $n \geq 2$.) Therefore the quotient $\Gamma(N)\backslash\mathbb{H}$ is a smooth manifold instead of being an orbifold such as $SL(2, \mathbb{Z})\backslash\mathbb{H}$. It represents the moduli space of elliptic curves with level $N$ structure. Any subgroup $\Gamma$ of $SL(2, \mathbb{Z})$ containing some $\Gamma(N)$ is called a congruence subgroup. There are other special congruence subgroups $\Gamma$ of $SL(2, \mathbb{Z})$ of finite index which are important for other moduli spaces of elliptic curves. For example,

$$\Gamma_1(N) = \left\{g = \begin{pmatrix} a & b \\ c & d \end{pmatrix} \in SL(2, \mathbb{Z}) \mid a, b \equiv 1, c \equiv 0 \mod N\right\},$$

$$\Gamma_0(N) = \left\{g = \begin{pmatrix} a & b \\ c & d \end{pmatrix} \in SL(2, \mathbb{Z}) \mid c \equiv 0 \mod N\right\}.$$

All these subgroups have finite index in $SL(2, \mathbb{Z})$. For explicit computations of the index of the above subgroups, see [Shi1].

More generally, a subgroup $\Gamma \subset SL(2, \mathbb{Z})$ of finite index is called an *arithmetic subgroup.* An important point to mention is that there are arithmetic subgroups of $SL(2,\mathbb{Z})$ which are not congruence subgroups. See [Mag]. This gives rise to the congruence subgroup problem.

For every arithmetic subgroup $\Gamma \subset SL(2,\mathbb{Z})$, if $\Omega$ is a fundamental domain of $SL(2,\mathbb{Z})$ and $\gamma_1, \ldots, \gamma_m$ are representatives of the quotient $\Gamma\backslash SL(2,\mathbb{Z})$, then $\gamma_1\Omega \cup \cdots \cup \gamma_m\Omega$ is a fundamental domain for $\Gamma$. The space $\Gamma\backslash\mathbb{H}$ is a covering space of $SL(2,\mathbb{Z})\backslash\mathbb{H}$.

See [Shi1] and also [Katok] for constructions and basic properties of other subgroups of $SL(2,\mathbb{R})$.

**3.3.10. Dirichlet fundamental domain.** As explained above, finding a good fundamental domain for $\Gamma$ on $\mathbb{H}$ is important and convenient for studying both geometry and topology of $\Gamma\backslash\mathbb{H}$ and group theoretical properties of $\Gamma$.

A direct method of obtaining a fundamental domain for a general subgroup $\Gamma$ of $SL(2,\mathbb{Z})$ is to use the *Dirichlet fundamental domain.* Specifically, for any point $x_0$ which is not fixed by any nontrivial element of $\Gamma$, the Dirichlet domain for $\Gamma$ with respect to the basepoint $x_0$ is defined by:

$$D_\Gamma(x_0) = \{x \in \mathbb{H} \mid d(x, x_0) \leq d(\gamma x, x_0), \text{ for every } \gamma \in \Gamma\}.$$

Such a base point $x_0$ clearly exists. This construction works for any discrete subgroup of $SL(2,\mathbb{R})$ (or rather $PSL(2,\mathbb{R})$). In fact, this method also works for every isometric and proper action of a discrete subgroup $\Gamma$ on a metric space $M$. One reason is that the existence of such a base point $x_0$ always holds if the $\Gamma$-action is effective, i.e., there is no nontrivial element of $\Gamma$ which fixes every point of $M$, and every proper action always induces an effective action after passing to a quotient by a finite subgroup.

For lattices $\Lambda \subset \mathbb{R}^n$, its Dirichlet domains with the center at points in $\Lambda$ are called Voronoi cells. Generalizations of Voronoi cells have found many applications. See [OkBSC] and the references there.

For finitely generated Fuchsian groups, their Dirichlet fundamental domains in $\mathbb{H}$ are bounded by finitely many geodesics. For Kleinian groups acting on the three dimensional real hyperbolic space, i.e., discrete subgroups of $SL(2,\mathbb{C})$ (or rather $PSL(2,\mathbb{C})$), though the Dirichlet fundamental domain is still bounded by totally geodesic surfaces, there could be infinitely many such boundary faces. Similarly, in higher dimensions, finite generation of a discrete groups does not imply that its Dirichlet fundamental domains are bounded by finitely many totally geodesic hypersurfaces. Fundamental domains with finitely many sides are often sought after, and a natural problem is to understand conditions under which a Dirichlet fundamental domain of $\Gamma$ is bounded by finitely many boundary faces.

A more natural condition imposed on such a discrete subgroup $\Gamma$ is the notion of geometric finiteness. These two finiteness conditions on discrete subgroups are not equivalent but closely related. See [Bea] [ViS] [BeneP] [Rat] [Th1] for more about fundamental domains. See [Bow1] [Ap2] for definitions and conditions related to geometric finiteness.

On the other hand, if $M$ is not a Riemannian manifold with constant curvature, the boundary faces of the Dirichlet fundamental domain are totally geodesic subspaces. Due to this, even for symmetric spaces of rank 1 but not of constant curvature, the Dirichlet domains are difficult to understand and may not so useful.

If the quotient $\Gamma\backslash M$ is noncompact, we often seek some fundamental domains whose structures at infinity can be described explicitly. For example, in the fundamental domain $\Omega$ of $SL(2,\mathbb{Z})$ in $\mathbb{H}$ discussed earlier in §3.3.1, the part $\Omega \cap \{x+iy \mid y>1\}$ near the infinity $\{i\infty\}$ is the product $[-\frac{1}{2},\frac{1}{2}] \times (1,+\infty)$. For general arithmetic subgroups, the so-called Siegel sets enjoy similar properties and are used to construct their fundamental domains or rather fundamental sets. See §4.13. (We note that the boundary of the fundamental domain $\Omega$ of $SL(2,\mathbb{Z})$ near the cusp point $\{i\infty\}$ consists of two geodesics converging to it. Such a property is not possible for general symmetric spaces.)

**3.3.11. Finite generation of discrete groups.** A natural and combinatorial way to describe a group with discrete topology is to find generators and relations among them. The most natural classes are those which are finitely generated and finitely presented. Understanding groups which are finitely generated and finitely presented is the basic problem in combinatorial group theory. For a historic account of this theory, see [ChaW]. For more detailed discussions, see [LyS] [MagKS]. See also [FiR].

One general method to find generators and relations of a group $\Gamma$ is to use actions of $\Gamma$ on some suitable spaces. These methods work for general actions of a discrete subgroup $\Gamma$ on a metric space $M$. Let $\Omega$ be a closed fundamental domain of $\Gamma$ on $M$. The idea is that elements $\gamma$ of $\Gamma$ which do not move $\Omega$ completely away, i.e., $\gamma\Omega \cap \Omega \neq \emptyset$, generate $\Gamma$. Therefore, if a fundamental domain $\Omega$ can be found such that

$$\{\gamma \in \Gamma \mid \gamma\Omega \cap \Omega \neq \emptyset\}$$

is a finite set, then $\Gamma$ is finitely generated.

This finiteness result is not easy to obtain. In fact, for arithmetic groups, this is one of the deepest parts of reduction theory and called the *Siegel finiteness property* (see §4.13). Finite presentation of $\Gamma$ can be similarly obtained. These conclusions hold for general actions of a discrete subgroup $\Gamma$ on a metric space $M$. See [PlR1, §4.4].

A corollary of the above discussions of fundamental domains of arithmetic subgroups $\Gamma$ is the following.

COROLLARY 3.3.2. *Every subgroup $\Gamma$ of $SL(2,\mathbb{Z})$ of finite index is finitely presented, in particular finitely generated.*

An interesting result to note is the following ([Si2] [Bea, Theorem 10.1.2] [Scot]).

PROPOSITION 3.3.3. *If $\Gamma$ is a discrete subgroup of $SL(2,\mathbb{R})$ and the area of the hyperbolic surface $\Gamma\backslash\mathbb{H}$ is finite, then its Dirichlet fundamental domain has finitely many sides and hence $\Gamma$ is finitely generated and also finitely presented.*

For relations and generators of Hilbert modular groups, see [KobN] and [KirW]. See also [FaP2] for generators and relations of the Eisenstein-Picard modular group.

**3.3.12. Arithmetic Fuchsian groups.** A discrete subgroup of $SL(2,\mathbb{R})$ (or rather $PSL(2,\mathbb{R})$, which is the identity component of the isometry group of $\mathbb{H}$) is called a Fuchsian group. We are mainly interested in cofinite Fuchsian groups $\Gamma$, i.e., those $\Gamma$ such that $\Gamma\backslash\mathbb{H}$ has finite area, which are Fuchsian groups of the first kind. If $\Gamma$ has a fundamental domain with finitely many geodesic sides of infinite area, then it is a Fuchsian group of second kind. Among Fuchsian groups of second kind, a particularly important class consists of convex-cocompact Fuchsian groups, which do not contain any nontrivial parabolic elements. To be precise, a discrete group $\Gamma$ acting on $\mathbb{H}$ is called a Fuchsian group of the first kind if its limit set is equal to the whole boundary $\partial\mathbb{H}$, and called a Fuchsian group of the second kind if its limit set is a proper subset of $\partial\mathbb{H}$. See [Bea, Chap. 8] for more details.

There are two classes of arithmetic Fuchsian subgroups of $SL(2,\mathbb{R})$. Subgroups of finite index of $SL(2,\mathbb{Z})$ are cofinite but not-cocompact, as explained above. More generally, any subgroup of $SL(2,\mathbb{Q})$ commensurable with $SL(2,\mathbb{Z})$ is called an arithmetic subgroup. Clearly there are some arithmetic subgroups of $SL(2,\mathbb{Q})$ which are not contained in $SL(2,\mathbb{Z})$, for example, a conjugate of $SL(2,\mathbb{Z})$ by a non-integral element of $SL(2,\mathbb{Q})$. There is an abstract definition of arithmetic subgroups of Lie groups (see §4.2 and §4.5 below). It can be shown that every non-cocompact arithmetic subgroup of $SL(2,\mathbb{R})$ is commensurable with $SL(2,\mathbb{Z})$.

Though it is less obvious, cocompact arithmetic discrete subgroups of $SL(2,\mathbb{R})$ do exist and can be constructed by quaternion algebras over $\mathbb{Q}$. See [Katok] [MaclR] [Vig] and references there. The basic point is to show that there are no parabolic elements in the constructed discrete groups. (This is a special case of the general compactness criterion in §4.11. The basic reason here is that a cusp of $\Gamma\backslash\mathbb{H}$ corresponds to a point (or rather an $\Gamma$-orbit) on the boundary $\partial\mathbb{H}$ and the parabolic elements fixing the boundary point produce the cusp neighborhood of $\Gamma\backslash\mathbb{H}$). These cocompact arithmetic subgroups of $SL(2,\mathbb{R})$ are important in the theory of Shimura curves. See [Shi1].

**3.3.13. Characterization of arithmetic Fuchsian groups.** The theory of uniformization of Riemann surfaces, or deformation of them, shows that there are many non-arithmetic Fuchsian groups in $SL(2,\mathbb{R})$. In fact, there are only countably infinitely many arithmetic Fuchsian subgroups of $SL(2, \mathbb{R})$, but the dimension of the moduli space of Riemann surfaces of negative Euler characteristic is positive and hence there are uncountably infinitely many non-arithmetic Fuchsian groups.

A natural problem is to characterize arithmetic Fuchsian groups. The classical paper [Ta] provides an *algebraic* characterization of arithmetic Fuchsian groups.

It was shown in [LuSa2] that the trace set (or, equivalently, the length set of closed geodesics) of a cofinite arithmetic Fuchsian group satisfies the bounded clustering property, i.e., there is a uniform upper bound on the number of lengths counted with multiplicity in every unit interval $[n, n+1]$, $n \geq 0$. In [Sar5], it was conjectured that the opposite is also true, namely, that if the trace set of a cofinite Fuchsian group satisfies the bounded clustering property, then it is arithmetic.

Schmutz [Schmu2] makes an even stronger conjecture: A cofinite Fuchsian group is arithmetic if and only if its trace set has linear growth. Unfortunately the "proof" of his conjecture, which he proposes in [Schmu2], contains a gap.

Based on ideas of Schmutz, Sarnak's conjecture is proved in [GenL] (under the additional assumption that the Fuchsian group contains parabolic elements.) In contrast to Takeuchi's result this yields a *geometric* characterization of arithmetic Fuchsian groups.

There is another characterization of arithmetic subgroups in terms of a condition on the factorizability of a Dirichlet series attached to a cusp form [JiaPS].

For a characterization of arithmetic subgroups of general semisimple Lie groups, see §5.5 below.

### 3.4. Spectral theory of $\Gamma\backslash\mathbb{H}$

If $\Gamma$ is a cocompact subgroup of $SL(2, \mathbb{R})$, then the Laplace operator $\Delta$ of $\Gamma\backslash\mathbb{H}$ acting on the Hilbert space $L^2(\Gamma\backslash\mathbb{H})$ has only the discrete spectrum, i.e., the spectrum consisting only of eigenvalues with finite multiplicity, and the counting function of the eigenvalues satisfy the Weyl law given in the next subsection. (See also the discussions of the Weyl law in §3.2.)

On the other hand, when $\Gamma\backslash\mathbb{H}$ is noncompact, the spectrum of the Laplace operator $\Delta$ contains both a continuous spectrum $[\frac{1}{4}, +\infty]$, which is in fact absolutely continuous and the generalized eigenfunctions can also be described in terms of Eisenstein series, and a discrete spectrum, which consists of the eigenvalues of square integrable eigenfunctions and can be embedded into the continuous spectrum. Understanding the location of the discrete spectrum and the spectral measure of the continuous spectrum are important but difficult. One reason is that the embedded eigenvalues are rather unstable and there is no simple way to detect them.

Several general references here are [Ku] [Bore9] [Sele1-2] [Hej1-2] [Ven] [Sar3] [Iw1] [Te1]. A closely related book for the three dimensional hyperbolic space is [ElsGM].

**3.4.1. Weyl law.** Assume that $\Gamma\backslash\mathbb{H}$ is compact. Let $\lambda_1, \dots, \lambda_n, \dots$ be the eigenvalues of the Laplace operator $\Delta$ of $\Gamma\backslash\mathbb{H}$. Then the counting function of the eigenvalues defined by

$$N(\lambda) = |\{\lambda_i \mid \lambda_i \leq \lambda\}|$$

satisfies the asymptotic formula: as $\lambda \to +\infty$,

$$N(\lambda) \sim 2\pi \operatorname{area}(\Gamma\backslash\mathbb{H})\lambda.$$

This is a special case of the general Weyl law for the counting function $N(\lambda)$ of eigenvalues of a compact Riemannian manifold $M^n$ of dimension $n \geq 2$: as $\lambda \to +\infty$,

$$N(\lambda) \sim \frac{\omega_n}{(2\pi)^n} \operatorname{vol}(M)\, \lambda^{\frac{n}{2}},$$

where $\omega_n$ is the volume of the unit sphere in $\mathbb{R}^n$. This Weyl law is related to the famous question *"Can one hear the shape of a drum?"* raised by Kac in [Kac]. This question is a basic or rather motivating question in spectral theory of compact Riemannian manifolds, i.e., how much geometry of a manifold is determined by the eigenvalues of the Riemannian manifolds. In fact, the Weyl law implies that one can hear the area of a drum. This result is part of the vast subject of spectral geometry, which studies various aspects of relations between geometry and eigenvalues of Riemannian manifolds. For locally symmetric spaces, much more is known. See the part of [Berg1] on spectral geometry and [Gord] for a survey on isospectral manifolds and related references. See [MiaR] [GordR] [Ze4] and their references for geometric data which can be determined from the isospectral property on differential forms.

The Weyl law above gives the leading term of the asymptotic expansion of $N(\lambda)$. A natural question concerns sharp bounds on the remainder. The Selberg trace formula is useful for this purpose (see [Hej1] and also [Sele1-2]). For general compact Riemannian manifolds, good bounds on the error terms are obtained in [DuG] [Ber] under certain conditions on the set of closed geodesics or the negative sectional curvature, and the book [Ivr] gives a systematical study of this question. On the other hand, as mentioned earlier in §3.2, this question on the remainder for the counting function of the eigenvalues of the flat torus $\mathbb{Z}^n\backslash\mathbb{R}^n$ is not completely understood yet.

**3.4.2. Spectral decomposition.** As mentioned above, when $\Gamma\backslash\mathbb{H}$ is noncompact, the spectrum of the Laplace operator $\Delta$ consists of two parts: the continuous part $[\frac{1}{4}, +\infty)$ and a discrete part, which consists of square integrable eigenvalues. The generalized eigenfunctions of the continuous spectrum are given by the Eisenstein series. It follows from this that the continuous spectrum is absolutely continuous. By generalized eigenfunctions, we mean functions that satisfy the eigen-equations but barely miss being square integrable in the sense each is not square integrable, but their superpositions by square integrable functions in the spectral variable are square integrable. These generalized eigenfunctions occur naturally in the spectral projection operators corresponding to the continuous spectrum, and hence are essential part of the spectral decomposition.

For each cusp point (or rather neighborhood) of $\Gamma\backslash\mathbb{H}$, there is an Eisenstein series depending on a complex parameter $s$. For the cusp point corresponding to $i\infty$ in the upper-half plane model, when $\mathrm{Re}(s) > 1$, the Eisenstein series is basically the sum of the translates of $y^s$ under the action of $\Gamma$ so that it becomes $\Gamma$-invariant. Since $y^s$ satisfies the equation $(\Delta - s(1-s))y^s = 0$ (recall that $\Delta = -y^2(\frac{\partial}{\partial x^2} + \frac{\partial}{\partial y^2})$), and the $\Gamma$-action commutes with $\Delta$, it follows that the Eisenstein series also satisfies the equation $(\Delta - s(1-s))f = 0$. In studying these Eisenstein series, the Riemann zeta function appears naturally. In fact, the zeta function appears in the constant term of the Eisenstein series, and the meromorphic continuation of the Eisenstein series to the whole complex plane of $s$ is related to the meromorphic continuation of the zeta function.

See [Iw1] for some discussions of the spectral decomposition and applications to analytic number theory, [Te1] for many applications to various fields. See also [Ven]. [Sar3] gives an updated summary.

**3.4.3. Discrete spectrum and Selberg's $\frac{1}{4}$-conjecture.** The existence of square integrable eigenfunctions is a difficult but fundamental problem. Since the area of $\Gamma\backslash\mathbb{H}$ is finite, it is certainly known that constant functions are square integrable eigenfunctions with the eigenvalue equal to 0. Therefore, the discrete part is nonempty and contains at least 0. Beyond this, the existence of other eigenvalues is not obvious but rather difficult. One reason for the difficulty is that almost all eigenvalues are embedded in the continuous spectrum $[\frac{1}{4}, +\infty)$, and it is not easy to detect them (under slight perturbations, these eigenvalues can be absorbed by the continuous spectrum). It is also not easy to construct them. For example, the naive method of constructing an approximating eigenfunction first and then perturb it into a genuine engenfunction will not work. (This naive method often works well for compact Riemannian manifolds. For example, it can be used to show easily that the eigenvalues of a smooth family of compact Riemannian manifolds depend continuously on the parameter. Note also that due to branching of eigenvalues, we do not have smooth dependence of each individual eigenvalue.)

Explicit computations of these eigenvalues even for $SL(2,\mathbb{Z})$ are complicated. One method to compute them is to use Hecke relations satisfied by the eigenfunctions to cut down the system of linear equations. See [Hej2-3] and references there.

A famous conjecture on small eigenvalues is the Selberg $\frac{1}{4}$-Conjecture, which says that the first positive eigenvalue of the hyperbolic surface $\Gamma(N)\backslash\mathbb{H}$ for the congruence subgroups $\Gamma(N)$ is at least $\frac{1}{4}$, i.e., it must be embedded in the continuous spectrum $[\frac{1}{4}, +\infty)$. This has many important applications and is one of the famous outstanding problems in analytic number theory.

In fact, it is a special case of the Ramanujan conjecture in §12.13. See [Sar4] for a survey and [LuRS] some important progress.

For some geometric approach to this problem, see [BroM] and the references there. The basic point is that arithmetic surfaces are full of symmetry, and it is difficult to support small positive eigenvalues. (If a surface is modeled after a big circle, then it will have small positive eigenvalues. Abundant symmetries prevent this from happening.) This might give a geometric way to characterize congruence subgroups. As mentioned before in §3.3.9, there are arithmetic subgroups of $SL(2,\mathbb{Z})$ which are not congruence subgroups.

**3.4.4. Selberg trace formula.** In order to show the existence of infinitely many eigenvalues for subgroups of finite index of $SL(2,\mathbb{Z})$, Selberg developed the trace formula for general Fuchsian groups $\Gamma$ of the first kind. This trace formula is a non-abelian generalization of the Poisson summation formula.

When $\Gamma\backslash\mathbb{H}$ is compact, it is obtained by computing the trace of certain integral operator on the diagonal in two different ways: directly in terms of the original function and in terms of the spectral decomposition of the function. This gives an equality between the spectral side involving eigenvalues and the geometric side involving lengths of closed geodesics of the hyperbolic surface $\Gamma\backslash\mathbb{H}$. (This gives one easy way to prove the Poisson summation formula for $\mathbb{Z}^n \subset \mathbb{R}^n$.) Hence, the Selberg trace formula can be used to study both the spectrum of the surface and also the distribution of the lengths of closed geodesics. For more general locally symmetric spaces, the geometric side involves orbital integrals parametrized by conjugacy classes of elements in $\Gamma$, instead of lengths of closed geodesics, which mainly happen in the constant sectional curvature case.

When the space $\Gamma\backslash\mathbb{H}$ is non-compact, the above trace of integral operators is not finite, i.e., the integral over a noncompact diagonal submanifold will not converge. Instead, we need to truncate the integral kernel suitably in order to get finite values, a sort of normalization, and two ways of computing this integral leads to an equality, which depends on the truncation parameter. It turns out that one can derive an equality independent of the truncation parameter. The spectral side of the trace formula also involves other terms coming from the continuous spectrum. For general arithmetic subgroups, this is the Arthur-Selberg trace formula. The original paper is [Sele1]. See [Art1] [Art3] [Bore9] [Ku] [Hej1-2] [Gelb2] [HejST] for other expositions.

A closely related formula is the Eichler-Selberg trace formula which computes the trace of Hecke operators on spaces of modular forms. For a self-contained discussion of this formula, see [KnL1].

For other related trace formulas, for example, the relative trace formula, the Kuznetsov formula, Petterson formula, see also [Iw1] [Goo] [Brug1] [BruM] [Jac] [LiuY] [KnL2] [MiaW2-3] and the references there. A wave equation generalization of the Kuznetsov formula to $\mathbb{Q}$-rank 1 locally symmetric spaces is given in [JiW2].

**3.4.5. Generalized Weyl law.** One corollary of the Selberg trace formula is a generalization of the Weyl law for eigenvalues when $\Gamma$ is noncocompact. Besides the counting function of the discrete spectrum, denoted by $N_d(\lambda)$, there is also a counting function of the continuous spectrum, denoted by $N_c(\lambda)$. The generalized Weyl law states that as $\lambda \to +\infty$,

$$N_d(\lambda) + N_c(\lambda) \sim 2\pi \text{ area}(\Gamma\backslash\mathbb{H})\lambda.$$

See [Sele1-2] [Hej2] [Ven]. (Note that the spectral counting function $N(\lambda)$ of a compact Riemannian manifold is given by integral over $[0, R]$ of the spectral measure. Similarly, the counting function of the continuous spectrum is also defined in terms of integrals of the spectral measure of the continuous spectrum, which involves the generalized eigenfunctions.)

If the counting function $N_c(\lambda)$ for the continuous spectrum is of smaller order than $\lambda$, then the Weyl law for the discrete spectrum holds. In particular, there are infinitely many eigenvalues.

When $\Gamma$ is a congruence subgroup, this is the case. In fact, it follows from some explicit computations of the constant terms of the Eisenstein series using the Riemann zeta function. See [Hej2].

Counting of the continuous spectrum is closely related to counting resonances, which are poles of the resolvent kernel and can be thought of as a discrete description of the continuous spectrum. Understanding resonances of noncompact Riemannian manifolds is important in mathematical physics and spectral geometry. See [GuiZ1-2] [Mul5] and the references there.

**3.4.6. Phillips-Sarnak conjecture on cusp Mass forms and spectral degeneration.** The spectral decomposition of $\Gamma\backslash\mathbb{H}$ is similarly described for every cofinite discrete subgroup of $SL(2,\mathbb{R})$, regardless whether $\Gamma$ is arithmetic or not. It is expected that for a general, non-arithmetic cofinite discrete subgroup $\Gamma$, there are only finitely many square integrable eigenfunctions. The reason is that except for finitely many of them, discrete eigenvalues are embedded in the continuous spectrum and are hence unstable, i.e., they will disappear under perturbations in the moduli spaces of complex structures on the surface. This is the content of the *Phillips and Sarnak conjecture*. See [PhS1-3] [Sar3].

It is well-known that Riemann surfaces of genus at least 2 can deform and degenerate into singular ones. Since compact hyperbolic surfaces have only discrete spectrum, i.e., discrete eigenvalues with finite multiplicity and going to infinity, but the limiting noncompact hyperbolic surfaces have the continuous spectrum $[\frac{1}{4}, +\infty)$, a natural question is to understand relations between the eigenvalues and the continuous spectrum during the degeneration process: for example, how the eigenvalues of compact hyperbolic surfaces become dense in the continuous spectrum of the limiting noncompact surfaces and how fast they become dense; and how eigenfunctions of compact surfaces converge to generalized eigenfunctions of the noncompact hyperbolic surfaces. Such a study on the spectral behavior of a degenerating family, called spectral degeneration, is related to the above conjecture of Phillips and Sarnak on eigenvalues. See [Wolp1]. For sharp bounds on the rate of spectral accumulation, convergence of eigenfunctions and related questions, see [JiZ3] [Wolp2] and the references there.

**3.4.7. Counting lengths of geodesics–the generalized prime number theorem.** The Weyl law describes the large scale distribution of eigenvalues of manifolds. The lengths of closed geodesics of a hyperbolic surface $\Gamma\backslash\mathbb{H}$ form an increasing sequence of positive numbers with finite multiplicity and are natural geometric invariants of the surface. It can be shown that this sequence is infinite and has no finite accumulation point, and hence its counting function is finite and goes to infinity as the counting parameter goes to infinity. A natural question is to understand its asymptotic behavior.

It turns out that instead of polynomial growth as in the Weyl law for eigenvalues, it grows exponentially. Specifically, let $L(x)$ be the number of closed geodesics in $\Gamma\backslash\mathbb{H}$ whose lengths are less than or equal to $x$. Then as $x \to +\infty$,

$$L(x) \sim const. \frac{e^{\frac{1}{4}x}}{x}.$$

This a natural generalization of the prime number theorem. In fact, it is similar to the counting of the logarithms of prime numbers. Therefore, such an asymptotic formula is called the generalized prime number theorem.

Since the Selberg trace formula relates the geometric and spectral data of $\Gamma\backslash\mathbb{H}$, it can be used to count the lengths of closed geodesics. In the prime number theorem, a refined version gives distribution of primes in arithmetic progressions. Similar results hold for counting of lengths of closed geodesics. See [PhS4] [Sun] [KatsS] [SarW] [Sharp2] [McSR] [McS].

Another naturally related problem is to distributions of simple closed geodesics as their lengths go to infinity. For compact Riemannian manifolds of constant negative sectional curvature, [Bowe] showed that they are uniformly distributed. A stronger, optimal version was proved in [Ze2] for compact hyperbolic surfaces.

Closed geodesics of $\Gamma\backslash\mathbb{H}$ are periodic orbits of $\Gamma\backslash\mathbb{H}$. Another flow for surfaces $\Gamma\backslash\mathbb{H}$ with cusps is the horocycle flow. See [Kai3] and the survey article [Stark].

The problem of counting lengths of closed geodesics makes sense also for closed manifolds of negative sectional curvature. In this case, geodesics are isolated, and their lengths form an increasing sequence with finite multiplicity and going to infinity. See [Marg4] for the first such general result. (The complete version of [Marg4] is now available in [Marg5]. Other related questions about distribution of lattices points and closed geodesics are also discussed in [Marg5].) Closed geodesics are periodic orbits of the geodesic flow, and the geodesic flow of negatively curved compact manifold is a particular case of Anosov flows, and a natural problem is to count lengths of periodic orbits of such flows. There is a similar generalized prime number theorem. See [ParP1-2] and the survey article by Sharp in [Marg5] and the references there for more details.

For closed manifolds of nonpositive sectional curvature, for example locally symmetric spaces which are quotients of symmetric spaces of higher rank, each homotopy class of simple closed curves may contain more than one simple closed geodesics. On the other hand, they are all of the same length, and the length counting function is still well-defined. A natural question is to understand their asymptotic behavior.

There is also a zeta function defined in terms of lengths of closed geodesics in compact manifolds of strictly negative sectional curvature. See [PolS3-4].

See also [ConzG1] [LedP1-2] [Led1] [PolS1-2] [BirS] for related results on distributions of orbits of linear groups. For distribution of periods of automorphic forms on closed geodesics, see [Sharp1]

As mentioned below in §3.4.10, if $\Gamma\backslash\mathbb{H}$ is noncompact, there is a special class of unbounded geodesics, called scattering geodesics, which have finite normalized lengths, called sojourn times. It was shown in [JiZ1] that they form an increasing sequence without finite accumulation points. A natural question is to find asymptotics of the counting function of sojourn times. See [Goo, §11] for results on this problem. As in the case of closed geodesics, another natural question is to understand how these scattering geodesics distribute when their sojourn times go to infinity.

Asymptotic behaviors of the counting function of the lengths of geodesics study their large scale behaviors. Another important aspect is to study individual lengths, in particular, the short ones. A systole on a Riemannian surface is a simple closed geodesic of shortest length. For hyperbolic surfaces $\Gamma\backslash\mathbb{H}$, bounds on the systoles are important. For example, if it is small and separates the surface, then the first positive eigenvalue is small. See [Schmu1-5] for discussions of many applications of systoles. For relations between geometry of numbers, in particular the Hermite invariants, and systoles, see [Bav1-3].

As mentioned earlier, a basic question in spectral geometry is to understand how much geometry of a Riemannian manifold is determined by its eigenvalues. For hyperbolic surfaces, see [Wolp5] [Mck]. The lengths of closed geodesics are important invariants of Riemannian manifolds. It is a natural problem to understand how much of geometry of a Riemannian manifold is determined by these lengths. For rigidity properties of hyperbolic surfaces in terms of length spectra, see [Wolp4] [Buse] and also [Kim1-3, 5] for three dimensional hyperbolic manifolds. For related results on rigidity of closed surfaces of strictly negative curvature with respect to the marked lengths of closed geodesics, see [Ot] and [Cro1] [Cro2].

**3.4.8. Modular symbols.** Assume that $\Gamma$ is an arithmetic subgroup of $SL(2,\mathbb{Z})$. Geodesics that connect cusps of $\Gamma\backslash\mathbb{H}$ are related to modular symbols. In fact, they give rise to rational homology cycles of degree 1 by integrating 1-forms on them. Let $\overline{\Gamma\backslash\mathbb{H}}$ be the compactification by filling in the cusp points, and $\partial\overline{\Gamma\backslash\mathbb{H}}$ the set of ideal points. Then these modular symbols generate the relative homology group $H_1(\overline{\Gamma\backslash\mathbb{H}}, \partial\overline{\Gamma\backslash\mathbb{H}}; \mathbb{Q})$. This representation of cycles by pairs of rational boundary points of $\mathbb{Q}\cup\{\infty\}\subset\partial\mathbb{H}$ is useful for various purposes. One reason is that they allow us to compute actions of Hecke operators on the cohomology groups of modular curves, which are an essential ingredient of arithmetic geometric properties of modular curves. The intersection of homology classes in $H_1(\Gamma\backslash\mathbb{H})$ with modular symbols $H_1(\overline{\Gamma\backslash\mathbb{H}}, \partial\overline{\Gamma\backslash\mathbb{H}}; \mathbb{Q})$ can also be described explicitly. See [Man] [Mer] and also [Vla].

Geodesics that connect cusps of $\Gamma\backslash\mathbb{H}$ are lifted to geodesics in $\mathbb{H}$ which have rational boundary points in the compactification $\mathbb{H}\cup\mathbb{H}(\infty)$. Instead of considering only these geodesics with end points in $\mathbb{Q}\cup\{\infty\}\subset\partial\mathbb{H}=\mathbb{H}(\infty)$, we can consider other geodesics. They lead to limiting modular symbols in [ManM], which are real homology classes, and patterns of the continued fractions of the boundary points determine vanishing properties of these cycles. They are related to the idea of blowing up each cusp point at infinity of $\Gamma\backslash\mathbb{H}$ by a non-commutative (or non-Hausdorff) boundary $SL(2,\mathbb{Z})\backslash\partial\mathbb{H}$ consisting of non-commutative tori and hence relating classical modular forms to $C^*$-algebras or rather non-commutative geometry. (Note that $SL(2,\mathbb{Z})$ does not act properly on $\partial\mathbb{H}$ and hence the quotient $SL(2,\mathbb{Z})\backslash\partial\mathbb{H}$ is non-Hausdorff. On the other hand, for many rigidity problems about discrete groups, ergodic actions of discrete subgroups on homogeneous spaces have played a fundamental role, and non-Hausdorff quotients are desirable for such applications.) See [ManM] [Marc1-2] and references there for details.

For modular symbols in other higher dimensional locally symmetric spaces, see [Gun1] for a survey and also [AsB] [Gun2-3] [As5] [SpV]. They are cycles arising from natural reductive subgroups, i.e., the Levi components of $\mathbb{Q}$-parabolic subgroups and related also to the cohomology of the Borel-Serre boundary. As above, a main point is that the modular symbol algorithm computes eigenvalues of Hecke operators on cohomology of arithmetic groups.

An important problem concerning modular symbols is to determine if they vanish or not. One method is to use unitary representations. More specifically, a vanishing theorem in [KobD] is proved by using the restriction of unitary representations in [Kob11], and a non-vanishing theorem [ToW] is proved by using the analysis on semisimple symmetric spaces from [Fle]. See [Kob12, Section 4] for an exposition of these two methods.

**3.4.9. Selberg zeta function.** For a number field such as $\mathbb{Q}$, properties of its primes, for example their distribution, are encoded in its zeta function $\zeta(s)$ through the Euler product (or factorization):

$$\zeta(s) = \prod_p (1 - p^{-s})^{-1},$$

where $p$ runs over all prime numbers.

For a hyperbolic surface $\Gamma\backslash\mathbb{H}$, the distribution of the lengths of closed geodesics is encoded in its the Selberg zeta function $\zeta_\Gamma(s)$. In fact, recall that a geodesic $\gamma$ in a Riemannian manifold $M$ is a parametrized curve $\gamma : \mathbb{R} \to M$ such that the covariant derivative of $\gamma'(t)$ along $\gamma(t)$ is zero, i.e., the vector field $\gamma'(t)$ is parallel on $\gamma(t)$. Assume that $\gamma(t)$ is parametrized by arclength. It is closed if there exists some $\ell > 0$ such $\gamma(t + \ell) = \gamma(t)$ for all $t \in \mathbb{R}$. Therefore, a closed geodesic is represented by a map $\mathbb{R}/\ell\mathbb{Z} = [0, \ell]/\{0 \sim \ell\} \to M$. Then $\gamma$ is called a *prime geodesic* if $\ell$ is equal to the length of the image of $\gamma(t)$, or equivalently, $\ell$ is equal to smallest number such that $\gamma(t + \ell) = \gamma(t)$ holds for all $t \in \mathbb{R}$, which says that when the parameter $t$ takes values in $[0, \ell]$, $\gamma(t)$ traces out the whole curve $\gamma$ only once. (If $t$ takes values in $[0, n\ell]$, $n \in \mathbb{N}$, then it will trace out the curve $n$-times.)

Denote the length of a geodesic $\gamma$ by $\ell(\gamma)$. Then the Selberg zeta function of $\Gamma\backslash\mathbb{H}$ is defined by

$$\zeta_\Gamma(s) = \prod_\gamma \prod_{k=0}^{\infty} (1 - e^{(s+k)\ell(\gamma)}),$$

where $\gamma$ runs over all prime closed geodesics of $\Gamma\backslash\mathbb{H}$.

Besides the lengths of closed geodesics, the function $\zeta_\Gamma(s)$ also contains information about the eigenvalues and the continuous spectrum of $\Gamma\backslash\mathbb{H}$. For example, if $\lambda = s_0(1 - s_0)$ is the eigenvalue of a Maass form on $\Gamma\backslash\mathbb{H}$, then $\zeta_\Gamma(s)$ has a zero at $s_0$ with multiplicity equal to the dimension of the space of Maass forms. The poles of $\zeta_\Gamma(s)$ are related to resonances of $\Gamma\backslash\mathbb{H}$. In fact, the study of the Selberg zeta function is closely related to the Selberg trace formula, as originally defined by Selberg. See [Sele1] [Hej1-2] and references there. Later, there are other dynamical approaches to the Selberg zeta function. See [Ju] [BuO2-5] [Gan1-2] [GanW] [ParP1-2] and the references there. See also [Pat4] [PatP] [Pe1-4] for results and questions on related Poisson formula, distribution of lengths and bounds on resonances.

The Selberg zeta function for higher rank locally symmetric spaces has not been understood as well as the rank one case. See [MoS] [Dei3-6] [DeiP] [Pav] and references there. For Selberg zeta functions of graphs and arithmetic groups acting on the Bruhat-Tits trees, see [Nag] [Iha] [StarT].

**3.4.10. Scattering geodesics and generalized Poisson relation.** As mentioned earlier in §3.4.8, geodesics connecting cusps of a noncompact

$\Gamma\backslash\mathbb{H}$ give rise to modular symbols. These infinite geodesics also have normalized lengths, the so-called sojourn times, which measures the times they spend around the compact core of the space $\Gamma\backslash\mathbb{H}$. They form an increasing sequence with finite multiplicity going to infinity.

The Selberg trace formula relates the spectral and geometric data of $\Gamma\backslash\mathbb{H}$. For a general compact Riemannian manifold, there is a Poisson relation relating the singularities of the Fourier transformation of the spectral measure to the length spectrum of the manifold in [DuG] (see also [PetS].)

A natural generalization of such a Poisson relation for noncompact locally symmetric spaces of $\mathbb{Q}$-rank 1, in particular for $\Gamma\backslash\mathbb{H}$, is given [JiZ1] [JiZ2]. In this formula, the sojourn times of scattering geodesics appear and are related to the singularities of the Fourier transformation of the spectral measure of the continuous spectrum. For a general locally symmetric space $\Gamma\backslash X$, unlike the case of $\Gamma\backslash\mathbb{H}$, scattering geodesics are not isolated, instead they form smooth families parametrized by locally symmetric spaces of smaller dimensions and their dimensions are reflected in the structure of the singularities.

The Poisson relation in [JiZ1] can be considered as a wave equation generalization of the Kuznetsov formula in automorphic forms [Goo] [Iw1] [Brug1] [BruM].

For general locally symmetric spaces $\Gamma\backslash X$ of higher rank, we can also define scattering flats, which are some particular immersed flat geodesic submanifolds with some shifted Weyl chambers embedded (similar to eventually distance minimizing property of scattering geodesics in both directions), and sojourn times, which are vector-valued. The generalized Poisson relation is expected to hold too, but the methods in [JiZ1] do not work directly (products of $\mathbb{Q}$-rank 1 locally symmetric spaces give special higher rank locally symmetric spaces and are suggestive for the results desired).

**3.4.11. Modular forms and Maass forms.** The classical *modular forms* with respect to $\Gamma$ are holomorphic functions on $\mathbb{H}$ that satisfy certain transformation rules under $\Gamma$ and suitable growth at infinity. They can be interpreted as holomorphic sections of some holomorphic line bundles on $\Gamma\backslash\mathbb{H}$.

The square integrable eigenfunctions of $\Gamma\backslash\mathbb{H}$ with respect to the Laplace operator are the so-called *Maass forms.* For the Mass forms $u$, the holomorphy condition for modular forms is replaced by satisfying an eigen-equation $\Delta u = \lambda u$, where $\Delta$ is the Laplace operator of $\mathbb{H}$ with respect to the hyperbolic metric.

One important aspect of modular forms for $SL(2,\mathbb{Z})$ and other commensurable arithmetic subgroups $\Gamma$ is their Fourier coefficients at the cusp $\{i\infty\}$. If all negative and zeroth Fourier coefficients vanish, they are called cuspidal modular forms. This implies that the modular forms decay rapidly near the infinity of the cusps of $\Gamma\backslash\mathbb{H}$. Similarly, Maass forms also admit Fourier expansions, where special functions such as K-Bessel functions appear, and cuspidal Maass forms can be defined similarly and characterized in terms of rapid decays near the infinity of the cusps. The constant functions are Maass forms but not cuspidal. An important problem is to understand cuspidal Maass forms, which are related to embedded eigenvalues of $\Gamma\backslash\mathbb{H}$. We can also defined bundle-valued Maass forms.

There are many books on modular and automorphic forms covering basics and many applications to other fields, for example [Iw2] [Kr] [Leh] [Maa] [Star] [Hej2] [Ven]. For a classical, historical account of automorphic forms initiated by Poincaré, see the book [Had]. For a historic book together with some discussions over function fields, see [Vla] (the most part of [Vla] studies modular forms over number fields and the theory of complex multiplication, but [Vla, Part III and Chap. 3] gives a concise introduction to elliptic modules, modular forms and modular varieties over function fields). For a through but technical treatment of spectral theory of automorphic forms over both number fields and functions fields for general arithmetic subgroups of semisimple linear algebraic groups, see [MoeW] and [Lang3].

**3.4.12. Automorphic representations.** Automorphic forms generate representations of $SL(2,\mathbb{R})$ on certain function spaces on $\Gamma\backslash SL(2,\mathbb{R})$, the so-called automorphic representations. For example, suppose $\varphi$ is a square integrable Mass form on $\Gamma\backslash\mathbb{H}$, and consider it as a right $SO(2)$-invariant function on $\Gamma\backslash SL(2,\mathbb{R})$. Then the right translates under elements of $SL(2,\mathbb{R})$ span a $SL(2,\mathbb{R})$-invariant subspace of $L^2(\Gamma\backslash SL(2,\mathbb{R}))$, equivalently a subrepresentation of the regular representation of $SL(2,\mathbb{R})$ on $L^2(\Gamma\backslash SL(2,\mathbb{R}))$. Similarly, holomorphic cuspidal modular forms also generate subrepresentations of $L^2(\Gamma\backslash SL(2,\mathbb{R}))$.

There are many reasons for considering this passage from functions to subrepresentations. Decomposing the regular representation of $SL(2,\mathbb{R})$ in $L^2(\Gamma\backslash SL(2,\mathbb{R}))$ (and regular representations of other Lie groups) into irreducible subrepresentations and understanding these constituents is a fundamental problem in representation theory and modern number theory. Since the Laplace operator commutes with the action of the Lie group $SL(2,\mathbb{R})$, the spectral decomposition of $\Gamma\backslash\mathbb{H}$ commutes with the regular representation, and each eigenspace is a subrepresentation of the regular representation. Therefore, the spectral decomposition is an important approximation to the above problem. For a general locally symmetric space $\Gamma\backslash X$, if the rank of $X$ is bigger than 1, there are other invariant differential operators besides the Laplace operator, and joint eigenfunctions of invariant differential operators are basically the general automorphic forms in §12.1.

There are many books discussing modular forms and automorphic representations and relations between them. See [Gelb1-3] [Bore9] [GelGP] [JacL] [Ku] [BernG] [Ts] and the references there for example.

**3.4.13. $L$-functions.** For each modular form and Mass form $\varphi$ with respect to an arithmetic subgroup $\Gamma \subset SL(2,\mathbb{Z})$, there is an associated $L$-function $L(\varphi, s)$. Denote the (suitably normalized) Fourier coefficients by $a_n$, then the $L$-function is defined by

$$L(s) = \sum_{n=1}^{\infty} \frac{a_n}{n^s}, \quad \mathrm{Re}(s) \gg 0.$$

If all $a_n = 1$, we get the Riemann zeta function. If $a_n$ is given by the Dirichlet character, we get the Dirichlet series (see [Kobl]). If the sequence of $a_n$ represents some other natural numbers, for example, related to the number of points of a variety reduced at primes, or the eigenvalues of the Frobenius operator of some Galois representations, or the eigenvalues of Hecke operators on cohomology groups of arithmetic subgroups, we get some other important $L$-functions.

The first basic problem is meromorphic continuation of $L(s)$ to the whole complex $s$-plane. It turns out that $L(s)$ is equal to the Mellin transform of the modular form, and the meromorphic continuation follows from the Poisson formula and the automorphy condition of the modular forms. It also follows that this $L$-function $L(s)$ satisfies a functional equation. See [CassF] [RamV] [KnL1] [Kobl].

If $\varphi$ is also an eigenfunction of all Hecke operators, then $L(\varphi, s)$ has an Euler product formula where the factors are over the prime numbers. As it is well-known, the Euler product for the Riemann zeta function is important in understanding the distribution of prime numbers.

See [Cog1] for a quick introduction to these and related topics. See also [Kobl] [Gelb1] [LiWi1].

**3.4.14. Automorphic forms on adele groups.** The natural language to explain the Euler product of $L$-functions of automorphic representations is to formulate the automorphic representations in terms of adeles, or rather as representations of adele groups such as $GL(2,\mathbb{A})$. As recalled earlier in §3.2.3, the ring of adeles $\mathbb{A}$ is the restricted product $\mathbb{R} \times \prod_p' \mathbb{Q}_p$ with respect to $\mathbb{Z}_p$, where the restricted product means that for every adele $a = (a_\infty, a_2, a_3, \dots, a_p, \dots)$, all but finitely many of $a_p$ belongs to $\mathbb{Z}_p$, the ring of $p$-adic integers. Similarly, the group $GL(2,\mathbb{A})$ is also a restricted product of $GL(2,\mathbb{Q}_p)$. In this adelic formulation, the automorphic representations of $GL(2,\mathbb{A})$ are tensor products of local representations of the local factors $GL(2,\mathbb{R})$ and $GL(2,\mathbb{Q}_p)$, and the Hecke correspondences acting on automorphic forms can be expressed in terms of multiplication by elements of the group $GL(2,\mathbb{Q}_p)$ at finite places (or primes) $p$. See [GelGP] [Gelb1] [RamV][ [KnL1] [CassF].

For construction and classification of locally compact fields such as the $p$-adic fields $\mathbb{Q}_p$, see [RaV, Chap. 4]. For adeles and ideles, see [RaV, Chap. 5].

Adeles are convenient in studying towers of covering locally symmetric spaces as well. See [Roh1] and also [BlFG].

**3.4.15. Converse theorems for $L$-functions.** The famous converse theorem of Weil says roughly that $L$-functions satisfying the expected conditions such as meromorphic continuation, functional equations and Euler products etc come from some modular forms. As mentioned before, a suitable (or natural) sequence of numbers $a_n$ defines an $L$-function. This provides a good way to recognize when the associated $L$-function comes from a modular form, i.e., when the numbers $a_n$ are related to Fourier coefficients of a modular form. If this is true, then we often say that $a_n$ has the modularity property. A famous example of an interpretation of the coefficients of modular forms as dimensions of representation spaces is the monster moonshine in [Borc].

It is perhaps also worthwhile to point out that the existence of cuspidal Maass forms is not easy and the Selberg trace formula was created for this purpose. Given their existence, the actual construction of Maass forms and modular forms is another whole different story. The converse theorem is useful for this purpose too.

This point of view has been generalized to other groups and had important applications to the Langlands functoriality conjecture. See [LiWi1] [Cog2] and the references there.

**3.4.16. Applications of modular forms.** One typical application of modular forms is to find numbers of ways to represent an integer by an integral quadratic form, for example, a sum of squares. Besides many other applications of modular forms and automorphic representations in number theory and arithmetic algebraic geometry such as the growth of the class number of a complex quadratic field as the discriminant of the field goes to infinity, division points of elliptic curves, the class field theory for abelian extensions of $\mathbb{Q}$ etc (see [Vla, Part III, Chap. 4] and also [Sar1] [Iw2] [CogKM] [Art2-5] [DiaT] [LiWi1] and references there), they also have applications in many fields through the modularity property,[8] for example, they are applied in the Monster Moonshine (see [Borc]), in topology [Liu1-2], in topological modular forms [Hop] [Luri] [AnHS], the mirror symmetry [Mey], differential equations [Hol1-2] [Yo1-2], and expanders in graph theory [Lub1] [LubPS1] [HoLW], constructions of algebraic error-correcting codes [Vla, Part III, Chap. 4], and other applications such as distribution of points on the sphere [LubPS2-3].

---

[8]In a conversation at IAS, 1995, P.Deligne explained that the essential point of a sequence having some regular properties is that they appear as the coefficients of a modular form. It is expected that many naturally occurring sequences satisfy some regular pattern. This is the reason that the modularity property arises in many different areas.

CHAPTER 4

# General arithmetic subgroups and locally symmetric spaces

The general reference of this section is [Bore4], and a summary was given in [Bore3], and also [Bore14-15]. Another book is [Hum2]. A more advanced book is [PlR1]. A preprint form of a book is [Morr1]. See also [Ji6]. A masterful introduction is [Serr1].

As emphasized earlier, to be considered as an arithmetic group, $\mathbb{Z}$ must be embedded into the ambient Lie group $\mathbb{R}$, and the pair $(\mathbb{Z}, \mathbb{R})$ rather than $\mathbb{Z}$ alone is considered as an arithemetic group.

For studying arithmetic groups, theory of linear algebraic groups is essential. Linear algebraic groups also provide a natural class of Lie groups and put real Lie groups and $p$-adic Lie groups on the same footing.

The reduction theory of arithmetic groups is closely related to geometry of lattices in $\mathbb{R}^n$. Arithmetic subgroups of semisimple linear algebraic groups are non-abelian lattices in related Lie groups. An important question is to characterize arithmetic subgroups among all lattices of Lie groups.

## 4.1. Algebraic groups

Two basic books on algebraic groups are [Bore13] [Hum3]. See also [Bore11] for historic essays on linear algebraic groups and relations to Lie groups. Other books with some emphasis on arithmetic of algebraic groups include [PlR1] [Vos].

A subgroup $\mathbf{G}$ of $GL(n, \mathbb{C})$ is called a linear algebraic group if it is a subvariety, i.e., defined by polynomial equations in the matrix entries and the inverse of the determinant, and the group operations are morphisms between varieties. (Recall that we need to embed $GL(n, \mathbb{C})$ into $\mathbb{C}^{n^2+1}$ by $g \mapsto (g, \det g)$ in order to get an affine variety.)

It is easy to see that

$$SL(n, \mathbb{C}) = \{A \in GL(n, \mathbb{C}) \mid \det A = 1\}$$

is a linear algebraic group, since the determinant is multiplicative, $\det AB = \det A \det B$ for all matrices in $GL(n, \mathbb{C})$. Perhaps it is not obvious in general that why polynomial conditions defining a subvariety of $GL(n, \mathbb{C})$ are preserved under multiplication in $GL(n, \mathbb{C})$. The point is to note that the conditions defining algebraic groups often come from constraints that some structures are preserved, i.e., algebraic groups are automorphism groups of some structures on the vector spaces $\mathbb{C}^n$. Then it becomes clear such conditions are preserved under multiplication. For example, for $SL(n, \mathbb{C})$, the top degree differential form on $\mathbb{C}^n$ is preserved under its action.

The algebraic group $\mathbf{G}$ is said to be defined over $\mathbb{Q}$ if it is a variety defined over $\mathbb{Q}$ and the morphisms are also defined over $\mathbb{Q}$. Assume from now on that $\mathbf{G}$ is defined over $\mathbb{Q}$.

Define $\mathbf{G}(\mathbb{Q}) = \mathbf{G} \cap GL(n, \mathbb{Q})$, the group of rational points; and $\mathbf{G}(\mathbb{Z}) = \mathbf{G} \cap GL(n, \mathbb{Z})$, the group of integral points.

The real locus $G = \mathbf{G}(\mathbb{R})$ is a real group with finitely many connected components. (Note that even if $\mathbf{G}$ is connected, $G$ is not necessarily connected.) If $\mathbf{G}$ is semisimple, then the center of $G$ is finite. Hence, semisimple Lie groups with infinite centers, for example, the universal covering of $SL(2, \mathbb{R})$, cannot be realized as the real locus of linear algebraic groups.

## 4.2. Definition of arithmetic subgroups

In the above notations, a subgroup $\Gamma$ of $\mathbf{G}(\mathbb{Q})$ is called an arithmetic subgroup if it is commensurable with $\mathbf{G}(\mathbb{Z})$, i.e., the intersection $\Gamma \cap \mathbf{G}(\mathbb{Z})$ has finite index in both $\Gamma$ and $\mathbf{G}(\mathbb{Z})$.

It should be emphasized that arithmetic subgroups $\Gamma$ need not to be contained in $\mathbf{G}(\mathbb{Z})$. For example, there are arithmetic subgroups of $GL(n, \mathbb{Q})$ which are not contained in $GL(n, \mathbb{Z})$.

Algebraic groups can be defined as abstract varieties such that the group operations are morphisms. If group operations are defined over $\mathbb{Q}$, $\mathbf{G}$ is said to be defined over $\mathbb{Q}$. Every projective algebraic group over $\mathbb{C}$ is a complex torus (see [GriH, Chap. 2, §6]). Since an abelian variety over $\mathbb{C}$ is defined to be a complex torus that is also a projective algebraic variety over $\mathbb{C}$, every projective algebraic group over $\mathbb{C}$ an abelian variety.

We are only interested in affine algebraic groups. Every affine algebraic group $\mathbf{G}$ can be embedded into some $GL(n, V)$, where $V$ is a vector space $\mathbb{C}$ and hence becomes a linear algebraic group when $V$ is identified with $\mathbb{C}^n$. If $\mathbf{G}$ is defined over $\mathbb{Q}$, then there exists a vector space $V$ with a $\mathbb{Q}$-structure, i.e., there exists a $\mathbb{Q}$-vector subspace $V(\mathbb{Q})$ with $V(\mathbb{Q}) \otimes \mathbb{C} = V$, such that the embedding of $\mathbf{G}$ into $GL(n, V) \cong GL(n, \mathbb{C})$ can be taken over $\mathbb{Q}$.

In general, such embeddings $\mathbf{G} \subset GL(n, \mathbb{C})$ are not unique, since they depend on the choice of a basis of $V$ in order to get the identification $V \cong \mathbb{C}^n$. If $\mathbf{G}$ is defined over $\mathbb{Q}$, each choice of a basis of $V(\mathbb{Q})$ gives the rational points $\mathbf{G}(\mathbb{Q}) = \mathbf{G}(\mathbb{C}) \cap GL(n, V(\mathbb{Q})) = \mathbf{G}(\mathbb{C}) \cap GL(n, \mathbb{Q})$, and the integral points $\mathbf{G}(\mathbb{Z}) = \mathbf{G}(\mathbb{C}) \cap GL(n, \mathbb{Z})$. If $\mathbf{G}$ is defined over $\mathbb{Q}$, it is known that different choices of bases of $V(\mathbb{Q})$ and hence different embeddings of $\mathbf{G}$ into $GL(n, \mathbb{C})$ over $\mathbb{Q}$ give rise to the same rational points $\mathbf{G}(\mathbb{Q})$, but these different $\mathbb{Q}$-bases often lead to different integral points $\mathbf{G}(\mathbb{Z})$, which depend on the identification $V(\mathbb{Q}) = \mathbb{Q}^n$. As a consequence, for every affine algebraic group $\mathbf{G}$ defined over $\mathbb{Q}$, there is a well-defined class of arithmetic subgroups of $\mathbf{G}(\mathbb{Q})$, but there is no well-defined integral points $\mathbf{G}(\mathbb{Z})$, which can only be defined up to finite index. (In fact, the sub-class of congruence subgroups is well-defined too.)

Another convenient way to define arithmetic subgroups of $GL(n, \mathbb{C})$ and hence of general linear algebraic groups $\mathbf{G}$ is as follows. Let $\Lambda \subset \mathbb{R}^n$ be a lattice. As mentioned before, this gives an integral structure of $\mathbb{R}^n$. (The resulting $\mathbb{Q}$-structure of $\mathbb{R}^n$ is given by the $\mathbb{Q}$-vector subspace $\Lambda \otimes_{\mathbb{Z}} \mathbb{Q}$.) Define

$$GL(\Lambda, \mathbb{C}) = \{g \in GL(n, \mathbb{C}) \mid g\Lambda = \Lambda\}.$$

Then a subgroup $\Gamma$ of $GL(n, \mathbb{C})$ is called arithmetic with respect to the rational structure defined by $\Lambda$ if it is commensurable with $GL(\Lambda, \mathbb{C})$. This point of view is convenient in dealing with algebraic groups defined over number fields and their arithmetic subgroups.

For precise definitions and details of the topics above, see the books [PlR1] [Bore4] [Hum2]. See also the preprint [Miln4] [Morr1].

## 4.3. Hilbert modular groups

The modular group $SL(2, \mathbb{Z})$ and its subgroups of finite index are basic and important. They act on a complete Riemannian manifold of constant negative curvature, which is a symmetric space of rank 1.

A particularly important example of arithmetic groups acting on a higher rank symmetric space is the class of Hilbert modular groups, and the associated locally symmetric spaces are Hilbert modular varieties. Besides the papers [Hirz1-2, 4], see also books [vand] [Fre] [Gat1]. See also [Od3] for relations between Hilbert modular surfaces and the Taniyama-Weil conjecture for elliptic curves over real quadratic fields.

The general definition of Hilbert modular groups is as follow. Let $k$ be a totally real number field, and $\mathcal{O}_k$ be its ring of integers. Then $SL(2, \mathcal{O}_k)$ is a full Hilbert modular group acting on a product of *copies* of $\mathbb{H}$ the upper half via the diagonal embedding

$$SL(2, \mathcal{O}_k) \to \prod SL(2, \mathbb{R}),$$

where the product ranges over all different embeddings of $k$ into $\mathbb{R}$. We can also replace $\mathcal{O}_k$ by its orders, i.e., subrings of finite index. For example, take $k = \mathbb{Q}(\sqrt{d})$, where $d$ is a positive square free integer, and $\Gamma = SL(2, \mathbb{Z}(\sqrt{d}))$ is a Hilbert modular group and is a discrete subgroup of $SL(2, \mathbb{R}) \times SL(2, \mathbb{R})$ via the two different embeddings of $\sqrt{d} \to \pm\sqrt{d}$. The quotient $\Gamma\backslash\mathbb{H} \times \mathbb{H}$ is a Hilbert modular surface. In this case, though the symmetric space $X = \mathbb{H} \times \mathbb{H}$ is reducible, i.e., an isometric product, the arithmetic group $\Gamma$ is irreducible in the sense that no finite cover of $\Gamma\backslash X$ is an isometric product, or equivalently that the image of $\Gamma$ in every factor $SL(2, \mathbb{R})$ is not a discrete subgroup.

It is in general difficult to find generators and relations for arithmetic subgroups. For generators and relations of Hilbert modular groups, see [KobN] [Kir].

In general, arithmetic groups acting on higher rank symmetric spaces are complicated due to many different reasons. For example, if the $\mathbb{Q}$-rank of $\Gamma\backslash X$ (or rather the associated algebraic group $\mathbf{G}$) is at least two, then the infinity of $\Gamma\backslash X$ is connected, i.e., the complement of every compact subset of $\Gamma\backslash X$ has only one unbounded connected component,[9] and the structure of the end is also complicated. On the other hand, if the $\mathbb{Q}$-rank is equal to 1, every end of $\Gamma\backslash X$ is topologically a cylinder. This is basically related to the fact that Siegel sets from different $\mathbb{Q}$-parabolic subgroups $\mathbf{P}$ are disjoint if the heights of the Siegel sets are sufficiently large, i.e., the $A_{\mathbf{P},t}$-component is shifted enough, and hence the reduction theory is simpler in this $\mathbb{Q}$-rank 1 case. Due to this, automorphic forms, the Selberg trace formula and other topics are also easier to be understood for $\mathbb{Q}$-rank 1 locally symmetric spaces.

These Hilbert modular varieties (or groups) are relatively simple to study since their $\mathbb{Q}$-rank is equal to 1. Even among locally symmetric spaces of $\mathbb{Q}$-rank 1, Hilbert modular varieties are easier and more explicit and hence are very important examples of locally symmetric spaces of $\mathbb{Q}$-rank 1.

For a detailed discussion of geometry and analysis, in particular the index theory, of locally symmetric spaces of $\mathbb{Q}$-rank 1 and more generalized manifolds (i.e., manifolds which are not locally symmetric spaces but their ends are modeled after ends of locally symmetric spaces of $\mathbb{Q}$-rank 1), see [Mul6].

By reduction theory, cusps of Hilbert modular surfaces are related to ideal class of the ring of integers of the real quadratic fields. In some cases, they are also related to Kac-Moody algebras. In fact, [LepM] showed that the data from a rank 2 hyperbolic Kac-Moody root system encode the quasiregular cusps on a Hilbert modular surface. Conversely, given the data from the quasiregular cusps on a Hilbert modular surface, one may construct a rank 2 hyperbolic Kac-Moody root system.

## 4.4. Congruence subgroups and the congruence kernel

A distinguished class of arithmetic subgroups consists of the congruence subgroups. Briefly, given a linear algebraic group $\mathbf{G} \subset GL(n, \mathbb{C})$ defined over $\mathbb{Q}$, a principal congruence subgroup of $\mathbf{G}(\mathbb{Z})$ is the kernel of the map

$$\mathbf{G}(\mathbb{Z}) \to GL(n, \mathbb{Z}/n\mathbb{Z}).$$

Clearly, a principal congruence subgroup of $\mathbf{G}(\mathbb{Z})$ is of finite index and hence is an arithmetic subgroup. Any arithmetic subgroup containing a principal congruence subgroup is called a congruence subgroup.

It is known that when $\mathbf{G}$ is the standard form of $SL(2, \mathbb{C})$ defined over $\mathbb{Q}$, there are arithmetic subgroups which are not congruence subgroups [Mag].

On the other hand, when $\mathbf{G}$ is the standard (or split) form of $SL(n, \mathbb{C})$, and $n \geq 3$, every arithmetic subgroup is a congruence subgroup [BasLS] [Men] [BasMS].

[9]Equivalently, $\Gamma\backslash X$ has only one end.

In general, the relation between arithmetic subgroups and congruence subgroups is described by the so-called congruence subgroup kernel. Briefly, denote the family of arithmetic subgroups of $\mathbf{G}(\mathbb{Q})$ by $\mathcal{A}$ and the family of congruence subgroups of $\mathbf{G}(\mathbb{Q})$ by $\mathcal{C}$. Then every arithmetic subgroup of $\mathbf{G}(\mathbb{Q})$ is a congruence subgroup if and only if these two families are cofinal.

If there are not cofinal, we can measure their differences through completions with respect to these systems. Specifically, take $\mathcal{A}$ as a system of neighborhoods of the identity element of $\mathbf{G}(\mathbb{Q})$ and define a topology on $\mathbf{G}(\mathbb{Q})$ by translation. The completion with respect to this topology is denoted by $\hat{G}(\mathcal{A})$. Similarly, there is a completion $\hat{G}(\mathcal{C})$ with respect to the family $\mathcal{C}$. Since every congruence subgroup is an arithmetic subgroup, $\mathcal{C} \subseteq \mathcal{A}$. This inclusion induces a continuous map $\hat{G}(\mathcal{A}) \to \hat{G}(\mathcal{C})$. It can be shown that it is surjective, i.e., there exists an exact sequence

$$1 \to C(G) \to \hat{G}(\mathcal{A}) \to \hat{G}(\mathcal{C}) \to 1;$$

and the kernel $C(G)$ is called the congruence subgroup kernel.

In the above case of $SL(2)$, $C(G)$ is an infinite group, and in the case of $SL(n)$ with $n \geq 3$, it is trivial. When the algebraic group $\mathbf{G}$ is simple over $\mathbb{Q}$ and the $\mathbb{Q}$-rank is at least 2, the congruence subgroup kernel $C(G)$ is a finite subgroup. For discussions on the congruence subgroup kernel and related problems, see [Ra1-4] [Serr5] [Pr3] [PrR1-3]. For a simpler proof of the finiteness of the congruence subgroup kernel, see [Pr5, §4-5].

The congruence subgroup kernel can be identified with the fundamental group of the reductive Borel-Serre compactification of a naturally associated locally symmetric space $\Gamma\backslash X$ when the $\mathbb{Q}$-rank is at least 2 (see [JiMSS]).

Congruence subgroups are often special among arithmetic subgroups, see, for example, [Bek] [Lub3].

## 4.5. Arithmetic subgroups as discrete subgroups of Lie groups

Let $\mathbf{G}$ be a linear algebraic group defined over $\mathbb{Q}$ as above. Let $G = \mathbf{G}(\mathbb{R})$ be the real locus of $\mathbf{G}$. Then it is a Lie group with finitely many connected components. If the algebraic group $\mathbf{G}$ is semisimple, then the center of the Lie group $G$ is finite. On the other hand, if $G$ is a semisimple Lie group, i.e., its Lie algebra is semisimple, its center is discrete but could be an infinite group (even under the assumption that $G$ is connected). Therefore, not all Lie groups can arise as $\mathbf{G}(\mathbb{R})$. But many natural, for example, classical Lie groups, arise this way.

As mentioned before, starting from a linear algebraic group $\mathbf{G}$, we can also get $p$-adic Lie groups $\mathbf{G}(\mathbb{Q}_p)$ for all primes $p$. So linear algebraic groups produce both real and $p$-adic Lie groups at the same time by the same completion procedure. Together with arithmetic subgroups and more general $S$-arithmetic subgroups introduced later, there are three different kinds of groups associated with a linear algebraic group defined over $\mathbb{Q}$ (or more

generally over a number field). It is a natural and important problem to understand relations between them.

It can be shown easily that every arithmetic subgroup $\Gamma$ of $\mathbf{G}(\mathbb{Q})$ is a discrete subgroup of $G$. In fact, $\mathbb{Z}$ is a discrete subgroup of $\mathbb{R}$, and hence $\mathbf{G}(\mathbb{Z})$ is a discrete subgroup of $G = \mathbf{G}(\mathbb{R})$. Discreteness is preserved under passage to subgroups of finite index, and hence is shared by commensurable groups.

On the other hand, given a Lie group, a natural question is to define a class of arithmetic subgroups. It can not always be done in the exact sense as we discussed above, since not every Lie group with finitely many connected components is the real locus of an algebraic group.

On the other hand, if a Lie group differs from the real locus of an algebraic group by a compact subgroup, then this can be done. This is the more general notion of arithmetic subgroups of Lie groups (see [Zi1, p.114] and also [Morr1]).

Briefly, let $G$ be a Lie group with finitely many connected components. Then a discrete subgroup $\Gamma$ of $G$ of finite co-volume (i.e., a lattice subgroup) is called an arithmetic subgroup if there exists an algebraic group $\mathbb{H} \subset GL(n, \mathbb{C})$ defined over $\mathbb{Q}$ such that there is a Lie group morphism $\varphi : G \to \mathbf{H}(\mathbb{R})$ with a compact kernel and a compact quotient $\mathbf{H}(\mathbb{R})/\varphi(G)$ such that the image $\varphi(\Gamma)$ is commensurable with $\mathbf{H}(\mathbb{Z})$. See [Zi, p. 105] and [Morr1, (6.16)].

The presence of the compact kernel is important. For example, the isometry (or the motion) group of $\mathbb{R}^n$ is the semidirect product $SO(n) \rtimes \mathbb{R}^n$ of $SO(n)$ and $\mathbb{R}^n$. If we only consider discrete subgroups of $\mathbb{R}^n$ with co-finite volume, we only get lattices in $\mathbb{R}^n$. But there are many other arithmetic subgroups in $SO(n) \rtimes \mathbb{R}^n$, the so-called crystallographic subgroups.

## 4.6. Zariski density of arithmetic subgroups

In many senses, arithmetic subgroups $\Gamma$ of semisimple Lie groups $G$ are of the same size as the Lie groups. This is, for example, reflected in the finiteness of the voulme of the quotient $\Gamma\backslash G$. As emphasized before, an arithmetic subgroup $\Gamma$ is a discrete subgroup of the Lie group $G$ and hence also of the algebraic group $\mathbf{G} = \mathbf{G}(\mathbb{C})$ when $G$ is equal to the real locus $\mathbf{G}(\mathbb{R})$ of an algebraic group $\mathbf{G}$. (An algebraic group $\mathbf{G}$ is the variety $\mathbf{G}(\mathbb{C})$ together with compatible group structures.)

Since $\mathbf{G}(\mathbb{C})$ is an algebraic variety, $G = \mathbf{G}(\mathbb{R})$ also has a Zariski topology. Another aspect of the non-smallness of $\Gamma$ is reflected in the Zariski density of arithmetic subgroups $\Gamma$ in $G$ when $G$ has no normal compact factors of positive dimension. See [Borel]. This result is called the Borel density theorem of arithmetic subgroups and is important for many applications. For example, it implies that the normalizer of an arithmetic subgroup is a discrete subgroup. This shows the importance of algebraic groups in studying arithmetic groups. See [Zi1] for more applications. See [Beno2]

[PrRa] [ConzG1] for other Zariski density subgroups and their applications and properties.

## 4.7. Symmetric spaces

Assume that $\mathbf{G}$ is a semisimple linear algebraic group defined over $\mathbb{Q}$. Then its real locus $G = \mathbf{G}(\mathbb{R})$ is a semisimple Lie group with finitely many connected components. Let $K \subset G$ be a maximal compact subgroup of $G$. Let $X = G/K$ be endowed with a $G$-invariant metric, usually the one induced from the Killing form. Then $X$ is a Riemannian symmetric space of non-compact type. (Note also that $X$ can be identified with the space of all maximal compact subgroups of $G$. This interpretation is useful for some applications.)

Symmetric spaces are important in many subjects, in particular to topics under discussion in this book. The basic philosophy is that to understand a group, an effective method is to find a good space for it to act on.

A basic reason that symmetric spaces are very useful in different fields is that they support of a lot of symmetry, and symmetry induces additional structures. By definition, a complete Riemannian manifold is *a symmetric space* if the local geodesic symmetry at every point is isometric and extends to a global isometry (abundance of symmetry). This characterization ties symmetric spaces immediately with Lie groups. For example, it is immediate from this definition that a symmetric space is a homogeneous space since its isometry group acts transitively on it. (Note that it is known that the isometry group of any Riemannian manifold is a Lie group.)

Products of symmetric spaces are also symmetric, and irreducible ones are those which can not be written as products. Irreducible symmetric spaces are classified by the signs of curvature into three types: compact type, flat, and noncompact type. More specifically, an irreducible symmetric space of noncompact type has non-positive sectional curvature and strictly negative Ricci curvature, and an irreducible symmetric space of compact type has curvatures of the opposite sign. The typical examples are the sphere $S^n$, the Euclidean space $\mathbb{R}^n$ and the real hyperbolic space $\mathbb{H}^n$ (i.e., the simply connected Riemannian manifold with constant sectional curvature $-1$). These are the so-called space forms.

An important fact is that symmetric spaces of noncompact type are nonpositively curved and diffeomorphic to $\mathbb{R}^n$, in particular contractible. They are important examples of the so-called Hadamard manifolds. They provide natural spaces for Lie groups (and also for discrete subgroups) to act upon, and the above fact has many consequences: for example, every two maximal compact subgroups of a semisimple Lie group $G$ with finite center are conjugate and the symmetric space $X$ can be identified with the space of all maximal compact subgroups of $G$, and furthermore, the symmetric space $X$ is a universal space for proper actions of every discrete subgroup of $G$.

On the other hand, symmetric spaces of compact type are compact and have finite fundamental group. They play an important role in topology. See [MimT] and [Adam].

A particularly important subclass of symmetric spaces consists of Hermitian symmetric spaces, which are symmetric spaces that admit an invariant complex structure. Bounded symmetric domains in $\mathbb{C}^n$ endowed with the Bergman metric are Hermitian symmetric spaces of noncompact type, and every Hermitian symmetric space of noncompact type can be realized as a bounded symmetric domain in the holomorphic tangent space at any point. Hermitian symmetric spaces of compact type are rational projective varieties, i.e., birational to $\mathbb{C}P^n$.

Furthermore, every Hermitian symmetric space of noncompact type is naturally embedded into a Hermitian symmetric space of compact type, as in the case that the unit disc is embedded in the sphere $S^2 = \mathbb{C}P^1$. This is stronger than the duality between symmetric spaces of compact type and noncompact type. This embedding allows one to pass from one type of geometry to another type. For example, there are a lot of similarities between the spherical geometry (sin and cos) and hyperbolic geometry (sinh and cosh.) Another important example is that the Baily-Borel compactification of a Hermitian symmetric space $X$ of noncompact type is the closure of $X$ under this embedding into the dual Hermitian symmetric space $\hat{X}$ of compact type. If $X$ is realized as a bounded symmetric domain in $\mathbb{C}^n$, the Baily-Borel compactification of $X$ is also the closure in $\mathbb{C}^n$. But this embedding has an important point: the holomorphic isometries of $X$ (or elements of $G$ if $X = G/K$) extends to holomorphic automorphism of $\hat{X}$, but they do not extend to holomorphic maps on $\mathbb{C}^n$. So the embedding of $X \hookrightarrow \hat{X}$ is equivariant, not $X \hookrightarrow \mathbb{C}^n$.

Though symmetric spaces are foundational to and important for many subjects, there are not many books available. The standard reference is [Hel1] which is comprehensive but is also quite long. The book [Borel0] is concise and covers fewer topics. The book [Eb] studies symmetric spaces from a more differential geometry point of view and contains less structure theory. The book [Wol1] contains a nice summary of symmetric spaces and many detailed information about spaces of constant sectional curvature. For detailed discussions about the special symmetric space $SL(n, \mathbb{R})/SO(n)$, see the book [Te2]. The article [Ji8] contains a crash course on symmetric spaces.

The book [Hel1] [Borel0] cover some basics of Hermitian symmetric spaces. The article [Wol2] deals with more refined structures of bounded symmetric domains, or equivalently Hermitian symmetric spaces of noncompact type. For a systematic study of bounded symmetric domains, see [Sat5]. For a nice summary of classical domains, see [Mok2]. See also [Si1] for the Siegel upper half planes.

## 4.8. Non-Riemannian symmetric spaces

Riemannian manifolds are certainly important in differential geometry, but non-Riemannian (or pseudo-Riemannian) manifolds are also important and natural for other purposes. More generally, a complete manifold $X$ with an affine connection is a *symmetric space* (or *affine symmetric space*) if the local geodesic symmetry at every point is affine and extends to a global affine diffeomorphism. In this context, $X$ is a Riemannian (more generally, pseudo-Riemannian) symmetric space if $X$ is symmetric and Riemannian (pseudo-Riemannian), and the affine connection is the Levi-Civita connection of a Riemannian metric.

In general, the group $G$ of all affine diffeomorphisms of an affine manifold $X$ becomes a Lie group. If $X$ is a symmetric space, then $G$ acts transitively on $X$, and the geodesic symmetry at each point induces an involutive automorphism of $G$ by conjugation.

Fix a point $o$ of $X$, and let $H$ be the isotropy subgroup at $o$ and $\sigma$ the involutive automorphism of $G$ induced from the geodesic symmetry at $o$. Then $X$ is diffeomorphic to $G/H$ and $H$ is an open (therefore, automatically closed) subgroup of the fixed point group

$$G^\sigma := \{g \in G : \sigma g = g\}.$$

Conversely, if $G$ is a Lie group, $\sigma$ is an involutive automorphism of $G$, and $H$ is an open subgroup of $G^\sigma$, then $G/H$ carries a $G$-invariant affine connection and becomes a symmetric space.

Any Lie group $G'$ is a (not necessarily Riemannian) symmetric space because it is expressed as the homogeneous space $G/H$ where $G = G' \times G'$, the involution $\sigma$ is defined by $\sigma(x, y) = (y, x)$ for $(x, y) \in G$, and $H$ is the diagonal subgroup of $G$ defined by

$$\Delta G' = \{(g, g) \in G' \times G' : g \in G'\}.$$

The infinitesimal classification, i.e., the classification at the level of Lie algebras, of all irreducible symmetric spaces was accomplished in [Berg2].

There is an important class of pseudo-symmetric spaces, the so-called *causal symmetric spaces*, see the book [HilO].

## 4.9. Locally symmetric spaces

By definition, a Riemannian manifold is locally symmetric if the local geodesic symmetry at every point is isometric, or equivalently that the curvature operator is parallel, i.e., their covariant derivative vanishes. The first characterization ties locally symmetric spaces immediately with Lie groups, and the second shows their special place among Riemannian manifolds. If $M$ is a complete locally symmetric space, then its universal covering space $\tilde{M}$ is a symmetric space, and hence $M$ is a quotient of a symmetric space, $M = \Gamma\backslash\tilde{M}$.

If $X$ is a symmetric space and $\Gamma$ is a group of isometry acting properly and fixed point freely, then the quotient $\Gamma\backslash X$ is a manifold and a locally symmetric space.

A good supply of locally symmetric spaces comes from arithmetic subgroups. Let $\mathbf{G}$ be a semisimple linear algebraic group defined over $\mathbb{Q}$, $G = \mathbf{G}(\mathbb{R})$, and $X = G/K$ the associated Riemannian symmetric space as above. Let $\Gamma$ be an arithmetic subgroup of $\mathbf{G}(\mathbb{Q})$. Then $\Gamma$ acts isometrically and properly on $X$. If $\Gamma$ is torsion-free, then the action is fixed point free, and hence $\Gamma\backslash X$ is a locally symmetric space. It is known by Selberg's lemma that every arithmetic subgroup contains a torsion-free subgroup of finite index (see [Bore4, §17] for example).

On the other hand, many natural arithmetic subgroups such as $SL(n, \mathbb{Z})$ are not torsion-free, and passing to a subgroup of finite index loses some valuable information. So it is natural to consider arithmetic subgroups $\Gamma$ containing nontrivial torsion elements and the resulting non-smooth quotients $\Gamma\backslash X$. They are orbifolds instead. These orbifolds $\Gamma\backslash X$ always admit finite smooth covers. For convenience, quotients $\Gamma\backslash X$ of symmetric spaces by discrete subgroups are also called locally symmetric spaces. It might be worthwhile to point out that the notion of orbifolds (or V-manifolds) was first introduced by Satake in [Sat9] in order to handle such non-smooth quotients. (The name orbifold was introduced by Thurston. See [Rat].)

As shown below, if $\Gamma$ is an arithmetic subgroup of a semisimple algebraic group $\mathbf{G}$, then the locally symmetric space $\Gamma\backslash X$ has finite volume. Due to these considerations and many examples provided through arithmetic subgroups, in many cases, a locally symmetric space implicitly means a locally symmetric space of finite volume.

There are not many books and expository references about locally symmetric spaces. For example, many books on symmetric spaces do not study locally symmetric spaces any further after giving the definition, and connections with arithmetic subgroups or discrete subgroups have not been discussed at all or emphasized.

The preprint book [Morr1] gives a more accessible introduction to locally symmetric spaces than the classical book [Bore4].

## 4.10. Space forms

A complete Riemannian manifold is called a space form if its sectional curvature is constant. For example, the hyperbolic surfaces $\Gamma\backslash\mathbb{H}$ we discussed earlier are all space forms. They are closely related to the uniformization of Riemann surfaces and more general Riemannian manifolds of constant sectional curvature.

In fact, if $M^n$ is a $n$-dimensional complete Riemannian manifold with constant sectional curvature. By a suitable scaling on the Riemannian metric, we can assume that the sectional curvature is equal to either 1, 0, or $-1$. Then its universal covering space is a simply connected Riemannian space of constant sectional curvature and equal to either the unit sphere $S^n \subset \mathbb{R}^{n+1}$, an Euclidean space $\mathbb{R}^n$, the real hyperbolic space $\mathbb{H}^n$. (It might be worthwhile to point out that every irreducible symmetric space has constant Ricci curvature. On the other hand, very few symmetric spaces have constant sectional curvature.)

Spaces form are particularly important but special examples of locally symmetric spaces. Even among locally symmetric spaces of rank 1, many do not have the constant sectional curvature.

Space forms which are quotients of $S^n$ and $\mathbb{R}^n$ can be basically classified. On the other hand, the complete classification of quotients of $\mathbb{H}^n$ might not be possible. For example, for $n = 3$, it looks quite difficult already. It is still not known which hyperbolic 3 dimensional manifold $\Gamma\backslash\mathbb{H}^3$ has the smallest volume. For some general and detailed discussions about spaces forms, see [Wol1].

## 4.11. Compactness criterion for locally symmetric spaces

To study the geometry and analysis of arithmetic locally symmetric spaces $\Gamma\backslash X$, an important tool is the so-called reduction theory. The main purpose is to find a good fundamental domain for the $\Gamma$-action on $X$. As seen in the example of $\Gamma = SL(2,\mathbb{Z})$ and $X = \mathbb{H}$, the existence of a good fundamental domain is useful for many purposes.

The first natural question asks when the quotient $\Gamma\backslash X$ is compact. This is equivalent to when there is a fundamental domain given by a compact subset of $X$. This question is answered by the famous conjecture of Godement, proved by Borel & Harish-Chandra [BoreHC], and Mostow & Tamagawa [MostT]. In the special case that $\mathbf{G}$ is a semisimple linear algebraic group, this conjecture states that the quotient $\Gamma\backslash X$ is compact if and only if the $\mathbb{Q}$-rank of $\mathbf{G}$ is 0.

The algebraic group $SL(n,\mathbb{C}) = \{g \in GL(n,\mathbb{C}) \mid \det g = 1\}$ is defined over $\mathbb{Q}$ and its $\mathbb{Q}$-rank is equal to $n-1$ and hence positive for $n \geq 2$. This explains why the quotient $SL(2,\mathbb{Z})\backslash\mathbb{H} = SL(2,\mathbb{Z})\backslash SL(2,\mathbb{R})/SO(2)$ is noncompact. For other classical algebraic groups in the split form such as $Sp(n,\mathbb{C})$, their $\mathbb{Q}$-ranks are also positive, and hence their arithmetic subgroups $\Gamma$ are not co-compact, i.e., $\Gamma\backslash X$ are not compact.

The existence of compact quotients of $X$, i.e., cocompact lattices acting on $X$, is guaranteed by a theorem of Borel in [Borel6] (see [Ra5] for an exposition). The construction is similar to the construction of compact arithmetic Fuchsian groups, and the above compactness criterion is used.

In this book, we concentrate on Riemannian symmetric spaces. On the other hand, there are also non-Riemannian symmetric spaces, for example, the Lorentz hyperbolic space. For such symmetric spaces, the existence of compact quotients is a nontrivial problem, which is still under intensive study [Kob1-3].

## 4.12. Siegel sets and fundamental sets

Assume that the $\mathbb{Q}$-rank of $\mathbf{G}$ is positive. Then $\Gamma\backslash X$ is noncompact, and there is no fundamental domain of $\Gamma$ given by a compact subset. The noncompactness of fundamental domains or the geometry at infinity of $\Gamma\backslash X$ is described by the Siegel sets.

Recall that for every $\mathbb{Q}$-parabolic subgroup $\mathbf{P}$, its real locus $P = \mathbf{P}(\mathbb{R})$ has a $\mathbb{Q}$-Langlands decomposition

$$P = N_P A_{\mathbf{P}} M_{\mathbf{P}} \cong N_P \times A_{\mathbf{P}} \times M_{\mathbf{P}}.$$

This induces a horospherical decomposition

$$X = N_P \times A_{\mathbf{P}} \times X_{\mathbf{P}},$$

where $X_{\mathbf{P}} = M_{\mathbf{P}}/M_{\mathbf{P}} \cap K$ is a symmetric space, called the boundary symmetric space associated with $\mathbf{P}$.

When $G = SL(2, \mathbb{R})$, $X = \mathbb{H}$. Let $P$ be the parabolic subgroup of upper triangle matrices. Then the horospherical decomposition of $X$ associated with $P$ corresponds to the decomposition of $\mathbb{H}$ by the $x$ and $y$ coordinates. When $y$ is fixed and $x$ changes, the curve traced out is a horocircle at infinity. This is the reason for the name of horospherical decomposition.

The horospherical decomposition is important for understanding structure of geodesics. In fact, it corresponds to a classificattion of geodesics in $X$ going to infinity in the direction determined by the parabolic subgroup $\mathbf{P}$. It is also important for harmonic analysis, for example, in computing eigenfunctions since the radial component of the invariant differential operators in the horopsherical decomposition can be rather easily computed. For more discussions about the horospherical decomposition and its roles in compactifications of symmetric spaces, see [GuiJT] and [BoreJ].

For every $t \in \mathbb{R}$, there is a cone $A_{\mathbf{P},t}$, a shift of the positive cone. Then for bounded sets $U \subset N_{\mathbf{P}}$ and $V \subset X_{\mathbf{P}}$, the subset $U \times A_{\mathbf{P},t} \times V$ of $X$ is called a *Siegel set* of $X$ associated with $\mathbf{P}$.

In the case of $X = \mathbb{H}$ and the above special parabolic subgroup $P$, a Siegel set is of the form $I \times [t, +\infty)$, where $I \subset \mathbb{R}$ is a bounded subset.

Recall that a fundamental domain $\Omega$ for the $\Gamma$-action on $X$ is a subset $\Omega \subset X$ that satisfies the following conditions:

(1) $\Gamma\Omega = X$,
(2) the interior of $\Omega$ is mapped 1-1 into $\Gamma\backslash X$.
(3) The restricted map $\Omega \to \Gamma\backslash X$ is finite-to-one.

On the other hand, a fundamental set for the $\Gamma$-action on $X$ is a subset $\Sigma \subset X$ satisfying the following conditions:

(1) $\Gamma\Sigma = X$,
(2) the restricted map $\Sigma \to \Gamma\backslash X$ is finite-to-one, and the further restriction to the interior of $\Sigma$ is not necessarily one-to-one into $\Gamma\backslash X$.

(The names here might be confusing. By a fundamental domain, we mean that it is basically mapped to the quotient in a one-to-one way. On the other hand, by a fundamental set, we allow it to be mapped to the quotient in a finite-to-one way. These are standard names in reduction theory of arithmetic groups.)

In the example of $X = \mathbb{H}$ and $P$ as above, if we take $I = [-\frac{1}{2}, \frac{1}{2}]$ and $t \geq 1$, then the Siegel set is contained in the fundamental domain $\Omega$ of $SL(2, \mathbb{Z})$ discussed earlier. On the other hand, for $t \leq \frac{\sqrt{3}}{2}$, it contains the fundamental domain $\Omega$. In this case, we can see clearly that one Siegel set never gives a fundamental domain of $SL(2, \mathbb{Z})$.

This example explains that it is often easier to get some nice fundamental sets, and Siegel sets naturally give rise to fundamental sets and describe the structure at infinity of arithmetic locally symmetric spaces in a rather precise way.

## 4.13. Reduction theory for arithmetic subgroups

There are many expositions of reduction theory. See [Bore3-4] [Bore14] [BoreHC] [Gra1-2] [Ji6] for example. For special arithmetic subgroups, more specific references will be given below in §4.14.

In general, it is difficult to find a fundamental domain for an arithmetic subgroup $\Gamma$. On the other hand, the existence and structure of a fundamental set can be described in terms of Siegel sets.

PROPOSITION 4.13.1. *There are finitely many $\Gamma$-conjugacy classes of $\mathbb{Q}$-parabolic subgroups of $\mathbf{G}$. Denote a finite set of representatives by $\mathbf{P}_0 = \mathbf{G}, \mathbf{P}_1, \dots, \mathbf{P}_m$. Then there are Siegel sets $\Sigma_1, \dots, \Sigma_m$ associated with them such that their union $\cup_{i=0}^m \Sigma_i$, denoted by $\Sigma$, satisfies the following conditions:*

1. *$\Sigma$ is mapped surjectively onto $\Gamma\backslash X$, i.e., $\Gamma\Sigma = X$,*
2. *for every $g \in \mathbf{G}(\mathbb{Q})$, the set $\{\gamma \in \Gamma \mid \gamma\Sigma \cap g\Sigma \neq \emptyset\}$ is finite,*
3. *in particular, the union $\Sigma$ is a fundamental set for $\Gamma$.*

In the above proposition, the group $\mathbf{G}$ itself is considered as a $\mathbb{Q}$-parabolic subgroup of $\mathbf{G}$. If the $\mathbb{Q}$-rank of $\mathbf{G}$ is equal to 0, then $\mathbf{G}$ is the only $\mathbb{Q}$-parabolic subgroup of $\mathbf{G}$. Its Siegel sets are bounded, since by assumption, $\mathbf{G}$ is semisimple. Therefore, the quotient $\Gamma\backslash X$ is compact. This gives one proof of the Godement conjecture on compactness criterion mentioned earlier. But there is a more direct proof of this compactness criterion of Godement without using the reduction theory for general $\mathbb{Q}$-rank [MosT].

In the above proposition, we can use only Siegel sets of minimal parabolic subgroups to get a fundamental set. This can be seen clearly in the example of $X = \mathbb{H}$. The basic point is that if $\mathbf{P}$ is contained in $\mathbf{P}'$, then every Siegel set of $\mathbf{P}'$ is contained in a Siegel set of $\mathbf{P}$, and that the union of two Siegel sets of the same parabolic subgroup $\mathbf{P}$ is contained in a Siegel set of $\mathbf{P}$. Therefore, $\Sigma$ is contained in union of finitely many Siegel sets of minimal $\mathbb{Q}$-parabolic subgroups.

The finiteness property in Proposition 4.13.1.(2) is called the *Siegel finiteness property* and is the heart of the reduction theory. The presence of $g \in \mathbf{G}(\mathbb{Q})$ is more than needed for $\Sigma$ to be a fundamental set of $\Gamma$. This also allows us to get fundamental sets for other discrete subgroups of $G$ commensurable with $\Gamma$. More importantly, this is crucial for showing that compactifications of $\Gamma\backslash X$ such as the Borel-Serre, the Baily-Borel compactifications are Hausdorff spaces.

From the above description of a fundamental set in terms of Siegel sets, it is clear that the noncompactness of $\Gamma\backslash X$ is essentially caused by the Weyl chamber cones $A_{\mathbf{P},t}$.

Together with an expression of the invariant metric in terms of the Langlands decomposition (see [B014]), an immediate corollary is that the volume of $\Gamma\backslash X$ is finite when $\mathbf{G}$ is semisimple as assumed before. Hence, arithmetic subgroups $\Gamma$ are lattices in $G$, and the constant functions are square integrable eigenfunctions of $\Gamma\backslash X$. In particular, the discrete spectrum of $\Gamma\backslash X$ is non-empty.

## 4.14. Precise reduction theory for arithmetic subgroups

Besides this classical reduction theory in Proposition 4.13.1, there is also a precise reduction theory, which gives a *fundamental domain* without overlap under action of arithmetic groups.

This theory has important application in the theory of Selberg trace formula and other problems. For a survey of the state of art of the Selberg trace formula, see [Art1].

For special arithmetic groups such as $SL(n, \mathbb{Z})$, much more is known about fundamental domains, see [Gren1-2] and also [Te2] [Cass1-2] [Gru] [GruL] [Si3-4]. They are related to the Minkowski reduction and geometry of numbers.

Identify the symmetric space $X = SL(n, \mathbb{R})/SO(n)$ with the space of positive definite quadratic forms. Then the Minkowski fundamental domain for $SL(n, \mathbb{Z})$ is defined by linear inequalities involving the entries of the quadratic forms. Hence, the faces of the fundamental domain are contained in linear subspaces. Though the infinitely many number of inequalities can be reduced to finitely many (see [Te2, p. 130]), it is not easy to determine these finitely many boundary faces. This limits the usefulness of this Minkowski fundamental domain.

There is another reduction theory, called the Voronoi reduction theory, building on concepts such as perfect quadratic forms. It is also related to the geometry of numbers. The basic idea is to pick out some special points, i.e., quadratic forms, by using intrinsic properties of the quadratic forms. Since these properties are intrinsic, these points are invariant under $SL(n, \mathbb{Z})$, and their induced polyhedral cone decompositions are also invariant under $SL(n, \mathbb{Z})$.

Besides the arithmetic subgroup $SL(n, \mathbb{Z})$, it also works for linear symmetric spaces and arithmetic subgroups acting on them. By definition, a symmetric space $X$ is called a linear symmetric space if it is a symmetric cone (i.e., a homogeneous and self-adjoint cone in some $\mathbb{R}^n$) or a homothety section of a symmetric cone. Though the existence of an exact fundamental domain for a general linear arithmetic subgroup can be found, it may not given explicitly [AsMRT] [As3]. This Voronoi reduction theory is important in the theory of toroidal compactifications of Hermitian locally symmetric spaces (see [AsMRT].)

For nonlinear symmetric spaces, finding exact fundamental domains using geometry of numbers or trying to induce from linear symmetric spaces is much more complicated and only works for few examples. See [MM1-2] [McC2] [Yas2].

For general symmetric spaces, there are several different ways to construct exact fundamental domains, either through the notion of Busemann function in Riemann geometry, or equivariant tiling of the symmetric spaces through the geodesic action of parabolic subgroups and the Borel-Serre partial compactification of $X$. References on this precise reduction theory for general arithmetic subgroups include [Leu4] [OsbW2] [Sap1] [Yas1-2].

The basic idea is roughly as follows. For each parabolic subgroup $\mathbf{P}$, there is a way to measure the "distance" from the infinity. Denote this "distance" by $d_{\mathbf{P}}$. The space $X$ is decomposed into pieces $\Delta_{\mathbf{P}}$ parametrized by $\mathbf{P}$ so that points $p \in \Delta_{\mathbf{P}}$ satisfy "$d_{\mathbf{P}}(p) \leq d_{\mathbf{P}'}(p)$" for every other $\mathbf{P}'$.

From this description, it is clear that both minimal and non-minimal $\mathbb{Q}$-parabolic subgroups are needed to get a fundamental domain. For example, in the example of $G = SL(2, \mathbb{R})$ and $X = \mathbb{H}$, the fundamental domain $\Omega$ in §3.3.1 is not equal to any Siegel set but is the union of a Siegel set associated with the parabolic subgroup of upper triangular matrices, and a bounded set which is actually a Siegel set of the whole group $\mathbf{G}$ considered as a parabolic subgroup.

## 4.15. Metric properties and $\mathbb{Q}$-rank of locally symmetric spaces

There are several points of views about metric properties of $\Gamma\backslash X$. Since unions of finitely many Siegel sets cover the whole space, it is natural to consider metrics restricted to Siegel sets. An example of problems from this point of view is the Siegel conjecture concerning comparison between two metrics on Siegel sets induced from the metrics of $X$ and $\Gamma\backslash X$. Another approach to understand the metric properties of $\Gamma\backslash X$ is to consider the whole space $\Gamma\backslash X$ directly. Some references are [Abe3] [AbeM] [Bore6] [Din] [Hatt2-3] [Ji4] [JiM] [Leu1, 2, 4].

In the study of symmetric spaces $X$ of noncompact type, flat and totally geodesic submanifolds, which are isometric to $\mathbb{R}^r$, are important. The maximal dimension of such flats is called the rank of $X$, which is also equal to the rank of the real Lie group $G$.

Motivated by this, it is natural to consider flat subspaces in $\Gamma\backslash X$. This is related to the notion of $\mathbb{Q}$-rank of $\Gamma\backslash X$ (or more appropriately the $\mathbb{Q}$-rank of the algebraic group $\mathbf{G}$.) In this case, if $\Gamma\backslash X$ is irreducible, there is in general no $\mathbb{R}^r$, with $r \geq 2$, which can be isometrically embedded into $\Gamma\backslash X$. (Note that if $\Gamma\backslash X$ has two ends, then there exist globally distance minimizing geodesics in it, i.e., isometric embedding of $\mathbb{R} \to \Gamma\backslash X$.) On the other, some can be properly and locally isometrically immersed into $\Gamma\backslash X$. For example, take $\Gamma\backslash X$ as the product of two noncompact quotients $\Gamma\backslash\mathbb{H}$, and $\mathbb{R}^2$ can be embedded via products of two geodesics connecting cusps of $\Gamma\backslash\mathbb{H}$. The maximal dimension of such properly immersed flat submanifolds is equal to the $\mathbb{Q}$-rank of $\Gamma\backslash X$. See [Weis] [ChaM]. The $\mathbb{Q}$-rank of $\Gamma\backslash X$ is also related to the dimension of families of rays which are distance minimizing and which are eventually isometric to rays in $\mathbb{R}^n$. It is also reflected to the dimension of the continuous spectrum of $\Gamma\backslash X$. See [JiM].

## 4.16. Volume spectrum of locally symmetric spaces

For each fixed symmetric space $X$, there are many locally symmetric spaces of the form $\Gamma\backslash X$. The problem of understanding the structure of the set of volumes of $\Gamma\backslash X$ for all $\Gamma$ has been studied by many people.

There are universal lower bounds for the volume of locally symmetric spaces. This is related to the Margulis Lemma in §5.2. See [Bore5] [KazM] and also [Ji9].

For the real hyperbolic spaces $\mathbb{H}^n$ of dimension $n$, the set of volumes of compact hyperbolic manifolds has the following structure:

(1) When $n = 2$, it forms an arithmetic sequence $\pi(6g - 6)$, $g \geq 2$, and each value has infinite multiplicity, i.e., taken by infinitely many hyperbolic manifolds.
(2) When $n = 3$, it has the structure of $\omega^\omega$, and each has finite multiplicity. See [Gro7].
(3) When $n \geq 4$, it is a discrete sequence, and each value has finite multiplicity. See [Wan3].

The above results imply that in dimension 2 and 3, the volumes of hyperbolic manifolds do not form an increasing sequence with finite multiplicity and going to infinity. On the other hand, if we restrict to arithmetic hyperbolic manifolds, i.e., quotients of $\mathbb{H}^n$ by arithmetic subgroups, their volumes form an increasing sequence with finite multiplicity. In fact, they are all integral multiples of a common volume (or rather a positive constant). See [Bore17] [Chi] [MargR] and also [Ji9] for more detailed information.

If the rank of $X$ is at least 2 and only the irreducible lattices $\Gamma$ (or irreducible locally symmetric spaces) are considered, then the set of volumes of $\Gamma\backslash X$ is a discrete subset, and each has finite multiplicity. See [Wan3]. For related finiteness results on $\Gamma\backslash X$, see [Gel].

Another closely related problem is to consider volumes of all locally symmetric spaces $\Gamma\backslash X$, where both $X$ and $\Gamma$ are allowed to change. To get well-defined volumes of many different locally symmetric spaces $\Gamma\backslash X$, normalization of the volume form on each $X$ is crucial. As pointed out earlier, non-arithmetic lattices cause problems for hyperbolic manifolds in dimension 2 and 3. So we have to restrict to arithmetic locally symmetric spaces $\Gamma\backslash X$, i.e., $\Gamma$ is an arithmetic subgroup. It turns out that suitable normalization of volume forms of symmetric spaces $X$ can be found, and volumes of all arithmetic locally symmetric spaces $\Gamma\backslash X$ form an increasing sequence with finite multiplicity and going to infinity. See [BoreP] for even more general results for covolumes of $S$-arithmetic subgroups of linear algebraic groups.

## 4.17. Maximal arithmetic subgroups and automorphism groups

Among all arithmetic subgroups, an important class consists of the so-called maximal arithmetic subgroups. Recall that an arithmetic subgroup is called maximal if it is not properly contained in any other arithmetic subgroup. If $G$ does not contain any normal compact factors of positive dimension, then the Borel density implies that every arithmetic subgroup is contained in some maximal arithmetic subgroup.

Some references on maximal arithmetic subgroups are [Belo1-2] [Gree] [GrosN] [All] [Bore2] [Bore17] [Roh3]. For maximal Kleinian reflection groups, see [Ag1] [AgBSW].

If $\Gamma$ contains torsion elements, then $\Gamma\backslash X$ is an orbifold. If $\Gamma' \subset \Gamma$, then $\Gamma'\backslash X$ is a covering space of $\Gamma\backslash X$, and hence the volume of $\Gamma'\backslash X$ is greater than the volume of $\Gamma\backslash X$.

This implies that if $\Gamma\backslash X$ is an orbifold of minimal volume, then $\Gamma$ is a maximal arithmetic subgroup. Hence, finding maximal arithmetic subgroups is basically equivalent to finding orbifolds of minimal volume. When $X = \mathbb{H}^3$, $\Gamma\backslash X$ is a three dimensional hyperbolic orbifold, and a lot of work has been done to find such orbifolds with small volume. See [MaclR] [GehM1] [Ag1] [AgBSW] [GabMM] and the references there.

For three dimensional hyperbolic manifolds and orbifolds with cusps of minimal volumes, see [Meye] [Ada]. See also [Par] and [Hw2] for related results.

For a discussion of maximal discrete subgroups in some special Lie groups, see [All] [Belo1-2] [Roh3] [Wan2] [MaclR].

If $\Gamma'$ is a normal subgroup of $\Gamma$ with finite index, then $\Gamma/\Gamma'$ acts on $\Gamma'\backslash X$; in fact, it is the deck transformation group of the covering map $\Gamma'\backslash X \to \Gamma\backslash X$ if $\Gamma$ is torsion-free. Hence lack of symmetry of $\Gamma\backslash X$ is related to the maximality of $\Gamma$. The problem of determining automorphism groups of hyperbolic manifolds is interesting in itself. It is known that the automorphism group of any compact hyperbolic manifold is finite, and the realization problem asks if every finite group can be realized as such an automorphism group. For this problem and related problems, see [Gree] [BeloL] and the references there.

## 4.18. Counting of volumes of hyperbolic manifolds

As mentioned in §4.16, the volumes of complete hyperbolic manifolds in dimension 2 and 3 do not form a sequence with finite multiplicity and going to infinity. Only for arithmetic hyperbolic manifolds, their volumes form an increasing sequence with finite multiplicity and diverging to infinity. Motivating by the counting of eigenvalues in the Weyl law and counting of lengths of closed geodesics in locally symmetric spaces, it is natural to count volumes of such manifolds.

For more discussions about volumes of arithmetic hyperbolic manifolds, see [Bore17] [Chi]. For a systematic treatment of arithmetic 3-dimensional hyperbolic manifolds, different definitions of arithmetic subgroups, such as construction of arithmetic subgroups via division algebras, estimates on volumes of hyperbolic manifolds, finite subgroups in arithmetic groups, see the book [MaclR]. For computations of volume, for example, the volume of ideal tetrahedrons, see [Milno3] [AleVS].

As mentioned in §4.16, the volumes of complete hyperbolic manifolds in dimension 4 and higher form an increasing sequence of finite multiplicity and going to infinity. Sharp bounds on their counting function are given in [BurGLM]. This is related to, but different from, counting lattice subgroups of the Lie group $\mathrm{Isom}(\mathbb{H}^n) = PO(n,1)$ to be discussed in the next subsection, since here we do not fix a rational structure on $PO(n,1)$. Construction of non-arithmetic lattices in [GrPS] and counting of subgroups of finite index of free groups are used crucially in [BurGLM] to obtain the lower bound, and the upper bound is easier and proved by patching together pieces of hyperbolic manifolds coming from the thick-thin decomposition via the Margulis Lemma.

For counting of volumes of complex hyperbolic locally symmetric spaces, see [Hw1], and for counting of volumes of higher rank locally symmetric spaces, see [Gel].

## 4.19. Counting of subgroups by index

One way to measure the internal structure of a group is to understand its subgroups. In studying arithmetic subgroups $\Gamma$, we often pass to subgroups of finite index. For example, a well-known theorem of Selberg says that every finitely generated linear group $\Gamma$ admits a torsion-free subgroup $\Gamma'$ of finite index.

Assume that $X$ is a symmetric space where $\Gamma$ acts. Then passing from $\Gamma$ to $\Gamma'$ is equivalent to passing from the space $\Gamma\backslash X$ to a finite cover $\Gamma'\backslash X$.

The problem of counting subgroups $\Gamma'$ of $\Gamma$ with finite index is to understand the counting function $N(I) = |\{\Gamma' \mid [\Gamma,\Gamma'] \leq I\}|$ when $I \to +\infty$.

If the volume of $\Gamma\backslash X$ is finite, this is the same as counting finite coverings of $\Gamma\backslash X$ according to their volume. But it is different from counting volumes of all locally symmetric spaces which are quotients of $X$. When $X$ is the real hyperbolic space $\mathbb{H}^n$, this was hinted in the previous subsection, where we counted all hyperbolic manifolds, but not only coverings of a fixed hyperbolic manifold.

There are also variations of counting subgroups by considering special subgroups, for example, counting only normal subgroups.

The numbers $N(I)$ can be used to define a zeta type function, and a natural question is to determine the growth rate of $N(I)$, in particular to characterize those groups such that $N(I)$ grows polynomially, and hence the corresponding zeta type function converges for parameters with sufficiently large real part. See the book [LubS] for a comprehensive introduction. Other references include [LubN] [GoldLP].

Another natural question asks if we can count finite subgroups of $G$. For every arithmetic subgroup $\Gamma$, there are only finitely many conjugacy classes of finite subgroups. Determining the precise structure of these finite subgroups is important. For example, if $\Gamma$ contains torsion elements, then $\Gamma\backslash X$ is not a manifold, but rather an orbifold, and finite subgroups describe the singular locus of $\Gamma\backslash X$.

In the case of integers, the counting of prime numbers can be studied by the Riemann zeta function. There is also a zeta function for counting subgroups of finite index. See [SauS] and the references there.

For counting subgroups over function fields, see [AbNS].

CHAPTER 5

# Discrete subgroups of Lie groups and arithmeticity of lattices in Lie groups

Let $G$ be a Lie group with finitely many connected components. Let $K \subset G$ a maximal compact subgroup. Then $X = G/K$ admits a $G$-invariant metric and is diffeomorphic to $\mathbb{R}^n$. (If $G$ is not semisimple or reductive, $X$ is not a symmetric space.)

Let $\Gamma \subset G$ be a discrete subgroup. Then $\Gamma$ acts properly on $X$. In fact, $X$ is the universal space for proper actions of $\Gamma$. In particular, the virtual cohomological dimension of $\Gamma$ is less than or equal to $\dim X$. This clearly affirms our point in this book that group actions are fruitful for both the groups and the spaces.

This fact that $X$ is the universal space for proper actions of $\Gamma$ is important for many applications in algebraic and geometric topological properties of $\Gamma$ such as the Novikov conjecture for $\Gamma$ and the Baum-Connes conjecture in the theory of $C^*$-algebras of $\Gamma$. See [BauCH] [LucR] [BarLR].

If $\Gamma$ is torsion-free, then $\Gamma\backslash X$ is a $B\Gamma$-space. Hence, $\Gamma\backslash X$ can be used effectively to compute the cohomology groups $H^*(\Gamma, \mathbb{Z})$. (Note that even though $H^*(\Gamma, \mathbb{Z})$ can be defined completely in terms of $\Gamma$ in an algebraic way, a good model of $B\Gamma$ is usually the effective way to compute them.)

If $G$ is a reductive Lie group, then $X$ is a symmetric space of nonpositive curvature. In this case, $X$ is obviously the universal space for proper actions.

Let $\Gamma \subset G$ be a discrete subgroup. If $\Gamma\backslash G$ has finite volume with respect to a left invariant metric, then $\Gamma$ is called a *lattice* (or *a co-finite subgroup*.) If $\Gamma\backslash G$ is compact, then $\Gamma$ is called a *co-compact lattice* (or *a uniform lattice*, or *a co-compact discrete subgroup*, or *a uniform discrete subgroup*). In the following we will use all these names, but mostly lattices.

General references of this section include [Bore4] [Ra5] [ViS] [Morr1] [Ohs] [Zi1] [Marg1] [Most1] [Ji6] [Miln4].

## 5.1. Crystallographic groups and Auslander conjecture

In the rest of this section, we assume that $\Gamma$ is a lattice subgroup of $G$, i.e., $\Gamma\backslash G$ has finite volume with respect to an invariant measure.

When $G = \mathbb{R}^n$, the structure of discrete subgroups is clear. In fact, every discrete subgroup is torsion-free and isomorphic to $\mathbb{Z}^r$, for some $r \leq n$. On the other hand, when $G = \mathrm{Isom}(\mathbb{R}^n)$ whose identity component is the semidirect product of $SO(n)$ with $\mathbb{R}^n$, then a discrete, co-finite (i.e., a lattice) subgroup is called a crystallographic group. It is known that any such group contains the abelian subgroup consisting of translations as a subgroup of finite index.

The classification, in particular, the finiteness of isomorphism classes of crystallographic groups is one of the famous list of problems by Hilbert. Though the complete classification is known in low dimension, it is a non-trivial problem in high dimension where the complete list is not known. See [Wol1] [ViS]. For a more elementary discussion, see [NiS] and [Arm]. For a historic description and relation to the Hilbert 18th problem, see [Milno1].

If $\mathrm{Isom}(\mathbb{R}^n)$ is replaced by the affine group $\mathrm{Aff}(\mathbb{R}^n)$, then a discrete subgroup $\Gamma$ of $\mathrm{Aff}(\mathbb{R}^n)$ is called an affine crystallographic group if it acts properly on $\mathbb{R}^n$ with a compact quotient $\Gamma\backslash\mathbb{R}^n$. A famous conjecture of Auslander [Aus] asks if every affine crystallographic group is virtually solvable, i.e., it contains a solvable subgroup of finite index. It is still open, except some low dimensional cases [AbeMS4]. As mentioned earlier, this is true for (Riemannian) crystallographic groups. But a stronger conjecture by Milnor [Milno4] requiring only that $\Gamma \subset \mathrm{Aff}(\mathbb{R}^n)$ act properly on $\mathbb{R}^n$ but not necessarily with compact quotient is false. See [Abe4] for a survey and references there. See also [AbeMS1-4] [Kaw] for some recent results.

A generalization of the Auslander conjecture to discrete groups acting on nilpotent simply connected, connected Lie groups is formulated and some cases are proved in [BurDD].

A closely related question of Milnor asks whether every torsion-free polycyclic by-finite group $\Gamma$ occurs as the fundamental group of a compact complete affinely flat manifold. This is equivalent to asking whether $\Gamma$ acts properly discontinuously by affine motions on $\mathbb{R}^h$ where $h$ is the Hirsch length of $\Gamma$. This was solved negatively in [Beno4]. For a positive related result on actions by polynomials of bounded degrees, see [DekI]. For related results on NIL-affine crystallographic actions, see [Dek].

### 5.2. Lattices in nilpotent Lie groups and Margulis Lemma

Lattices in the abelian Lie group $\mathbb{R}^n$ have been discussed earlier in §3. They are also special case of crystallographic groups. Lattices in $\mathbb{R}^n$ are exactly co-compact discrete subgroups of $\mathbb{R}^n$. The next class to consider consists of nilpotent Lie groups and their lattices. Besides being interesting in themselves, they also occur naturally in locally symmetric spaces, for example, in the boundary of Borel-Serre compactification, and in the Margulis Lemma for thick-thin decomposition.

If $G$ is a nilpotent Lie group, then it is known that the existence of a lattice subgroup $\Gamma$ implies that $G$ has a $\mathbb{Q}$-structure, i.e., is equal to the real locus of an algebraic group $\mathbf{G}$ defined over $\mathbb{Q}$, and $\Gamma$ is an arithmetic subgroup with respect to this $\mathbb{Q}$-structure.

Note that if $\Gamma$ is any discrete subgroup of a simply connected nilpotent Lie group $G$, then $\Gamma$ is torsion-free and also nilpotent as an abstract group. The converse is also true: A group is isomorphic to a lattice in a simply connected nilpotent Lie group if it is finitely generated, nilpotent, and torsion-free. (See [Mal] [Ra5].)

Nilpotent Lie groups and their discrete subgroups occur naturally in studying semisimple Lie groups and arithmetic subgroups of linear semisimple Lie groups $\mathbf{G}$. In fact, for every $\mathbb{Q}$-parabolic subgroup $\mathbf{P}$ of $\mathbf{G}$, its unipotent radical $N_{\mathbf{P}}$ is a nilpotent subgroup, and an arithmetic subgroup $\Gamma$ induces a lattice $\Gamma \cap N_{\mathbf{P}}$ in $N_{\mathbf{P}}$. They play an important role in studying $\Gamma$ and the associated locally symmetric space $\Gamma\backslash X$. For example, the results on cohomology of nilpotent lattices in [vanE] [Kos] are crucial for the work in [GorHM] on cohomology of semisimple arithmetic groups. Quotients of nilpotent manifolds by lattices are called nilmanifolds. They occur naturally in the boundary of the Borel-Serre compactification of locally symmetric spaces $\Gamma\backslash X$. In fact, the Borel-Serre boundary components are nilmanifold bundles over the boundary components of the reductive Borel-Serre compactification.

Nilpotent discrete groups $\Gamma$ also have a clean characterization in terms of growth in [Gro5]. In fact, with respect to any word metric of $\Gamma$, the number of elements of $\Gamma$ inside a ball of radius $R$ has a polynomial growth in $R$ if and only if $\Gamma$ is nilpotent.

Nilmanifolds are also closely related to collapsing phenomenon in Riemannian geometry. See [Rong] for a survey and related references. A concrete instance in Riemannian geometry where nilpotent discrete subgroups arise naturally is in the Margulis Lemma. Let $M$ be a complete Riemannian manifold. For any point $M$, let $i_x$ be the injectivity radius of $M$ at $x$. For any $\varepsilon > 0$, define the thick part $M_{\geq\varepsilon}$ by

$$M_{\geq\varepsilon} = \{x \in M \mid i_x \geq \varepsilon\},$$

and the thin part $M_{\leq\varepsilon}$ by

$$M_{\leq\varepsilon} = \{x \in M \mid i_x \leq \varepsilon\}.$$

Then $M$ admits the thick-thin decomposition:

$$M = M_{\geq\varepsilon} \cup M_{\leq\varepsilon}.$$

This decomposition allows us to localize geometric problems and has played a fundamental role in Riemannian geometry. On the thick part $M_{\geq\varepsilon}$, the injectivity radius is uniformly bounded away from 0, and the geometry of $M$ is uniformly controlled and relatively easy to understand. Hence, it is important to understand the thin part $M_{\leq\varepsilon}$.

If $M$ is a hyperbolic surface $\Gamma\backslash\mathbb{H}$ of finite topology, i.e., homotopic to a finite CW-complex, then when $\varepsilon$ is small enough, the thin part consists of either cusp neighborhoods or thin necks. Similar results hold for higher dimensional hyperbolic manifolds $\Gamma\backslash\mathbb{H}^n$. It can be seen easily that the sections of these cusp ends are flat manifolds. The converse question that asks if flat manifolds can be realized as sections of cusp ends of hyperbolic manifolds has been studied by several people. See [McR] [Kami] and the references there. See also [BeleK] for related results.

For general manifolds, this thin part is complicated. Assume that $M$ is compact or more generally of finite topology, which could mean that it is either homotopic to a finite CW-complex or homeomorphic to the interior of a compact manifold with boundary. If the sectional curvature of $M$ is strictly negative, a well-known result of Margulis, called *Margulis Lemma*, says that the structure of the thin part $M_{\leq\varepsilon}$ is similar (see [Ji9] for more details about the history, statement, and applications of the Margulis Lemma). In particular, the fundamental group of each connected component of $M_{\leq\varepsilon}$ is virtually nilpotent. Note that this result does not hold for noncompact finite-volume locally symmetric spaces $\Gamma\backslash X$ if the rank of $X$ is greater than or equal to 2. For simplicity, assume that the $\mathbb{Q}$-rank of $\Gamma\backslash X$ is equal to 1. Then for every $\mathbb{Q}$-parabolic subgroup $\mathbf{P}$, its boundary symmetric space $X_{\mathbf{P}}$ has positive dimension, and $\Gamma$ induces a discrete subgroup $\Gamma_{X_{\mathbf{P}}}$ acting on $X_{\mathbf{P}}$. Then each connected component of the thin part of $\Gamma\backslash X$ is a nilmanifold bundle over $[a, +\infty) \times \Gamma_{X_{\mathbf{P}}}\backslash X_{\mathbf{P}}$. Clearly, its fundamental group is not virtually nilpotent if $\Gamma_{X_{\mathbf{P}}}$ is not.

This problem is solved by another version of the Margulis lemma, which works for all discrete groups $\Gamma$ acting properly and isometrically on symmetric spaces $X$ of of noncompact type. It says roughly that when $\varepsilon$ is sufficiently small, for every point $x \in X$, the subgroup of $\Gamma$ generated by $\{\gamma \in \Gamma \mid d(\gamma x, x) \leq \varepsilon\}$ is virtually nilpotent. (Note that non-torsion elements in $\Gamma_{X_{\mathbf{P}}}$ above do not belong to this subgroup if $\varepsilon$ is small enough since $\Gamma_{X_{\mathbf{P}}}\backslash X_{\mathbf{P}}$ is compact.)

See [Fuk] for a general survey about the role of the Margulis Lemma in Riemannian geometry, and [BeneP] for hyperbolic manifolds [BalGS, p. 107] for symmetric spaces of noncompact type. See also [FukY] for a generalization.

## 5.3. Lattices in solvable Lie groups

The next class of lattices to consider after the class of nilpotent groups is the class of solvable Lie groups and their lattices. Solvable Lie groups also occur naturally in studying semisimple Lie groups. In fact, if $G$ is a noncompact semisimple Lie group, $X = G/K$ the associated symmetric space. Then the Iwasawa decomposition (see [Hel1] for example) shows that a solvable Lie group $S = NA$ acts simply transitively on $X$. (It should be pointed out that if $\Gamma$ is a lattice acting on $X$, $\Gamma$ can not be realized as a lattice in $S$.)

Quotients of solvable Lie groups by lattices are called solvmanifolds. The rigidity of solvmanifolds in [Most3] motivated the famous Borel rigidity of aspherical manifolds in §11.12 (also §14.6).

The structure of discrete subgroups of solvable Lie groups is more complicated than nilpotent but can also be understood quite well. For example, there are no above results on $\mathbb{Q}$-algebraic structures on the Lie groups, and no characterization in terms of growth in word metrics.

Recall that a group $\Gamma$ is called *polycyclic* if there exists a finite sequence of subgroups

$$\Gamma = \Gamma_0 \supset \Gamma_1 \supset \cdots \supset \Gamma_k = \{e\}$$

such that $\Gamma_i$ is normal in $\Gamma_{i-1}$ and $\Gamma_{i-1}/\Gamma_i$ is cyclic. If each quotient $\Gamma_{i-1}/\Gamma_i$ is isomorphic to $\mathbb{Z}$, $\Gamma$ is called *strongly polycyclic.* A group $\Gamma$ is called *virtually polycyclic* (resp. *virtually strongly polycyclic*) if it contains a ploycyclic (resp. strongly ploycyclic) group $\Gamma'$ as a subgroup of finite index.

It is known [Ra5, Theorem 4.28] that any lattice (which is automatically cocompact) in a simply connected solvable Lie group is strongly polycyclic, and every polycyclic group $\Gamma$ admits a normal subgroup $\Gamma'$ of finite index that is isomorphic to a lattice in a solvable simply connected Lie group. See [Ra5] [Wi2] and the references there for various results about solvable groups and solvamanifolds.

A particularly interesting result on one type of rigidity of polycyclic groups states that there are only infinitely many nonisomorphic polycyclic groups with isomorphic finite quotients [GruPiS]. Arithmetic groups $GL(n, \mathbb{Z})$ play an important role in such study.

## 5.4. Lattices in semisimple Lie groups

The basic references here are [Marg1] [Ra5] [Zi1] [Bore4] [Morr1] [ViS].

As it is known, general Lie groups can be written as semidirect products of semisimple Lie groups and solvable Lie groups through the Levi decomposition. Though there is no such decomposition for lattices of general Lie groups, study of lattices in general Lie groups can be reduced two cases: lattices in solvable Lie groups, and lattices in semisimple Lie groups.

Solvable lattices were discussed in the previous subsection. The remaining and also the most interesting case concerns lattices of semisimple Lie groups. Assume for the rest of this section that $G$ is a semisimple Lie group. As mentioned earlier, if $\Gamma$ is an arithmetic subgroup of $G$, then the volume of $\Gamma\backslash X$ is finite. Therefore, lattices are natural generalizations of arithmetic subgroups.

A natural question is when a lattice of a semisimple Lie group is an arithmetic subgroup. A closely related problem is to understand special features of arithmetic quotients $\Gamma\backslash X$ among all locally symmetric spaces covered by $X$ both in terms of their spectrum and geometric properties. These questions will be discussed in more detail in the next several subsections together with specific references. Margulis has made fundamental contributions to many questions discussed here. See [Ji9] for a summary of many deep results of Margulis.

Locally symmetric spaces $\Gamma\backslash X$ provide many important examples of Riemannian manifolds. They also arise as moduli spaces in algebraic geometry, and one basic problem there is to give moduli interpretation of locally symmetric spaces. See [Ji6] [DolK] [Yo1-2] [AllCT1] [KugI] and the references there.

## 5.5. Characterization of arithmetic groups

For any subgroup $\Gamma$ of $G$, define its commensurability subgroup by

$$C(\Gamma) = \{g \in G \mid g\Gamma g^{-1} \text{ is commensurable with } \Gamma\}.$$

Clearly, $\Gamma \subset C(\Gamma)$. A characterization of an arithmetic subgroup $\Gamma$ is that its commensurability subgroup $C(\Gamma)$ is not a discrete subgroup of $G$ [Zi1, Theorem 6.2.5]. More specifically, assume that $G$ is a connected semisimple Lie group with the trivial center and without any nontrivial compact factors, and $\Gamma$ is an irreducible lattice in $G$. Then $\Gamma$ is arithmetic if and only if $C(\Gamma)$ is dense in $G$, which is also equivalent to that $C(\Gamma)$ contains $\Gamma$ as a subgroup of infinite index.

Alternatively, it says that arithmetic locally symmetric space $\Gamma\backslash X$ has infinitely many symmetries. Specifically, $\Gamma\backslash X$ admit infinitely many correspondences. (It is known that the isometry group of $\Gamma\backslash X$ is finite.) Another instance of rich symmetry is given in [FarbW1]. For a characterization of locally symmetric spaces in terms of non-discrete group of symmetries, see [FarbW2].

These symmetries are important for spectral theory of automorphic forms on $\Gamma\backslash X$. For example, Hecke operators on automorphic forms and cohomology of $\Gamma$ and various versions of cohomology groups of $\Gamma\backslash X$ are induced from these correspondences. They are also the reason behind a conjecture of Sarnak [Sar6] that the cuspidal eigenvalues of $\Gamma\backslash X$ satisfy the Weyl law. They might also be the reason for the validity of the Selberg-$\frac{1}{4}$ conjecture on the first positive eigenvalue of $\Gamma\backslash\mathbb{H}$ for congruence subgroups $\Gamma$ mentioned in §3.4.3.

Another important property of lattice subgroups is the following normal subgroup property: Assume that $G$ is a connected semisimple Lie group with the trivial center and without any nontrivial compact factors, and $\Gamma$ is an irreducible lattice in $G$. Then any normal subgroup of $\Gamma$ is either finite or has finite index in $\Gamma$. It says that such a lattice subgroup is simple modulo finite subgroups. See [Marg1] [BadS] and also [Ji9, pp. 19–21] for some applications.

A famous result of Margulis (see [Marg1] [Zi1]) is that irreducible lattices in semisimple Lie groups of rank at least 2 is arithmetic. This follows from his super-rigidity (see §11.4 below.) This result is useful for many reason. For example, if $\Gamma$ is a discrete subgroup acting on a symmetric space $X$. Assume that $\Gamma\backslash X$ is noncompact but the volume of $\Gamma\backslash X$ is finite. In order to develop the spectral theory of the locally symmetric space $\Gamma\backslash X$, for example, determination of the continuous spectrum and their generalized eigenfunctions, we need to understand shapes of the space $\Gamma\backslash X$ near infinity well. For hyperbolic surfaces $\Gamma\backslash\mathbb{H}$, each end is a cuspidal end, and it follows from this that the continuous spectrum of $\Gamma\backslash\mathbb{H}$ is equal to $[\frac{1}{4}, +\infty)$. The reduction theory in [Bore4] only works for arithmetic subgroups. In the book [Lang3], several properties on $\Gamma$, motivated by the reduction theory of arithmetic subgroups, were assumed before the spectral theory was developed. By this arithmeticity result of Margulis, these assumptions are satisfied for irreducible higher rank irreducible lattices. The arithmeticity result does not hold for lattices in rank one Lie groups, but the reduction theory for them has been developed in [GaR]; in particular, there is a satisfying spectral theory of automorphic forms on $\Gamma\backslash X$ when $\Gamma\backslash X$ is a locally symmetric space of finite volume and when the rank of $X$ is equal to 1.

For a more detailed discussion about the history, evolution and motivations of this arithmeticity result of Margulis and related results, see the survey paper [Ji9].

For characterization of arithmetic Fuchsian subgroups of $SL(2, \mathbb{R})$, see the discussion in §3.3.13. See [JiaPS] for related results.

## 5.6. Non-arithmeticity of lattices in rank 1 cases

It is well-known that there are many non-arithmetic lattice subgroups in $SL(2, \mathbb{R})$. The reason is that if $\Sigma_g$ is a compact Riemann surface of genus at least 2, then by the uniformization theorem for Riemann surfaces, $\Sigma_g = \Gamma\backslash\mathbb{H}$ for some lattice subgroup $\Gamma \subset SL(2, \mathbb{R})$. It is well-known that the moduli space of complex structures of a closed surface of genus $g$ is of dimension $6g - 6$, but clearly there are only countably infinitely many arithmetic subgroups. In fact, in the moduli space of hyperbolic surfaces $\Gamma\backslash\mathbb{H}$ of a fixed genus $g \geq 2$, there are only finitely many arithmetic hyperbolic surfaces, which follows from the result in [Chi] that the areas of arithmetic hyperbolic surfaces form an increasing sequence of finite multiplicity. As a consequence, most (or rather generic) Riemann surfaces are not arithmetic.

For the other real hyperbolic spaces $\mathbb{H}^n$, $n \geq 3$, the construction of non-arithmetic subgroups of $SO(n, 1)$ or $PO(n, 1)$ is more complicated. See [Vin] [GrPS] [ViS].

The complex hyperbolic spaces form an important class of symmetric spaces of rank 1. There exist non-arithmetic lattice subgroups. In fact, $PU(2, 1)$ [Most2] and $PU(3, 1)$ [DeM] contains non-arithmetic lattices. But for $n \geq 4$, it is not known if they exist.

Some general reference on discrete subgroups acting on the real hyperbolic spaces and related hyperbolic manifolds are [Bea] [BeneP] [ViS] [Ohs] [Rat] [Th1].

## 5.7. Arithmeticity of lattices in rank 1 cases

Besides the real and complex hyperbolic hyperbolic spaces, there are two remaining types of symmetric spaces of rank 1: the quaternionic hyperbolic spaces and the octonionic plane. In these cases, all lattice subgroups are arithmetic, by combining the super-rigidity type results of Corlette, Gromov-Schoen [Cor1] [GrS]. In these papers, harmonic maps into CAT(0)-spaces, i.e., nonpositively curved metric spaces, play an important role. See [Ji5] for explanations of how harmonic maps into Bruhat-Tits buildings are used to prove the super-rigidity results. The methods of proofs using harmonic maps motivated a lot of work on analysis on metric spaces, see [Jos] [Bal1] and references there.

Though there are non-arithmetic lattice subgroups for the real and lower dimensional complex hyperbolic spaces, they satisfy some weaker conditions, the so-called integrality, which says roughly that some subgroups of finite index are contained in some arithmetic subgroups but not necessarily of finite index. See [Kligl] [Ye1-2]. For some related applications to fake projective spaces, see [PYe1-2].

## 5.8. Linear discrete subgroups and Tits alternative

A group $\Gamma$ is said to satisfies the *Tits alternative* if it is either virtually solvable, i.e., contains a solvable subgroup of finite index, or contains a free group $F_n$ on $n$ generators for some $n \geq 2$, and hence contains $F_n$ for all $n$.

It is known that every finitely generated linear group over a field $k$, i.e., a subgroup of $GL(n, k)$, satisfies the Tits alternative [Tit3]. This is an important property of linear groups. For example, it implies that for every such group, its growth, i.e, the number of elements in metric balls with respect to every word metric, is either of polynomial growth or of exponential growth. (This is another alternative.) See [Sha7] for a simpler proof. See [BrG1-2] for related results. A different proof of the Tits alternative using the Oseledets multiplicative ergodic theorem was later given by Guivarch [Guiv].

The Tits alternative has many important applications, for example, it is an important ingredient in Gromov's proof that finitely generated groups with polynomial growth are nilpotent, and it is also used to prove that a famous conjecture of von Neumann that non-amenable groups must contain free groups is true for linear groups.

This Tits alternative property has been studied for many other classes of groups, for example, mapping class groups and outer automorphism group of free groups. See [Iv5] [Mc] [BesFH2-3] [NoV]. See also [Ji9] for more related results and references.

Arithmetic subgroups are linear groups. But there are many other linear groups which are not arithmetic subgroups. For example, an arbitrary subgroup of $GL(n, \mathbb{Q})$ generated by finitely many elements is in general not, or even contained in, an arithmetic subgroup. For the Zariski density property of linear groups which are not arithmetic subgroups and related results, see [Beno2] [PrRa] [GuiS] [AbeMS1] [DalK].

For linear groups which divide a convex domain, see [Benol]. See [ConzG1] for distribution of orbits of linear groups. See also [Sof].

## 5.9. Reflection groups

For symmetric spaces of constant curvature, reflections with respect to geodesic hypersurfaces are isometries. Suitable collections generate discrete subgroups. This is a basic method of generating Kleinian groups acting on the hyperbolic spaces $\mathbb{H}^n$, but cocompact lattices can only be constructed in dimension less than 18 (see [ViS][Bea] [Th1]. See also [Mas2] [Mas1] for related method of generating discrete Fuchsian groups from a fundamental domain with geodesic sides).

The most basic example is the Coxeter group associated with a root system. They are fundamental in studying structures of Lie algebras, Lie groups and many related topics. see [Hum1] and references there.

For real hyperbolic spaces, reflections can be used to construct non-arithmetic lattice subgroups. They are some of the first non-arithemtic subgroups constructed. See [ViS] [Vin] [Nik].

Reflections and Weyl group type reflection groups occur naturally in algebraic geometry. For a comprehensive discussion of the roles of reflection groups in algebraic geometry, see [Dol1]. One basic point is that lattices occur naturally in cohomology groups of varieties, and some exceptional rational curves naturally give rise to reflections.

For other references on groups related to Coxeter groups and applications to geometric topology, see [Dav1-3] [JaS3] and the references there. They are related to CAT(0)-groups.

A simple but important remark is that compositions of reflections give rise to other symmetries, such as translations. These two operations can be combined together in basic groups such as crystallographic groups and affine Weyl groups.

## 5.10. Discrete groups related to Kac-Moody groups and algebras

Arithmetic groups and discrete subgroups of Lie groups are related to finite dimensional linear algebraic group and Lie groups. There are also natural discrete groups associated with infinite dimensional Lie groups or Lie algebras. In particular, two important examples are $SL(n, \mathbb{Z}[t])$ and $SL(n, \mathbb{F}_q[t])$, where $t$ is a free variable, and $\mathbb{F}_q$ is a finite field.

As it is well-known, the structure of a simple finite dimensional complex Lie algebra $\mathfrak{g}$ is determined by its root system, in particular, its Cartan matrix, which is an integral positive definite matrix. The generalized Cartan matrices are obtained by relaxing the positivity condition and naturally lead to the Kac-Moody algebras, which are usually infinite dimensional Lie algebras [Kacv]. It is also well-known that there are correspondences between Lie algebras and Lie groups. Groups corresponding to Kac-Moody algebras are called Kac-Moody groups and constructed in [Tit5], which are usually not finite dimensional Lie groups. For each simple finite dimensional complex Lie algebra, there is an associated affine Kac-Moody algebra, and their Kac-Moody groups contain discrete subgroups such as $SL(n, \mathbb{Z}[t])$.

Kac-Moody algebras and Kac-Moody groups can be defined and constructed over other fields as well. They give rise to different kind of discrete subgroups. Specifically, let $G$ be a completion of Tits' Kac-Moody group functor over a finite field [Tit5]. Then $G$ is a locally compact, totally disconnected topological group. Such completions have been constructed in [CarG1-2] and [ReR]. A reference for the comparison between these constructions is [CarER]. Let $B^-$ be the subgroup of $G$ generated by the 'diagonal subgroup' $H$ and the subgroup $U^-$ generated by all negative real root groups. The group $B^-$ is the minimal parabolic subgroup of the negative $BN$-pair for $G$. It was shown independently in [CarG1-2] and [Re2] that $B^-$ is a non-uniform lattice subgroup of $G$.

The group $G$ admits a cocompact action on its corresponding Bruhat-Tits building, which is an affine or hyperbolic building in rank greater than 2, and a homogeneous tree in rank 2. The non-uniform lattices of [CarG1-2] are constructed in analogy with the 'arithmetic' method for Lie groups using the Tits system for such groups.

[CarG1] also constructed an uncountably infinite family of both uniform and non-uniform lattices in the rank 2 Kac-Moody case by generalizing Lubotzky's construction of lattices in $SL_2$ over a Laurent series field using the Bruhat-Tits tree ([Lub2].) Thus one can deduce that if one discovers a natural notion of 'arithmeticity' for Kac-Moody groups, then in rank 2, there should be 'non-arithmetic' lattices as well.

When $G$ has rank 2, [Car3] defines *congruence subgroups* of $B^-$ as a natural generalization of the corresponding notion for lattices in Lie groups and constructs congruence subgroups of $B^-$ using the Tits building of $G$, which is a tree. The technique involves determining a graph of groups presentation for $B^-$. Though the Kac-Moody group $G$ has no apparent arithmetic or algebraic structure, these results reveal strong analogies between lattices in $G$ and arithmetic lattices in the Lie group $SL_2(\mathbb{F}_q(t^{-1}))$.

## 5.11. Infinite dimensional Lie groups and discrete groups associated with them

In the previous subsection, we have considered some discrete groups associated with infinite dimensional Kac-Moody groups. In this subsection, we consider other infinite dimensional Lie groups and their related groups.

The group $SL(n, \mathbb{Z}[t])$ mentioned in the previous subsection is a discrete subgroup of the loop group $L(SL(n, \mathbb{C}))$, which is defined to be the group of continuous maps

$$S^1 \to SL(n, \mathbb{C}).$$

We can also consider discrete subgroups of loop groups of other finite dimensional Lie groups and algebraic groups. A natural question is whether some results for arithmetic subgroups such as $SL(n, \mathbb{Z})$ can be generalized to them. For example, the integral Novikov conjecture is known for all arithmetic subgroups (see [Ji1]). It is natural to expect that it also holds for $SL(n, \mathbb{Z}[t])$. See [Sta3] for some related discussions.

We can also consider loop groups of algebraic groups over finite fields. For the reduction theory and the theory of Eisenstein series on such loop groups, see [Gar1-3]. It seems that topological and geometric properties of such discrete (or arithmetic) subgroups have not been understood yet.

There are also other infinite dimensional Lie groups arising from geometry and mechanics. It is known that if $M$ is a Riemannian manifold, then its isometry group is a finite dimensional Lie group. For various applications, it is natural to relax the condition of preserving the metric and consider the group $\mathrm{Diff}(M)$ of all diffeomorphisms of $M$ and subgroups of $\mathrm{Diff}(M)$ preserving some other structures, for example, the subgroup $\mathrm{SDiff}(M)$ of special diffeomorphisms, i.e., diffeomorphisms preserving the volume form of $M$. These groups are in general infinite dimensional and have important applications in mechanics. For example, geodesics in these Lie groups with respect to left invariant metrics often describe trajectories of motion. When $M = S^1$, the group $\mathrm{Diff}(M)$ is infinite dimensional and has been intensively studied. See the book [ArnK] and also the papers [Arn] [EbM] for discussions of geometry of infinite dimensional diffeomorphism groups and applications to dynamics such as fluid dynamics and other related problems. For some general results on infinite dimensional Lie transformation groups, see [Om].

For a finite dimensional semisimple Lie group $G$, an important discrete group associated with it is the Weyl group $W$, which is the quotient $N(T)/T$, where $T$ is a Cartan subgroup of $G$, and $N(T)$ the normalizer of $T$. The Weyl group acts simply transitively on the set of Weyl chambers of the Cartan subalgebra $\mathfrak{t}$, the Lie algebra of $T$. The Weyl chambers induce a simplicial cone decomposition of $\mathfrak{t}$, which is dual to the convex hull of the Weyl group orbit of a generic point in $\mathfrak{t}$. These convex polytopes are related to several celebrated convexity properties of the images of the moment map and of the projection map to the Cartan subgroups.

In general, infinite dimensional Lie groups have not been understood as well as their finite dimensional analogues. On the other hand, some important results have been obtained. For example, for the infinite dimensional Lie group of area preserving diffeomorphisms of the annulus, the Weyl group can be defined and related convexity results hold. See [BloFR1-2] and the references there.

In connection with discrete groups, several natural questions occur. The first question is what are natural analogues of lattices or more specially arithmetic subgroups of $\mathrm{Diff}(M)$ and $\mathrm{SDiff}(M)$.

By [Ji7], it is known that the integral Novikov conjecture holds for every finitely generated subgroup of $GL(n, \overline{\mathbb{Q}})$, where $\overline{\mathbb{Q}}$ is the algebraic closure of $\mathbb{Q}$. The second question asks if the integral Novikov conjecture holds for finitely generated subgroups of Diff($M$). One problem is to get an analogue of algebraic elements as those in $GL(n, \overline{\mathbb{Q}})$. If $M$ is a projective algebraic variety defined over $\mathbb{Q}$, $M \subset \mathbb{C}P^n$, then it can be shown any subgroup of Diff($M$) generated by finitely many diffeomorphisms which are restrictions of rational elements of the automorphism group $\mathrm{Aut}^0(\mathbb{C}P^n) = PSL(n+1, \mathbb{C})$ satisfies the integral Novikov conjecture. In fact, it follows from the result in [Ji7]. But there are few such groups. A natural problem is to remove such algebraic conditions on the elements of Diff($M$) and prove the integral Novikov conjecture for larger classes of finitely generated subgroups of Diff($M$) and SDiff($M$).

Finitely generated subgroups of Diff($M$) and SDiff($M$) certainly give rise to natural groups in the theory of combinatorial group theory and should share similar properties as those finitely generated subgroups of the classical Lie group $GL(n, \mathbb{R})$.

CHAPTER 6

# Different completions of $\mathbb{Q}$ and $S$-arithmetic groups over number fields

A general reference on algebraic number theory is [CassF]. For a comprehensive discussion of harmonic analysis on locally compact fields, completions of global fields, and classification of locally compact fields, see [RamV]. For linear algebraic groups over number fields and related number theoretic questions, see [PlR1].

## 6.1. $p$-adic completions

It is known that with respect to any nontrivial absolute value of $\mathbb{Q}$, the field $\mathbb{Q}$ is not complete or locally compact. For the purpose of harmonic analysis such as the Fourier transformation, the local compactness is important. A natural idea is to complete $\mathbb{Q}$. There are many completions depending on the absolute values used.

For each prime $p$, there is a valuation of $\mathbb{Q}$ and an associated norm on it. The associated completion of $\mathbb{Q}$ in this norm is the field of $p$-adic numbers $\mathbb{Q}_p$. This is similar to that $\mathbb{R}$ is the completion of $\mathbb{Q}$ with respect to the usual absolute value, which corresponds to the archimedean valuation of $\mathbb{Q}$.

Then $\mathbb{Q}_p$ is a locally compact field. Every locally compact field containing $\mathbb{Q}$ as a dense subset is equal to either $\mathbb{Q}_p$ for some prime $p$, or $\mathbb{R}$.

Similarly we can get completions of other number fields $k$, which are finite extensions of $\mathbb{Q}$. Each completion of a number field $k$ corresponds to a place of $k$, which is defined to be an equivalence classes of valuations of the field $k$.

Every locally compact field of characteristic 0 is one of the above completions of a number field. See [RaV, Chap. 4].

## 6.2. $S$-integers

The idea of localization is important in algebra and number theory. Let $S$ be the union of finitely many primes (or places) $\{p_1, \dots, p_m\}$ and the infinite place of $\mathbb{Q}$. Then there is an associated ring of $S$-integers $\mathbb{Z}_S = \mathbb{Z}[\frac{1}{p_1}, \dots, \frac{1}{p_m}]$, which consists of all rational numbers containing only powers of $p_1, \dots, p_m$ in their denominators.

In general, let $k$ be a number field, and $\mathcal{O}_k$ be the ring of integers. Let $S$ be a finite set of places containing all infinite places of $k$. Then we can also define the ring of $S$-integers $\mathcal{O}_{k,S}$. When $S$ consists of exactly the infinite places, then the ring $\mathcal{O}_{k,S}$ is equal to the ring of integers $\mathcal{O}_k$ of the number field $k$. See [PR, Chap. 1].

$S$-integers have been studied from various points of view. See [Burg] [GunMM] for example.

## 6.3. $S$-arithmetic subgroups

Let $\mathbf{G}$ be an algebraic group defined over $k$. Then an $S$-arithmetic subgroup $\Gamma$ of $\mathbf{G}(k)$ is a subgroup commensurable with $\mathbf{G}(\mathcal{O}_{S,k})$.

An important example of $S$-arithmetic subgroups is $SL(n, \mathbb{Z}[\frac{1}{p_1}, \dots, \frac{1}{p_m}])$, where $p_1, \dots, p_m$ are primes. In this case, the algebraic group is the standard form of $SL(n, \mathbb{C}) \subset GL(n, \mathbb{C})$, the number field is taken to be $\mathbb{Q}$, and $S = \{\infty, p_1, \dots, p_m\}$.

One important reason for considering $S$-arithmetic subgroups is that a general finitely generated subgroup of $SL(n, \mathbb{Q})$ is not contained in an arithmetic subgroup, but is always contained in some $S$-arithmetic subgroup, where the set $S$ contains the prime factors in the denominators of the finitely many generators. Therefore, $S$-arithmetic subgroups provide a natural class of groups which strictly contains the class of arithmetic subgroups. Another important reason is that they share many properties with arithmetic subgroups.

For striking applications of $S$-arithmetic subgroups of $SL(2)$ to the construction of expander (or Ramanujan) graphs and distributions of points on the sphere, see [LubPS1-3].

## 6.4. $S$-arithmetic subgroups as discrete subgroups of Lie groups

As discussed before, the embedding of an arithmetic subgroup into a Lie group (with finitely many connected components) is crucial for understanding structures of the arithmetic subgroup. Many important locally symmetric spaces are constructed this way. For example, such an embedding induces a proper action of the arithmetic subgroup on a suitable homogeneous space.

$S$-arithmetic subgroups are not discrete subgroups of real Lie groups if $S$ contains finite places. For example, $SL(n, \mathbb{Z}[\frac{1}{p}])$ is not a discrete subgroup of $SL(n, \mathbb{R})$. It is not a discrete subgroup of the $p$-adic Lie group $SL(n, \mathbb{Q}_p)$ either.

Let $k$ be a number field as before. For each place $\nu \in S$, there is a locally compact field $k_\nu$, the completion of $k$ with respect to a norm associated with $\nu$. And the locus $\mathbf{G}(k_\nu)$ is a locally compact group (in fact, either a real Lie group or a $p$-adic Lie group.) Under the diagonal embedding,

$$\Gamma \hookrightarrow \prod_{\nu \in S} \mathbf{G}(k_\nu),$$

the $S$-arithmetic subgroup $\Gamma$ becomes a discrete subgroup of the group $\prod_{\nu \in S} \mathbf{G}(k_\nu)$. In the definition of $S$-integers, it might not be clear why we needed to include all the archimedean places in $S$. Now this embedding makes it clear. If $S$ does not contain all the infinite places, the image of the above embedding of $\Gamma$ will not be a discrete subgroup. For example, as mentioned before, $SL(n, \mathbb{Z}[\frac{1}{p}])$ is a discrete subgroup of $SL(n, \mathbb{R}) \times SL(n, \mathbb{Q}_p)$, but not a discrete subgroup of $SL(n, \mathbb{Q}_p)$.

In the super-rigidity of lattices, it is important to treat both real and $p$-adic Lie groups simultaneously. In fact, from page 1 on in [Marg1], rigidity properties are discussed for $S$-arithmetic subgroups.

Basic references on $S$-arithmetic subgroups include [Marg1] [PlR1] [Borel4] [Serr1-3] [BoreS2]. For a characterization of $S$-arithmetic subgroups, see [LubV].

CHAPTER 7

# Global fields and $S$-arithmetic groups over function fields

## 7.1. Function fields

By a function field, we mean the field of rational functions on a projective curve $C$ defined over a finite field $\mathbb{F}_q$, usually denoted by $\mathbb{F}_q(C)$. For example, when the curve is the projective curve $\mathbb{P}^1$, then the function field is $\mathbb{F}_q(t)$, where $t$ is a variable.

Places of a function field $\mathbb{F}_q(C)$ correspond to points of the curve $C$. We can also define completions of function fields. The completions are also locally compact fields, or local fields. For example, the quotient field of formal power series $\mathbb{F}_p[[t]]$ is a completion of $\mathbb{F}_p(t)$ corresponding to the point of the origin in $\mathbb{P}^1 = \mathbb{F}_q(t) \cup \{\infty\}$.

For an elementary and comprehensive introduction to function fields, see [SalD]. For discussion of modular forms over function fields together with some history on complex multiplication, see [Vla, Part III and Chap. 3].

## 7.2. Global fields

By a global field $k$, we mean either a number field or the function field of a projective curve over a finite field. One basic thing they have in common is that for any place, the completion of the field is a locally compact field such as $\mathbb{R}$ and $\mathbb{Q}_p$. This is one of the reasons why they are called global fields.

They share other properties. For example, for every non-zero number $x \in k$, for every place $\nu$ of $k$, denote the corresponding normalized norm by $|x|_\nu$, where the norm is normalized to coincide with the modular function of the additive Haar measure. Then both types of global fields $k$ satisfy the following product formula:

$$\prod_{\nu} |x|_\nu = 1.$$

Local compactness is important for various reasons, for example, for harmonic analysis on the fields, such as the existence of Haar measure. Problems over global fields are often what is really interesting, for example, the existence of $\mathbb{Q}$-points of varieties. On the other hand, the corresponding problems over the completed local fields are easier. For example, it is not obvious whether a quadratic form $Q(x)$, $x \in \mathbb{R}^n$, with rational coefficients, represents 0 over $\mathbb{Q}$, i.e., if there exists $x \in \mathbb{Q}^n$ such that $Q(x) = 0$. But over $\mathbb{R}$, the archmedean completion of $\mathbb{Q}$, the corresponding problem is easy. Passing from solutions of the local fields back to the global fields is important and nontrivial. There are often obstructions associated with individual problems.

A foundational book showing similarities between these two types of fields regarding both results proved and methods used is the book [Wei].

Among many similarities between them, the most important one to us here is that they both give rise to $S$-arithemtic subgroups, and Bruhat-Tits buildings and symmetric spaces are natural spaces for them to act upon.

For some overview of relations between number fields and function fields, see [VadMS]. For discussions of modular forms over both number fields and function fields, see [Vla, Part III and Chap. 3]. For another instance of interplay between number fields and function fields, see [KatzS1-2].

## 7.3. $S$-arithmetic subgroups over function fields

For number fields, $S$-arithmetic subgroups were defined above. $S$-arithmetic groups over function fields can be defined similarly, except that function fields do not have any infinite place, and hence $S$ consists of only finite places.

Important examples include $SL(n, \mathbb{F}_p[t])$, $SL(n, \mathbb{F}_p[t, \frac{1}{t}])$. For a nice and short summary, see [Bro1].

For function fields, there is no analogue of arithmetic subgroups, since there is no infinite place. From this point of view, $S$-arithmetic subgroups are the only analogue of arithmetic subgroups of linear real Lie groups over function fields, and are particularly important for automorphic forms and automorphic representations over function fields.

For some discussions of spectral theory and applications of $S$-arithmetic subgroups of rank 1, see §12.14 below. Other references include [Abe1-2] [Gra3] [Serr1] [Bux] [Abr1-2] [BuW1-2] [Stu1-3].

CHAPTER 8

# Finiteness properties of arithmetic and $S$-arithmetic groups

For general discrete groups, there are many finiteness conditions. Several natural ones are related to cohomological properties of the groups and are hence called cohomological finiteness results. One reason why these conditions are important is that they give important invariants of the groups under discussion.

The general references for the next several subsections are [Bro2] [Serr1-2] [Abel] [Bier1].

## 8.1. Finite generation

A natural way to describe a discrete group is to find generators. Then other elements can be expressed as words in terms of them. For example, the group $\mathbb{Z}^n$ is generated by $e_1 = (1, 0, \dots, 0), \dots, e_n = (0, \dots, 0, 1)$. This is the free abelian group on $n$ generators. If the field $\mathbb{Q}$ of rational numbers is considered as a group under addition, then it is not finitely generated. In fact, elements of any finitely generated subgroup has bounded denominators.

Given a discrete group $\Gamma$, the first basic finiteness question asks if the group is finitely generated. Finite generation is important for studying the group $\Gamma$ as metric spaces, in particular from the point of view of large scale geometry.

For example, given a finite generating set $S$ of $\Gamma$ which is symmetric in the sense that for any element $\gamma$ in $S$, its inverse $\gamma^{-1}$ also belongs to $S$, we can define a word metric $d_S$ on $\Gamma$.

Briefly, let $e$ be the identity element of $\Gamma$. Then for every $\gamma \in \Gamma$, $d_S(\gamma, e)$ is defined to be the minimum length of words in $S$ which express $\gamma$. For any two elements $\gamma_1, \gamma_2$, define the distance between them by

$$d_S(\gamma_1, \gamma_2) = d_S(\gamma_1^{-1}\gamma_2, e).$$

This defines a left invariant metric on $\Gamma$. Then $\Gamma$ can be studied in terms of the metric space $(\Gamma, d_S)$. Clearly, the word metric $d_S$ depends on the choice of the set $S$ of generators. On the other hand, any two word metrics $d_{S_1}$ and $d_{S_2}$ are quasi-isometric in the sense that there exists a positive constant $k > 1$ such that for every pairs of elements $\gamma_1, \gamma_1 \in \Gamma$,

$$k^{-1} d_{S_1}(\gamma_1, \gamma_2) \leq d_{S_2}(\gamma_1, \gamma_2) \leq k d_{S_1}(\gamma_1, \gamma_2).$$

Consequently, the large scale geometry of the metric space $(\Gamma, d_S)$ is independent of $S$.

We can also define a Cayley graph of $\Gamma$, whose vertices are elements of $\Gamma$, and two vertices are connected by an edge if and only if they are related by an element in $S$. If every edge is assigned the unit length, the Cayley graph becomes a geodesic length space, i.e., a metric space where distances between points are realized by geodesics connecting them. Then the metric space $(\Gamma, d_S)$ is isometrically embedded into the Cayley graph. If $G$ is the free group $F_2$ on two generators $a, b$ with the set $S = \{a, a^{-1}, b, b^{-1}\}$, then its Cayley graph is the standard binary tree (without any root or terminating branch.)

Finite generation can be expressed algebraically in terms of maps from free groups as well. If $F_n$ denotes the free group on $n$ generators, then the finite generation of $\Gamma$ is equivalent to the existence of a surjective homomorphism $F_n \to \Gamma$ for some $n$.

## 8.2. Bounded generation

An important subclass of the class of finitely generated groups consists of boundedly generated groups.

A finitely generated group $G$ is called boundedly generated if there exists a finite set $S$ of generators and a number $k$ depending only on $G$ and $S$ such that every element $\gamma \in \Gamma$ can be written as a product:

$$\gamma = \gamma_1^{m_1} \dots \gamma_k^{m_k},$$

where $\gamma_1, \dots, \gamma_k \in S$, and $m_1, \dots, m_k \in \mathbb{Z}$.

This notion was first introduced in [CartK]. In fact, [CartK] proved that the arithmetic subgroup $SL(n, \mathcal{O}_k)$ is boundedly generated, where $n \geq 3$ and $\mathcal{O}_k$ is the ring of integers of a number field $k$. (See [AdM] for a simpler proof.) This notion was motivated by the positive solution of the congruence subgroup problem for $SL(n, \mathcal{O}_k)$. For more connections with the congruence subgroup problem, see [Rap1] [Rap4] [Kum] [PlR2] and the references there.

For bounded generation of arithmetic subgroups of higher rank algebraic groups, see [Tav] and also [SharV] [Mur]. For $S$-arithmetic subgroups, see [ErR] [AbLP]. For non-bounded generation of discrete subgroups of rank 1 simple Lie groups, see [Fuj1].

Besides connections with the congruence subgroup problems, it is also closely related to property (T) and used in understanding the latter. In fact, explicit Kazhdan constants, which characterizes the property (T), of arithmetic subgroups are computed in [Shal4] using methods of the bounded generation.

For relations on bounded generation and growth rate of the counting function of subgroups, see [Py].

Bounded generation is stronger than finite generation. But it does not imply finite presentation of a group, which will be discussed in the next subsection. See [Sur].

## 8.3. Finite presentations

Once it is known that a group $\Gamma$ is finitely generated, it is important to understand relations between these generators. Two simplest examples are the free abelian group $\mathbb{Z}^n$, in which every two elements commute and there is no other restriction, and the free group $F_n$, in which there is no relation at all between any two elements.

Most groups satisfy some relations among their generators. Given a finitely generated group $\Gamma$, a natural question asks if $\Gamma$ is finitely presented, i.e., if there are only finitely many non-redundant relations among the finitely many generators. This is equivalent to the existence of an exact sequence

$$F_m \to F_n \to \Gamma \to \{1\}.$$

Finite relations are important for computations in the group. As mentioned in §3.3.11, finite generation and finite presentation of $\Gamma$ often follow from existence of a nice fundamental domain of $\Gamma$ acting on a topological space.

Understanding groups which are finitely generated and finitely presented is the basic problem in combinatorial group theory. For a historic account of this theory, see [ChaW]. For more detailed discussions, see [LyS] [MagKS]. See also [FiS] for groups defined essentially by one relation.

It is known that arithmetic subgroups of linear algebraic groups are finitely generated and presented. On the other hand, finding explicit generators and relations is often difficult. For some results on generators of arithmetic and $S$-arithmetic subgroups, see [GruS1] [GruS2] [GruS3] [Ra8] [Ve5].

If a group is finitely generated, a natural question is the word problem, i.e, to decide if a given word in generators is trivial or not by following an algorithm. (Note that to store generators and relations of a group on a machine, their finiteness is important.) For general groups, this problem is not solvable. Automatic groups form a class of groups for which this problem can be solved efficiently. They also describe in a combinatorial way fundamental groups of compact manifolds of non-positive curvature, similar to that hyperbolic groups describe combinatorially fundamental groups of compact manifolds with strictly negative curvature. See [EepCHL].

Another generalization of the fundamental group of compact nonpositively curved Riemannian manifolds is the class of CAT(0) groups in §9.4. A natural question is to understand relations between automatic groups and CAT(0)-groups. See [El] and the references there for results on this question.

## 8.4. Finiteness properties such as $FP_\infty$

There are other ways to formulate finiteness properties of $\Gamma$. Let $\mathbb{Z}[\Gamma]$ be the group ring of $\Gamma$ with coefficients in $\mathbb{Z}$. Then $\mathbb{Z}$ can be considered as a trivial $\mathbb{Z}[\Gamma]$-module. Some finiteness properties of $\Gamma$ can be expressed in terms of resolution of $\mathbb{Z}$ by projective or free $\mathbb{Z}[\Gamma]$-modules. Here, several typical properties are $FP_n$, $FP_\infty$, $FL$ etc. For example, $\Gamma$ is said to be of type $FP_\infty$ if it admits a projective resolution by finitely generated $\mathbb{Z}[\Gamma]$-module. It is of type $FP_n$ if only the first $n$ projective modules are finitely generated. If it admits a resolution of finite length by finitely generated free modules, then it is said to be of type $FL$. See the book [Bro2] for detailed discussions on these conditions.

Recall that for a group $\Gamma$, a classifying space of $\Gamma$, i.e., a $B\Gamma$-space (or a $K(\Gamma,1)$-space), is a space satisfying the conditions:

$$\pi_1(B\Gamma)=\Gamma,\quad \pi_i(B\Gamma)=\{1\}\ \text{ for }\ i\geq 2.$$

For later purpose, we point out that the universal covering space $E\Gamma=\widetilde{B\Gamma}$ is the universal space for proper and fixed free actions of $\Gamma$.

A very strong finiteness condition on $\Gamma$ is that there exists a $B\Gamma$-space given by a finite CW-complex, a so-called finite $B\Gamma$-space. This implies all the previous finiteness properties such as $FP_\infty$ and $FL$.

Usually $B\Gamma$-spaces are CW-complexes. There are also some weaker conditions, for example, the existence of a finite dimensional $B\Gamma$, or a space $B\Gamma$ such that in each dimension, there are only finite many cells of that dimension. The former is related to finiteness of cohomological dimension of $\Gamma$, and the latter is related to finite generation of the cohomology groups $H^*(\Gamma,\mathbb{Z})$ of $\Gamma$ in that degree.

By definition, the cohomological dimension of a group $\Gamma$ is the largest integer $i$ such that $H^i(\Gamma,M)\neq 0$ for some $\Gamma$-module. If $\Gamma$ contains a nontrivial torsion element, then its cohomological dimension is equal to $+\infty$. The virtually cohomological dimension of a group containing torsion elements is the cohomological dimension of its torsion-free subgroups of finite index if they exist. (It is independent of the choice of the torsion-free subgroups of finite index.) Since many natural arithmetic subgroups such as $SL(n,\mathbb{Z})$ contain nontrivial torsion-elements and admit torsion-free subgroups of finite index, this virtual definition is important. If there exists a $B\Gamma$-space of dimension $n$, then the cohomological dimension of $\Gamma$ is less than or equal to $n$.

See §13 below about cohomology of groups and related invariants. A general reference on finiteness properties is the book [Bro2]. See also [Serr1-2]. Some other references on these topics are [Abe1-2] [Abr1-2] [Beh1-2] [BS1-2] [Bro1-4] [Bux] [BuW1-2] [Ra6] [Tie].

## 8.5. Cofinite universal space for proper actions and arithmetic groups

It is known that a finite group does not admit a finite dimensional, in particular a finite, $B\Gamma$-space. Hence, a group containing torsion elements does not admit a finite $B\Gamma$-space. For such groups, a more natural condition is the existence of cofinite space $\underline{E}\Gamma$ for proper actions of $\Gamma$. Roughly, $\underline{E}\Gamma$ is characterized by the following conditions:

(1) $\underline{E}\Gamma$ is a $\Gamma$-CW-complex, and $\Gamma$ acts properly on it.
(2) For every finite subgroup $H \subset \Gamma$, the fixed point set $(\underline{E}\Gamma)^H$ is nonempty and contractible. In particular, $\underline{E}\Gamma$ is contractible.

If the quotient $\Gamma\backslash\underline{E}\Gamma$ is a finite CW-complex, then $\underline{E}\Gamma$ is called a cofinite $\underline{E}\Gamma$-space. See [Luc2] for more precise definitions.

If $\Gamma$ is torsion-free, then $\underline{E}\Gamma$ is equal to $E\Gamma$, which is the universal covering space of $B\Gamma$ and is the universal space for proper and *fixed point free* actions of $\Gamma$.

If $\Gamma$ is a finite group, then $\underline{E}\Gamma$ can be given by the trivial space consisting of only one point. If $G$ is a connected Lie group, then for any maximal compact subgroup $K$ of $G$, the quotient $X = G/K$ is diffeomorphic to $\mathbb{R}^n$, where $n = \dim X$ (see [Most4] [Hoc]. This is related to the general statement that all the topology of $G$ is contained in a maximal compact subgroup $K$.). If $\Gamma \subset G$ is a discrete subgroup, then $\Gamma$ acts properly on $X$, and $X$ is a $\underline{E}\Gamma$-space. In particular, for arithmetic subgroups $\Gamma$, there are natural $\underline{E}\Gamma$ spaces given by symmetric spaces.

Similarly, $S$-arithmetic subgroups of linear algebraic groups also admit explicit finite dimensional models of $\underline{E}\Gamma$. These considerations make $\underline{E}\Gamma$ a natural and relatively simple object to study. See [BauCH] [Bro2] [Luc2] [Ji7] for detailed definitions and discussions.

On the other hand, it is not easy to show the existence of cofinite $\underline{E}\Gamma$-spaces. By the reduction theory for arithmetic subgroups, it is known that every arithmetic subgroup $\Gamma$ is finitely generated and presented. Using either the Borel-Serre partial compactification of symmetric spaces (or equivalently the Borel-Serre compactification of locally symmetric spaces) in [BoreS1], it can be shown that arithmetic subgroups admits cofinite $\underline{E}\Gamma$-spaces. (For some applications to $K$-groups of the Borel-Serre compactification as a cofinite $\underline{E}\Gamma$-space, see [AdeR, Remark 5.8].) An easier way to obtain a cofinite $\underline{E}\Gamma$-space for arithmetic subgroups is to use a truncation of the associated symmetric spaces induced from the precise reduction theory and the related equivariant deformation retraction of the symmetric spaces in [Sap1]. See [Ji10] for more detail and references. A similar construction can be applied to show that the mapping class groups admit cofinite spaces $\underline{E}\Gamma$. As a simple corollary, it follows that the rational Novikov conjecture in algebraic $K$-theory holds for mapping class groups.

On the other hand, $S$-arithmetic groups over function fields do not enjoy all these finiteness conditions. In fact, some are not even finitely generated. See [Br2-4] [Bux] [BuW1-2].

For finite classifying spaces $B\Gamma$ for Artin groups, see [ChaD1]. For finite $B\Gamma$ spaces for Coxeter groups, see [ChaD2].

## 8.6. Finiteness properties of arithmetic subgroups

After discussing definitions of several finiteness properties of discrete groups, we summarize most of the known finiteness properties of arithmetic and $S$-arithmetic subgroups of semisimple linear algebraic groups defined over $\mathbb{Q}$.

Let $\mathbf{G}$ be a semisimple linear algebraic group defined over $\mathbb{Q}$. Denote its $\mathbb{Q}$-rank by $r$. Let $X = G/K$ be the symmetric space associated with the real locus $G = \mathbf{G}(\mathbb{R})$, and $\Gamma \subset \mathbf{G}(\mathbb{Q})$ an arithmetic subgroup as above. Then $\Gamma$ has the following finiteness properties:

(1) $\Gamma$ is finitely generated.
(2) $\Gamma$ is finitely presented.
(3) $\Gamma$ is virtually torsion-free, i.e., contains a torsion-free subgroup $\Gamma'$ of finite index.
(4) The cohomology $H^i(\Gamma, \mathbb{Z})$ and homology groups $H_i(\Gamma, \mathbb{Z})$ of $\Gamma$ are finitely generated in every degree $i$.
(5) $\Gamma$ is of type $FP_\infty$. If $\Gamma$ is torsion-free, then it is also of type $FL$.
(6) $\Gamma$ admits a cofinite $\underline{E}\Gamma$-space. If $\Gamma$ is torsion-free, it admits a finite $B\Gamma$-space.
(7) If $\Gamma$ is torsion-free, then $\Gamma$ is of finite cohomological dimension, which is equal to $\dim X - r$.
(8) If $\Gamma$ is torsion-free, then $\Gamma$ is a duality group of dimension $\dim X - r$. If $r > 0$, it is not a Poincaré duality group.
(9) $\Gamma$ is residually finite.
(10) $\Gamma$ contains only finitely many conjugacy classes of finite subgroups.

$S$-arithmetic subgroups of semisimple linear also enjoy similar properties. Similarly, mapping class groups of surfaces and outer automorphism groups of free groups also enjoy many of the above finiteness properties. For more details about definitions of the above properties and references for their proofs, see [Ji10].

CHAPTER 9

# Symmetric spaces, Bruhat-Tits buildings and their arithmetic quotients

Assume that $G$ is a semisimple Lie group. Let $K \subset G$ be a maximal compact subgroup. Then $X = G/K$ with a $G$-invariant metric is a Riemannian symmetric space of noncompact type. In particular, it is a simply connected nonpositively curved Riemannian manifold.

The Lie group $G$ acts isometrically and properly on $X$. This action can be used to understand $G$. For example, the nonpositive curvarure and simply connectedness of $X$ imply that every compact subgroup of $G$ has at least one fixed point (*the Cartan fixed point theorem*), and hence every compact subgroup of $G$ is contained in a maximal compact subgroup; and the transitivity of the action of $G$ on $X$ implies that there is only one conjugacy class of maximal compact subgroups of $G$. (It might be helpful to point out that this last conclusion does not hold for semisimple $p$-adic groups such as $SL(n, \mathbb{Q}_p)$ since they do not act transitively on their associated Bruhat-Tits buildings. On the other hand, there are only finitely many conjugacy classes of maximal compact subgroups. Though we have tried to emphasize similarities between symmetric spaces of noncompact type and Bruhat-Tits buildings, this is an important difference.)

Some basic books on symmetric spaces are [Bore10] [Eb] [Hel1] [Wol1] (see [Ji8] for a quick summary). For bounded symmetric domains, see [Sat5]. For more refined structures of Hermitian symmetric spaces, see [Wol2] and also [Mok2].

For a detailed discussion of the special symmetric spaces $SL(n, \mathbb{R})/SO(n)$ and $GL(n, \mathbb{R})/O(n)$, see the book [Te2]. For a more geometric approach to symmetric spaces, see the book [Eb].

## 9.1. Flats in symmetric spaces and the spherical Tits building

The original reference on spherical Tits buildings is [Tit1]. See also [Tit2] for a concise survey of basic definitions and applications of Tits buildings. One general reference on buildings is [Bro1] (see also [Gat2] [Ron1-2].) Other references concerning relations between spherical Tits buildings and symmetric spaces are [BalGS] [GuiJT]. (See also [Ji5] and the references there.)

The richness of structure of $X$ is reflected by the existence of flat subspaces. Recall that a flat in $X$ is an isometric embedding $\mathbb{R}^r \to X$, i.e., a flat totally geodesic submanifold. All maximal flats of $X$ have the same dimension, which is called the rank of $X$ (also the $\mathbb{R}$-rank of $\mathbf{G}$ if $G = \mathbf{G}(\mathbb{R})$). The rank of $X$ is equal to 1 if and only if the sectional curvature of $X$ is strictly negative.

The relations between these flat subspaces are described by the spherical Tits building of $X$, which is also related to the asymptotic geometry of $X$ at infinity. Briefly, the spherical Tits building of $X$ is an infinite simplicial complex of dimension equal to the rank of $X$ minus 1 such that its simplexes are parametrized by proper parabolic subgroups of $G$ satisfying the following conditions:

(1) the face inclusion relation between the simplexes is opposite to the inclusion relation of the corresponding parabolic subgroups,
(2) the minimal parabolic subgroups of $G$ correspond to the top dimensional simplexes, and the proper maximal parabolic subgroups of $G$ correspond to simplexes of dimension 0.

On the other hand, the sphere at infinity $X(\infty)$ of $X$ is defined to be the set of equivalence classes of unit speed oriented geodesics in $X$, where two geodesics $\gamma_1(t)$ and $\gamma_2(t)$ are called equivalent if the distance $d(\gamma_1(t), \gamma_2(t))$ is bounded for $t \geq 0$. The sphere $X(\infty)$ can be canonically identified with the underlying topological space of the spherical Tits building of $X$. In particular, $X(\infty)$ has a natural simplicial structure given by the Tits building. It turns out that this simplicial complex structure can be defined in terms of intersection of flats in $X$.

For example, if the rank of $X$ is equal to 1, then the Tits building of $X$ a zero dimensional complex with one vertex for every point of $X(\infty)$, i.e., it can be identified with the set $X(\infty)$ endowed with the discrete topology. On the other hand, if $X_1$ and $X_2$ are both rank 1 symmetric spaces, then the rank of $X_1 \times X_2$ is equal to 2, and the Tits building of $X_1 \times X_2$ is an one-dimensional simplicial complex with one simplex for every pair of points $p, q$, where $p \in X_1(\infty)$ and $q \in X_2(\infty)$.

The above realization of the spherical Tits building in $X(\infty)$ also gives the topological Tits building of $X$ (or $G$), where a non-discrete topology is put on the set of simplexes of the spherical Tits building of every type. For example, if the rank of $X$ is equal to 1, then its topological Tits building is homeomorphic to $X(\infty)$ with its usual topology. This notion of topological buildings is important in proving rigidity of locally symmetric spaces. See [Ji5]. In fact, it was used implicitly in the proof of the Mostow strong rigidity of locally symmetric spaces.

It can be shown that $G$ acts transitively on the set of geodesics of $X$ if and only if the rank of $X$ is equal to 1. This is equivalent to that $G$ acts transitively on $X(\infty)$ if and only if the rank of $X$ is equal to 1. Hence, if the rank of $X$ is at least two, there are different types of geodesics. The spherical Tits building also classifies different types of geodesics in $X$. See [GuiJT] [BoreJ] and [Ji5].

If $\Gamma$ is a discrete isometric group acting on $X$, then it also acts on the boundary $X(\infty)$. This action is important for understanding various properties of the group $\Gamma$.

## 9.2. Bruhat-Tits buildings

A brief introduction to Euclidean buildings is [Bro1]. Other books include [Gat2] [Ron1]. See also [Ron2-3] for a survey on related topics. Another introduction with references on many applications of buildings in geometry and topology is [Ji5], where other types of buildings and buildings with non-simplicial topology are also considered: spherical buildings, Euclidean buildings, $\mathbb{R}$-buildings, topological buildings.

If $\mathbf{G}$ is a semisimple algebraic group defined over a number field $k$, and $\nu$ is a non-Archimedean place, then $\mathbf{G}(k_\nu)$ is a locally compact and totally disconnected group. The analogue of the symmetric space is the Bruhat-Tits building $X_\nu$ associated with $\mathbf{G}(k_\nu)$.

Briefly, it is an infinite simplicial complex with simplexes corresponding roughly to compact and open subgroups of $\mathbf{G}(k_\nu)$. It is a simply connected metric space of nonpositive curvature, a so-called CAT(0)-space (see below), and contains many flats of dimension equal to the $k_\nu$-rank $r_\nu$ of $\mathbf{G}$, called the apartments of $X_\nu$, which are isometric to $\mathbb{R}^{r_\nu}$ with the standard Euclidean metric.

$\mathbf{G}(k_\nu)$ acts isometrically and properly on $X_\nu$, but not transitively. (Note that a semisimple real Lie group $G$ acts transitively on its associated symmetric space $X$.) On the other hand, the quotient of $\Delta_\nu$ by $\mathbf{G}(k_\nu)$ is equal to the union of finitely many simplexes and hence compact. (Therefore, in a certain sense, $\mathbf{G}(k_\nu)$ acts transitively on $X_\nu$ up to compact subsets.) This action is important to understand the internal structure of $\mathbf{G}(k_\nu)$. For example, it follows from the Cartan fixed point theorem and the CAT(0)-space property of $X_\nu$ that there are only finitely many conjugacy classes of maximal compact subgroups of $\mathbf{G}(k_\nu)$.

The same construction works if $k$ is a function field and $k_\nu$ is a completion of $k$ and gives a Bruhat-Tits building for $\mathbf{G}(k_\nu)$.

One difference between number fields and function fields is that for the former, there are also Riemannian symmetric spaces associated with the infinite places of them. In fact, if $\nu$ is an infinite place of $k$, then $\mathbf{G}(k_\nu)$ is a Lie group. Denote the associated Riemannian symmetric space by $X_\nu$ as well. This will be convenient for later discussions on $S$-arithmetic subgroups.

## 9.3. Action of $S$-arithmetic subgroups on products of symmetric spaces and buildings

As mentioned earlier in §6.4, if $\Gamma$ is an $S$-arithmetic subgroup of $\mathbf{G}(k)$, where $k$ is a global field, then $\Gamma$ is a discrete subgroup of $\prod_{\nu\in S}\mathbf{G}(k_\nu)$.

This implies that $\Gamma$ acts isometrically and properly on the product space

$$X_S = \prod_{\nu \in S} X_\nu.$$

If $k$ is a number field and $S$ contains non-archimedean places, i.e., finite places, then $X_S$ is the product of Riemannian symmetric spaces and Bruhat-Tits buildings, in particular, it is not a manifold. On the other hand, if $k$ is a function field, then $X_S$ is product of buildings, hence has a natural CW-complex structure. Since Riemannian symmetric spaces of noncompact type and Euclidean buildings are CAT(0)-spaces [BridH] (see the next subsection), their products, in particular, $X_S$, are also CAT(0)-spaces.

The action of the $S$-arithmetic subgroup $\Gamma$ on $X_S$ is important for understanding properties of $\Gamma$. For example, the fact that $X_S = \prod_{\nu \in S} X_\nu$ is a CAT(0)-space implies that $X_S$ is a finite dimensional $\underline{E}\Gamma$-space. (Note that in the number field case, $X_S$ does not admit a canonical structure of simplicial complex. On the other hand, for any $S$-arithmetic subgroup $\Gamma$, $X_S$ admits an $\Gamma$-equivariant simplicial complex structure. This is needed in discussing $\underline{E}\Gamma$-spaces.) This explicit model of $\underline{E}\Gamma$ is crucial for the solution of the integral Novikov conjectures for $S$-arithmetic subgroups in [Ji7]. See [Bro1] [Abr1-2] and references there for other applications.

For volumes of quotients $\Gamma \backslash \prod_{\nu \in S} \mathbf{G}(k_\nu)$ by $S$-arithmetic subgroups, see [Pr1]. Computation of such volumes is useful for various purposes, for example, in determining the fake projective spaces in [PYe1-2].

## 9.4. CAT(0)-spaces and CAT(0)-groups

A comprehensive introduction to CAT(0)-spaces is [BridH]. Other references include [Bal1-2] [Jos] [BalGS]. CAT(0)-spaces are geodesic metric spaces with nonpositive curvarure, which is suitably interpreted.

As it is well-known, if $M$ is a Hadamard manifold, i.e., a simply connected and nonpositively curved Riemannian manifold, then the exponential map based at every point is a diffeomorphism, and hence every two different points in $M$ are connected by a unique geodesic segment. More importantly, for every three points, the triangle in $M$ with vertices on them is thinner than the corresponding triangle in $\mathbb{R}^2$ of the same side lengths. Symmetric spaces of non-compact type are important examples of Hadamard manifolds.

Recall that a geodesic in a metric space $M$ is an isometric embedding of some interval $[a, b]$ into $M$. Motivated by geodesic properties of Hadamard manifolds, a metric space $M$ is a CAT(0)-space if it satisfies the following conditions:

(1) Every two different points are connected by a unique geodesic.
(2) Every triangle in $M$ is thinner than the corresponding triangle in $\mathbb{R}^2$ of the same side lengths. (Note that this thinness is measured by the distance from points on one side to other two sides.)

If $M$ is a CAT(0)-space, then it is contractible by contracting rays issued from any fixed basepoint. If $M$ is a proper space, then $M$ also admits a compactification by equivalence classes of geodesic rays. This compactification is useful for the integral Novikov conjecture for groups acting cocompactly on CAT(0)-spaces. In particular, it can be used to prove this conjecture for some $S$-arithmetic subgroups. See the references of [Ji7].

Groups that act cocompactly on CAT(0)-spaces are called CAT(0)-groups. They include fundamental groups of closed and nonpositively curved Riemannian manifolds and cocompact $S$-arithmetic subgroups of linear semisimple algebraic groups over global fields. There are many others, for example, from the simplicial complexes with simplicial nonpositive curvature in [JaS2] and hyperbolization of polyhedra in [DavJ] (see surveys [Da1-3] [Jan] [DavM] and also [BrC] [Cr] [KapK] for example). For other results about actions of groups on CAT(0)-spaces, see [BiG1-3] [ConMT].

CAT(0)-spaces are important because the Bruhat-Tits buildings are not manifolds, but are CAT(0)-spaces when endowed with the Tits metric. Products of symmetric spaces and buildings are also CAT(0)-spaces that are not manifolds, as pointed out in the previous subsection. They are important for studying discrete subgroups over local fields in [Lub2] [Pra2] and $S$-arithmetic groups.

Another reason why they are important is that limits of Riemannian manifolds are not necessarily Riemannian manifolds. But limits of simply connected and nonpositively curved Riemannian manifolds are CAT(0)-spaces. For example, the limit of scaled down hyperbolic space $\mathbb{H}^n$ is an $\mathbb{R}$-tree, a particular example of CAT(0)-spaces.

An important property of discrete groups, called the the property of rapid decay, is related to CAT(0)-spaces. This property is important for applications to the Baum-Connes conjecture and the Novikov conjecture. See [ChaR] and its references for details of this property and relations to these conjectures.

A combinatorial characterization of the fundamental group of compact nonpositively curved Riemannian manifolds is the class of automorphic groups in §8.3. See [El] and the references there for results on relations between automatic groups and CAT(0)-groups.

## 9.5. Reduction theory for $S$-arithmetic subgroups

Let $k$ be a global field, and $\Gamma$ an $S$-arithmetic subgroup of a linear semisimple algebraic group $\mathbf{G}$ defined over $k$ as above.

If $k$ is a function field, then $X_S$ is the product of finitely many Bruhat-Tits buildings and hence has a natural structure of a simplicial complex structure. If the quotient $\Gamma\backslash X_S$ is compact, then it is a finite complex. On the other hand, if $\Gamma\backslash X_S$ is non-compact, it is an infinite simplicial complex. When $X_S$ is a tree, then $\Gamma\backslash X_S$ is the union of a finite graph together with finitely many rays. In general, when $\dim X_S \geq 2$, the quotient $\Gamma\backslash X_S$ will contain chambers of the Bruhat-Tits buildings. See [Abr1] and references there.

Assume that $k$ is a number field. In this case, as for arithmetic subgroups, fundamental sets for $S$-arithmetic subgroups $\Gamma$ acting on $X_S = \prod_{\nu\in S} X_\nu$ can be constructed. See [BoreS2] [Borel4-15]. Briefly, the action on $X_\nu$ for a finite place $\nu$ is cocompact, and the noncompactness of the quotient $\Gamma\backslash X_S$ comes from the cusp neighborhoods of the Siegel sets associated with arithmetic subgroups for the archimedean places in $S$.

More precisely, let $S_f$ be the subset of $S$ consisting of all the finite places. Let $X_{S_f}$ be the product of Bruhat-Tits buildings $X_\nu$, $\nu \in S_f$. Then $X_{S_f}$ has a natural simplicial complex structure. It can be shown that the quotient $\Gamma\backslash X_{S_f}$ is a finite simplicial complex. Let $X_{S_\infty}$ be the product of symmetric spaces $X_\nu$, where $\nu$ runs over all infinite places of $k$. For every simplex $\sigma$ in $X_{S_f}$, its stabilizer in $\Gamma$, denoted by $\Gamma_\sigma$, is an arithmetic subgroup acting on $X_{S_\infty}$. Then the reduction theory for arithmetic subgroups allows us to construct fundamental sets for $\Gamma_\sigma$. These can be combined to construct a fundamental set for the $\Gamma$-action on $X_S$.

This reduction theory can also be used to define and study compactifications of the quotient $\Gamma\backslash X_S$ and cohomology of $\Gamma$. See [BoreS1] [Serr1-3].

CHAPTER 10

# Compactifications of locally symmetric spaces

There is a long history of compactifications of locally symmetric spaces motivated by different applications in analysis, geometry, topology and number theory. See the introduction of [BoreJ] for more details. They are closely related to compactifications of symmetric spaces. That's why compactifications of both symmetric spaces and locally symmetric spaces are considered together here and also in the book [BoreJ].

## 10.1. Why locally symmetric spaces are often noncompact

As pointed out earlier in §4.11, if $\mathbf{G}$ is an algebraic group defined over $\mathbb{Q}$ and the $\mathbb{Q}$-rank of $\mathbf{G}$ is positive, then for every arithmetic subgroup $\Gamma \subset \mathbf{G}(\mathbb{Q})$, the locally symmetric space $\Gamma\backslash X$ is noncompact. So compact locally symmetric spaces correspond to the exceptional rank 0 case.

Many natural locally symmetric spaces are non-compact. For example, if a locally symmetric space is the moduli space and the objects in this moduli space can degenerate, then this locally symmetric space is noncompact. For example, $SL(2,\mathbb{Z})\backslash\mathbb{H}$ is the moduli space of elliptic curves (or rather closed Riemann surfaces of genus 1.) Clearly, such curves can degenerate into singular Riemann surfaces (or curves with nodal points), and hence $SL(2,\mathbb{Z})\backslash\mathbb{H}$ is noncompact. For other instances where locally symmetric spaces arise, for example, in algebraic geometry as moduli spaces, see [Hol2] [HuKW] [Hun] [DolGK] [DolK] [AllCT1] [Yo1-2].

Therefore, a natural problem is to compactify such noncompact arithmetic quotients $\Gamma\backslash X$. If $\Gamma\backslash X$ has a moduli interpretation, then it is important to relate the boundary points to some degenerate objects. An important example here is that when $SL(2,\mathbb{Z})\backslash\mathbb{H}$ is identified with the moduli space of elliptic curves, then the ideal point in the compactification obtained by filling in the cusp point corresponds to a rational curve $\mathbb{C}P^1$ with two points identified.

## 10.2. Compactifications of symmetric spaces

Compactifications of locally symmetric spaces are closely related to compactifications of the symmetric spaces. The first papers giving systematical study of compactifications of symmetric spaces are [Sat1-4]. For a summary, see [Sat8]. The main motivation of these papers was to use compactifications of symmetric spaces to get the desired *rational* boundary components so that they can be used to compactify locally symmetric spaces, which in turn might be used to study boundary behaviors (or behaviors at infinity) of automorphic forms.

There are many reasons to compactify noncompact symmetric spaces. We are mainly interested in symmetric spaces of noncompact type. This is not surprising. According to the classification of symmetric spaces, there are three types of (irreducible) symmetric spaces. Symmetric spaces of compact type are compact and no compactification is needed for them; and symmetric spaces of flat type are just $\mathbb{R}^n$, and their structures at infinity and hence their compactifications, are relatively well-understood. Besides, symmetric spaces of noncompact type give rise to most interesting locally symmetric spaces and hence their compactifications are needed for compactifications of these locally symmetric spaces. For example, in dimension 2, most complete Riemannian manifolds of constant curvature are quotients of the Poincaré upper half-plane $\mathbb{H}$.

The simplest symmetric space of noncompact type is the Poincaré disc $D = \{z \in \mathbb{C} \mid |z| < 1\}$ with the Poincaré metric $ds^2 = \frac{4|dz|^2}{(1-|z|^2)^2}$, which is the unique simply connected complete surface of constant curvature $-1$ (another model of it was given by $\mathbb{H}$ with the Poincaré metric in §3.3). It has a natural compactification by adding the unit circle $S^1$. The classical Poisson integral formula expresses harmonic functions on $D$ which are continuous up to the boundary $S^1$ in terms of their boundary values on $S^1$. It can be generalized to represent all positive harmonic functions on $D$ in terms of nonnegative functions (or rather measures) on $S^1$. From this point of view, $S^1$ becomes the Martin boundary of the Poincaré disc, and $D \cup S^1$ can be identified with the Martin compactification of $D$.

For a general symmetric space $X$ of noncompact type, the space of positive harmonic functions is nonempty and a natural question is to seek a formula analogous to the Poisson integral formula over the Martin boundary. There is a such boundary, called the Martin boundary, for every complete Riemannian manifold, which is the boundary of the Martin compactification of the manifold constructed purely in terms of asymptotic behaviors of the Green function of the Riemannian manifold. Geometrically determining the Martin compactification and hence the Martin boundary of symmetric spaces $X$ is the main motivation in [GuiJT]. For general noncompact and complete Riemannian manifolds, for example, even for simply connected and nonpositively curved Riemannian manifolds, their Martin compactifications have not been completely understood.

The compactification $D \cup S^1$ can also be used to classify the isometries of the Poincaré metric into elliptic, parabolic and hyperbolic types in terms of the fixed points in $D \cup S^1$. Specifically, a nontrivial elliptic isometry has a fixed point in $D$, a parabolic isometry has exactly one fixed point on the boundary $S^1$, and a hyperbolic isometry has exactly two fixed points on the boundary $S^1$. Similar classifications hold for isometries of rank one symmetric spaces of non-compact type by using the geodesic compactification $X \cup X(\infty)$, where $X(\infty)$ is the set of equivalence classes of oriented unit speed geodesics in $X$ and is called the sphere at infinity.

This geodesic compactification can be defined for any so-called Hadamard manifold, i.e., a simply connected and nonpositively curved complete Riemannian manifold. In particular, any symmetric space $X$ of noncompact type admits such a geodesic compactification $X \cup X(\infty)$.

Probably one of the most important compactifications of symmetric spaces is the maximal Satake-Furstenberg compactification. Its boundary contains a distinguished subset, called the maximal Furstenberg boundary. In fact, if $X = G/K$, then the $G$-action on $X$ extends to the compactification, and the unique closed $G$-orbit in the boundary is the maximal Furstenberg boundary.

The maximal Furstenberg boundary is used crucially in the Mostow strong rigidity of higher rank lattices in [Most1] and an explanation is given in [Ji5]. Two books on compactifications of symmetric and related spaces are [GuiJT] and [BoreJ]. See the introduction of [BoreJ] for more motivations of compactifications of symmetric spaces.

As mentioned before, the motivation in [Sat1-2] was to get some boundary components of compactifications of symmetric spaces in order to use them to compactify locally symmetric spaces. To pass from a compactification of symmetric space to a compactification of an arithmetic quotient, i.e., a locally symmetric space given by an arithmetic subgroup, is not easy. One basic question is to decide which boundary components of compactifications of symmetric spaces are needed. These are the so-called rational boundary components. To understand them often requires some rationality of the compactifications of symmetric spaces. A simple and important example to keep in mind is when $X = \mathbb{H}$. In this case, by adding only the rational boundary points $\mathbb{Q} \cup \{i\infty\}$ of $\mathbb{R} \cup \{i\infty\} = \mathbb{H}(\infty)$, we get a partial compactification of $\mathbb{H}$ with the topology described by horodiscs, called the Satake topology, which can be used to compactify every arithmetic quotient $\Gamma\backslash\mathbb{H}$. (This example suggested the terminology of rational boundary components for compactifications of general symmetric spaces. It is easy to imagine that it is more subtle to define rational boundary components of general symmetric spaces.) See [Cass] [Sap2] for discussions about rational boundary components of the Satake compactifications of symmetric spaces.

## 10.3. Limit sets of Kleinian groups and Patterson-Sullivan theory

The limit set of Fuchsian groups acting on the upper half plane has been intensively studied by many people.

Let $\overline{\mathbb{H}} = \mathbb{H} \cup \mathbb{R} \cup \{i\infty\}$ be the compactification of the upper half plane by adding the circle at infinity $\partial\mathbb{H} = \mathbb{R} \cup \{i\infty\} \cong S^1$. Then for any discrete subgroup $\Gamma$ acting isometrically on $\mathbb{H}$ and any point $p \in \mathbb{H}$, the set of accumulation points of the orbit $\Gamma \cdot p$ in the boundary $\partial\mathbb{H}$ is independent of the choice of $p$ and is called the limit set of $\Gamma$, denoted by $\Lambda_\Gamma$. (This follows from the fact that if two unbounded sequences in $\mathbb{H}$ converge to boundary points in $\overline{\mathbb{H}}$ and the distance between the two sequences is bounded, then they converge to the same point.) Let $\Omega_\Gamma = \partial\mathbb{H} - \Lambda_\Gamma$, called the domain of discontinuity. Then $\Gamma$ acts properly on $\mathbb{H} \cup \Omega_\Gamma$ and the quotient $\Gamma\backslash\mathbb{H} \cup \Omega_\Gamma$ is a surface with boundary if $\Gamma$ is torsion-free. If $\Gamma$ is a Fuchsian group of first kind, i.e., the area of $\Gamma\backslash\mathbb{H}$ is finite, then $\Lambda_\Gamma = \partial\mathbb{H}$. On the other hand, if $\Gamma$ is of the second kind, then $\Lambda_\Gamma$ is in general a proper subset of $\partial\mathbb{H}$. Similar results hold for Kleinian groups acting on the real hyperbolic space $\mathbb{H}^n$ of dimension $n$.

A lot of work has been done to understand the Hausdorff dimension of $\Lambda_\Gamma$ and relations with other invariants of $\Gamma$, for example, the bottom of the spectrum of $\Gamma\backslash\mathbb{H}$. See [Bea] [Mas1] [Kra] [BisP] [Bis1-2] [BuO7] and references there.

If $X = G/K$ is a rank one symmetric space of noncompact type, i.e., the sectional curvature is strictly negative but not necessarily of constant sectional curvature (the complex hyperbolic spaces are such examples), then as for the real hyperbolic space $\mathbb{H}^n$, $X$ essentially admits only one nontrivial compactification by adding the sphere $X(\infty)$. The compactification $X \cup X(\infty)$ is a real analytic manifold with boundary. See §12.2 below. (Note that if $X$ is a symmetric space of rank at least 2, then $X \cup X(\infty)$ is not a differential manifold. On the other hand, the maximal Satake compactification is a real analytic manifold with corners. See [BoreJ].) For any discrete subgroup $\Gamma$ of $G$, we can also define its limit set $\Lambda_\Gamma$ and the domain of discontinuity $\Omega_\Gamma$.

An important concept connected with the limit set $\Lambda_\Gamma$ is the notion of Patterson-Sullivan measure. It is supported on the limit set and related to the Hausdorff measure of the limit set, and the bottom of the spectrum of $\Gamma\backslash X$. This was first introduced for Kleinian groups acting on the real hyperbolic space $\mathbb{H}^n$. The theory is similar for rank one symmetric spaces. See [Sul2] [Pat1-2] [Cor2] [New] [Ham3] [Nay] [Leu6] and the references there. For relations between the Hausdorff dimension of the limit set and the cohomological dimension of the convex co-compact discrete subgroup $\Gamma$, see [Iz] [Ol3, §7.1].

For generalizations to simply connected Riemannian manifolds of strictly negative sectional curvature and related results, see [Led2] [Yue1-3] [Con2-3].

Now let $X$ be a symmetric space of rank at least two. In this case, $X$ admits many different nontrivial compactifications, and the notion of limit sets depends on choice of a compactification of $X$. The theory of Patterson-Sullivan measure can also be generalized to higher rank symmetric spaces. See [Alb] [Quin1] [Link1] [Hatt1] [Kor2] [Leu6].

Another natural generalization of cocompact lattices acting on symmetric spaces of rank 1 is the class of Gromov hyperbolic groups. The theory of Patterson-Sullivan measure has also been developed for them. See [Coo].

## 10.4. Compactifications of locally symmetric spaces

A general reference on compactifications of locally symmetric spaces is [BoreJ]. This book does not discuss in detail the important toroidal compactifications of Hermitian locally symmetric spaces, which were first constructed in [AsMRT]. See [Mum1] for a summary and [Mum3] for an application to a generalization of Hirzebruch's proportionality theorem to the noncompact case.

Among many compactifications mentioned in [BoreJ], several are particularly important. The Borel-Serre compactification in [BoreS1] is fundamental for many applications in topology, for example in determining the virtual cohomological dimension of arithmetic subgroups of reductive groups $\mathbf{G}$ over $\mathbb{Q}$ and showing that such subgroups are duality groups. As mentioned in the introduction of this book, various finiteness properties of arithmetic groups can be proved without using the Borel-Serre compactification. On the other hand, to determine their virtual cohomological dimensions and to show that they are virtual duality groups of dimension $\dim X - r$, where $r$ is the $\mathbb{Q}$-rank of $\mathbf{G}$, the structure of the boundary of the Borel-Serre partial compactification of $X$, in particular that the boundary of the Borel-Serre partial compactification is homotopic to the spherical Tits building is crucial. In fact, the boundary components are contractible and correspond to simplexes of the Tits building of $\mathbf{G}$. Furthermore, it also implies that arithmetic groups are not virtual Poincaré duality groups if they are not uniform lattices (See [Serr1-2] [Bro1-2] for summaries of these and related results.)

In the construction of the Borel-Serre compactification, the method or rather idea of attaching only rational boundary components directly without using (real) compactifications of symmetric spaces avoids the difficulties about defining *rational* boundary components of compactifications of symmetric spaces, and it has played a fundamental role in uniformly constructing other compactifications in [BoreJ]. Furthermore, the Borel-Serre partial compactification is also used in the equivariant tiling of symmetric spaces and the resulting deformation retraction in [Sap1] and later results [Yas1-2].

The Baily-Borel compactification in [BaiB] is a normal projective variety defined over a number fields and has many applications in algebraic geometry and number theory. Connections with automorphic forms are also clearly visible in its construction. In some sense, it is the canonical compactification of $\Gamma\backslash X$ when $\Gamma\backslash X$ is considered as a quasi-projective variety.

Another important compactification is the reductive Borel-Serre compactification, since it occurs naturally in many different contexts, for example, as an intermediate compactification between the Baily-Borel compactification and the Borel-Serre compactification for studying the $L^2$-cohomology groups in [Zu1] [GorT], as a natural space for the weighted cohomology groups [GorHM] and related conjectures [Sap3-5], as the topological space to realize the $L^p$-cohomology groups of $\Gamma\backslash X$ in [Zu3] for $p \gg 0$.

One basic reason is that the Baily-Borel compactification is singular. The toroidal compactifications are resolutions in the category of algebraic varieties, and the Borel-Serre compactification can be thought of as a topological resolution. But the Borel-Serre compactification is too large in the sense that it does not support partitions of unity for $L^2$-analysis. The reductive Borel-Serre compactification is the largest quotient (or equivalently the least blow-down) of the Borel-Serre compactification which supports such partitions of unity, and its singularities (for example, the links of singular strata) are relatively easy to be described and understood.

A survey on some compactifications important for applications to cohomology groups is [Gore1]. Other references on compactifications include [Ki] [KK1-2] [KobO] [Loo1-3] [PiaS1] [Ra6] [Sap3] [Zu2]. See [Ji3] for relations between the Baily-Borel, the Borel-Serre, the reductive Borel-Serre, and the toroidal compactifications of Hermitian locally symmetric spaces. The introduction of the book [BoreJ] gives a historic survey of most of the compactifications of both symmetric and locally symmetric spaces and their motivations and applications.

For toroidal compactifications through examples and applications to moduli spaces of abelian varieties in algebraic geometry, see [Nam1-4] [Alex] [She]. For applications to moduli spaces of abelian surfaces, see [HuKW].

For compactifications of quotients of the complex balls and applications to moduli spaces in algebraic geometry, see [Loo1-3] [DolK] [DolGK] [LooS] [AllCT1] [Yo1-2].

For applications of Hermitian locally symmetric spaces to number theory, in particular, the theory of Shimura varieties, see [Miln2] [Vas] and the references there. For applications of compactifications of Hermitian locally symmetric spaces to number theory and arithmetic geometry, we need arithmetic compactifications. See [Ch1-4] [Lar] [Miln1-3] [NoR] [Pin1] [HarrT].

For compactifications of more general, non-locally symmetric complex manifolds and spaces, see [MokZ] [Nad] [NadT] [SiuY1] [Ye3]. These papers also give alternative constructions of the Baily-Borel compactification of Hermitian locally symmetric spaces using complex analysis instead of the usual group theory. Therefore, they can be applied to more general spaces.

## 10.5. Compactifications of Bruhat-Tits buildings

The Bruhat-Tits buildings are natural analogues of the symmetric spaces, and their compactifications are important for some problems involving $S$-arithmetic subgroups, for example, cohomological groups of, virtual duality properties, and the integral Novikov conjecture for $S$-arithmetic subgroups. Since they are CAT(0)-spaces, they admit the geodesic compactification by adding the set of equivalence classes of geodesics. This set at infinity turns out to be the spherical Tits buildings, and the geodesic compactification is exactly the compactification in [BoreS2]. This was used to show that $S$-arithmetic subgroups are virtual duality group.

For symmetric spaces, the Satake compactifications are inductive in the sense that the boundary can be decomposed into boundary components, which are also symmetric spaces of similar nature but lower dimension. For example, for the symmetric space associated with $SL(n, \mathbb{R})$, the boundary symmetric spaces are associated with $SL(k, \mathbb{R})$, where $k \leq n-1$.

There is a compactification of Bruhat-Tits buildings in [Land], which is similar to and corresponds to the maximal Satake compactification of symmetric spaces. The basic idea of the construction is to start with compactifications of each apartment, and then glue them together. This procedure also allows one to construct more compactifications of Bruhat-Tits buildings by choosing different compactifications of apartments. A crucial point is to get some compatibility between these compactifications of apartments along their intersections. This idea was used in constructing the dual cell compactification of symmetric spaces in [GuiJT]. The name of dual-cell comes from the fact that the compactification of every flat is obtained by adding the cell complex dual to the Weyl chamber decomposition. For compactifications of $\mathbb{C}^n$ in algebraic geometry, see [Furu] and references there.

Similarly, we can also use compactifications of Bruhat-Tits buildings to study compactifications of quotients of Bruhat-Tits buildings. When the dimension of the building is equal to 1, it is a tree, and the quotient is union of finitely many rays and a finite graph, and the compactification is clear and obtained by adding one point for each ray. In general, we have some infinite simplicial complexes which form some cones in apartments, or rather in $\mathbb{R}^r$, where $r$ is the rank of the algebraic group.

## 10.6. Compactifications for $S$-arithmetic groups

For some problems involving $S$-arithmetic subgroups, we also need to study compactifications of quotients of products of symmetric spaces and buildings under $S$-arithmetic subgroups.

For example, the congruence subgroup kernel for $S$-arithmetic subgroups can be identified with the fundamental group of such a compactification of $\Gamma \backslash X_S$ for suitable $\Gamma$.

Since the spectral theory of automorphic forms with respect to arithmetic subgroups has had many applications and is closely related to compactifications or structure at infinity of locally symmetric spaces, it is natural and important to develop similar spectral theory of automorphic forms for $S$-arithmetic subgroups. One difficulty is that $X_S$ is not a manifold, but a suitable combination of differential forms together with some more combinatorial theory for the Bruhat-Tits buildings $X_\nu$ might work, where $\nu$ are finite places in $S$. Then the compactifications hinted above will be useful.

An interesting feature of these compactifications of $\Gamma \backslash X_S$ is that they involve both symmetric spaces and Bruhat-Tits buildings.

## 10.7. Geometry and topology of compactifications

In §10.4, we mentioned several applications of compactifications of locally symmetric spaces. In those results, the underlying locally symmetric space $\Gamma\backslash X$ as a Riemannian manifold has not been emphasized. But the invariant Riemannian metric of a symmetric space is an important ingredient of the space, and it is a natural question to relate the geometry of $\Gamma\backslash X$ to boundaries of compactifications of $\Gamma\backslash X$. Such a study was carried out in [JiM]. It studies and classifies the set of geodesic rays, where a ray means an isometric embedding $[0, +\infty) \to \Gamma\backslash X$, or equivalently the set of geodesics which are eventually distance minimizing, and relates them (or different equivalence classes of such rays) to boundaries of various compactifications of $\Gamma\backslash X$. These relations also connect the spectral theory of $\Gamma\backslash X$ to the structure of boundaries of compactifications of $\Gamma\backslash X$. In [Ji4], the sizes of compactifications are also considered from the point of view of metric spaces. These results can be used to obtain rather simply the Borel extension theorem in [Bore6] and a proof of a Siegel conjecture. (See also [Din] [Leu1] for different proofs of the Siegel conjecture.)

Another natural question is to understand the global topology of the spaces. As mentioned earlier, when $\Gamma$ is torsion-free, then $\Gamma\backslash X$ is a $B\Gamma$-space, and hence the cohomology groups of $\Gamma\backslash X$ are equal to the cohomology groups of $\Gamma$. Since for most compactifications of $\Gamma\backslash X$, the inclusion of $\Gamma\backslash X$ is not a homotopy equivalence, relations between cohomology groups of compactifications of $\Gamma\backslash X$ and cohomology groups of $\Gamma$ are not obvious, it is an important and natural problem to understand them. Besides relation to the cohomology of locally symmetric spaces or arithmetic groups, for example, the relation between the intersection cohomology of the Baily-Borel compactification and the $L^2$-cohomology (see §13.3), structure of the homotopy groups of these compactifications is also interesting. After all, locally symmetric spaces and their compactifications are often moduli spaces of some natural objects. Understanding homotopy groups of them are certainly important.

For the reductive Borel-Serre compactifications and the Satake compactifications of locally symmetric spaces, it turns out that their fundamental groups are related to the congruence subgroup kernel. Some references are [BoreS1] [ChaL] [LeeW1-2] [JiMSS] [McC1] [San].

## 10.8. Truncation of locally symmetric spaces

One important reason for introducing the Borel-Serre compactification of a locally symmetric space $\Gamma\backslash X$ is that it is homotopic to the interior $\Gamma\backslash X$, and hence gives a finite $B\Gamma$-space when $\Gamma$ is torsion-feee. In fact, since a symmetric space $X$ of noncompact type is simply connected and non-positively curved, for any torsion-free group $\Gamma$ of $G$, $\Gamma\backslash X$ is a $B\Gamma$-space. The fact that the inclusion $\Gamma\backslash X \to \overline{\Gamma\backslash X}^{BS}$ is an homotopy equivalence implies that

$\overline{\Gamma\backslash X}^{BS}$ is also a $B\Gamma$-space. (Note that $\overline{\Gamma\backslash X}^{BS}$ is a compact real analytic manifold with corners, and its set of interior points is equal to $\Gamma\backslash X$. For a Hermitian locally symmetric space $\Gamma\backslash X$, there exist smooth compactifications which are clearly not homotopic to $\Gamma\backslash X$. Therefore, the condition that the set of interior points of $\overline{\Gamma\backslash X}^{BS}$ is equal to $\Gamma\backslash X$ is important.)

The boundary of the Borel-Serre compactification $\overline{\Gamma\backslash X}^{BS}$ is also important for the purpose to obtain more refined properties of arithmetic subgroups such as virtual cohomological dimension and virtual duality property, and to $L^2$-analysis on $\Gamma\backslash X$, and behaviors at infinity of eigenfunctions.

Another natural method for similar applications in topology is to find a compact subspace $\Gamma\backslash X_T$ of $\Gamma\backslash X$ so that $\Gamma\backslash X$ deformation retracts to it. This is a truncated subspace and the subscript $T$ stands both for truncation and also a truncation parameter. This has one advantage that the invariant Riemannian metric induces a Riemannian metric on this subspace.

On the other hand, the existence of a good and natural metric on a compactification of $\Gamma\backslash X$ as the metric on the truncated subspace $\Gamma\backslash X_T$ in the previous paragraph is not clear. See [Ji4] for considering sizes of various compactifications from several metric points of view.

When $\Gamma\backslash X$ is a surface with a hyperbolic metric, the existence of such a compact submanifold $\Gamma\backslash X_T$ is intuitively clear and is obtained by cutting off the cusps. The resulting space is a manifold with boundary. The same method works for $\mathbb{Q}$-rank one locally symmetric spaces.

In higher rank, it is more complicated. One reason is that the infinity of $\Gamma\backslash X$ is connected, and Siegel sets associated with different $\mathbb{Q}$-parabolic subgroups have non-empty intersection. The resulting submanifold is a compact submanifold with corners. See [Leu4] [Sap1] and also [Bav3]. One way to see why the corner structure arises is to consider a reducible locally symmetric space $\Gamma\backslash X = \Gamma_1\backslash X_1 \times \Gamma_2\backslash X_2$, where each factor gives a manifold with boundary and the product gives a manifold with corners. (The same construction works for Teichmüller spaces, but the Borel-Serre partial compactification of Teichmüller spaces is more difficult to construct.)

If $\Gamma$ is torsion-free and $\Gamma\backslash X$ is noncompact, then the cohomological dimension of $\Gamma$ is equal to $\dim X - r$, where $r$ is the $\mathbb{Q}$-rank of $\mathbf{G}$. A natural question is whether there exists a subspace $Y$ of $\Gamma\backslash X$ and a deformation retraction to it such that $\dim Y = \dim X - r$. This is the best dimension. The submanifold $\Gamma\backslash X_T$ is of the dimension $\dim X$.

On the other hand, when $X$ is linear in the sense that it is a homothety section of a symmetric cone, there is such a subspace given by a simplicial complex. In fact, there is a simplicial complex in $X$ which is invariant under any arithmetic subgroup $\Gamma$, not necessarily torsion-free, and there is a $\Gamma$-equivariant deformation retraction from $X$ to it. See [As1-4] and also [AsM]. Such minimal dimensional simplicial complex is also called spines. For construction of spines for non-linear symmetric spaces, see [MacPM1-2] and [Yas1-2]. Perhaps, it should be mentioned that for general nonlinear symmetric spaces $X$, the existence of such an equivariant deformation retraction of $X$ to a subspace of the minimal dimension is still open.

Once a spine is constructed, it can be used to construct an exact fundamental domain for arithmetic groups. The converse is also true in some sense [Yas1-2].

CHAPTER 11

# Rigidity of locally symmetric spaces

The basic problem here is understand how the topology of locally symmetric spaces determines the locally symmetric spaces, or rather the geometry of the spaces.

In general, the homotopy type of a Riemannian manifold does not determine the manifold up to isometry. For example, we could slightly modify the Riemannian metric in small regions. Even for special metrics, for example, of constant sectional curvature $-1$, this is still not true. For example, there are many non-isometric closed surfaces with hyperbolic metrics of the same genus. Since they have the same genus, then they are homotopic (or even diffeomorphic).

The Mostow strong rigidity says roughly that for locally symmetric spaces of finite volume with the invariant metric, their fundamental groups determine the metrics up to a scalar multiple, with the only exception of the above hyperbolic surfaces.

A different formulation of the Mostow strong rigidity is as follows: Let $G_1, G_2$ be two connected semisimple Lie groups with trivial center and no compact normal factors of positive dimension, and $\Gamma_1 \subset G_1, \Gamma_2 \subset G_2$ two irreducible lattices. Assume that one of $G_1, G_2$ is not isomorphic to $PSL(2, \mathbb{R})$, then any isomorphism between $\Gamma_1$ and $\Gamma_2$ extends to an isomorphism between $G_1$ and $G_2$. This formulation leads naturally to the super-rigidity of Margulis [Marg1]. (Note that the reason for the above assumption that $G_i$ has trivial center and no nontrivial compact factor is that for a symmetric space $X$ of noncompact type, then the identity component of the isometry group of $X$ is such a semisimple Lie group.)

A detailed exposition of the Mostow strong rigidity for compact locally symmetric spaces is given in [Most1]. Extension to finite volumes were proved by Margulis and Prasad (see [Ji9] [Pr4]). There are many expositions of the strong rigidity in the rank one case (see for example [GrP] [Spa1-2] and also [BesCG2] [Iv1] [Mun] [HaaM] for different proofs). For a discussion of the higher rank case, see [Ji5], where Tits buildings and topological Tits buildings play a crucial role (in fact, the rigidity follows from the rigidity of buildings). For related results, see [Klei] [KlL1-3] [BonK2-3]. For an extension to infinite volume spaces, see [Yue1]. For generalizations to infinite volume three dimensional hyperbolic manifolds, see [Mins] and the references there, and also [Bro]. (See also §16.2.)

Selberg's work and conjectures on strong rigidity and arithmeticity of lattices have played a crucial role in the whole development [Sele2] (see [Ji9] for an overview of the work of Margulis and related rigidity results of lattices). His work [Sele1-2] also motivated the modern theory of spectral theory of automorphic forms and the Selberg trace formula.

There are many different notions of rigidities. The general references here are [Most1] [Marg1] [Zi1]. More specific ones will be given in subsections below. For more references about rigidity problems, see the references in [Ji5] (and also [Ji9]) and the description there about relations to buildings. See [Bor22] for a nice summary of some results on rigidity of lattices. See also [BisS] for different notions of rigidity of lattices. For some analogous rigidity problems for von Neumann algebras associated with lattices of semisimple Lie groups having property (T), see [Jon, Problems 7, 8 and 9] and [Pop1, p. 455].

## 11.1. As special Riemannian manifolds

Locally symmetric spaces are special Riemannian manifolds. For example, one definition says that a Riemannian manifold is locally symmetric space if the covariant derivative of the curvature tensor is equal to zero. Another definition is that the local geodesic symmetry is an isometry. Clearly, these conditions are not preserved under any perturbation.

Many results in differential geometry are often tested out on them first (see [Berg1]). One such example is the Margulis Lemma in §5.2. Symmetric and locally symmetric spaces provide important examples of Einstein manifolds. See [Bes].

Locally symmetric spaces naturally arise in different contexts and from different sources. For example, a bounded symmetric domain in $\mathbb{C}^n$ with the Bergman metric is a Hermitian symmetric space (see [Hel1] [Mok2]), and their quotients are related to Shimura varieties.

Many natural locally symmetric spaces can be interpreted as moduli spaces of algebraic varieties and other objects (see [DolK] [DolGK] [Yo1-2] and their references). This is certainly very special among Riemannian manifolds. In this way, geometry at infinity (or compactifications) of locally symmetric spaces are related to degenerations of natural objects.

## 11.2. Local and infinitesimal rigidity of locally symmetric spaces

There are many different versions or notions of rigidity of locally symmetric spaces. The first version concerns deformation of the embedding of a discrete subgroup into an ambient Lie group and the local rigidity. Given an embedding of $\Gamma \subset G$, the conjugation by an element of $g \in G$ gives a different embedding, which should be considered as an equivalent one (in fact, they give rise to the same locally symmetric space). The local rigidity says that any small deformation is of this form, and hence the lattice $\Gamma$ can not be deformed in $G$ in a nontrivial way. It is equivalent to that character (or reduced representation) variety of $\Gamma$ in $\mathbf{G}$ is zero dimensional. See [Fis1] for a survey and references there. See [BergeG] for simpler proofs. See also [GoldM].

A closely related notion is the infinitesimal rigidity. Roughly the tangent space of the character variety is given by some cohomology groups, and the infinitesimal rigidity amounts to the vanishing of such cohomology groups. See [Sele2] [Ra5] [Ra7] there (see [Ji5] for more references).

For a related paper on the character variety of the modular group, see [FaP1]. For other relater results, see [FaK] [GuP1-2]. See also [Gh].

## 11.3. Global (strong) rigidity of locally symmetric spaces

The global rigidity asks if two locally symmetric spaces are isometric if they share some common weaker properties. Since they are not necessarily connected by a family of objects of similar properties as in the case of character varieties and are not close by in suitable senses, this type of rigidity property is called global rigidity in comparison to the local and deformation rigidity properties. The Mostow strong rigidity is a distinguished example of this phenomenon. It states that if two locally symmetric spaces of finite volume are not hyperbolic surfaces or products of them up to finite coverings, then their fundamental groups determine them as Riemannian manifolds up to suitable scaling.

Besides the book [Most1], there are also many other papers, for example [GrP] [GruP1] [GruP2] [Iv1] [Most1] [Pan1] [Pr2] [Pr4] [Sha1-2] [Spa1-2] [MonS1-2] [MinMS], giving expositions of the proof in [Most1] or new proofs of the Mostow strong rigidity of mainly rank 1 case, and generalizations to non-locally symmetric spaces or solvable groups [GruP3]. For an exposition of the Mostow strong rigidity of higher rank case together with many applications of spherical buildings, see [Ji5], where it was emphasized that the underlying Tits building in the proof of the Mostow strong rigidity is a topological building, but not the simplicial Tits building.

Locally symmetric spaces $\Gamma\backslash X$ are complete. For a generalization of the Mostow strong rigidity to a particular class of incomplete hyperbolic manifolds, hyperbolic cone manifolds, see [Koj] [HodK]. For a generalization to some singular spaces, see [Laf]. For yet another generalization, see [Klig2]. For a generalization to manifolds of finite volume with variable negative curvature, see [Bele3].

For other types of rigidity of locally symmetric spaces related to the length spectrum, see [DalK] [Kim1-4]. See also [Wolp4]. For marked length spectrum rigidity of hyperbolic surfaces, see [Ot] [Cro1] [Cro2].

## 11.4. Super-rigidity of lattices

The super-rigidity of irreducible lattices $\Gamma$ in semisimple Lie groups $G$ of rank at least 2 was motivated by and used to answer the question if such lattices $\Gamma$ in semisimple Lie groups $G$ are arithmetic with respect to suitable $\mathbb{Q}$-structures on $G$. This is the question of arithmeticity of lattices.

Given a lattice subgroup $\Gamma$ of a semisimple Lie group $G$, and another (real) Lie group or a $p$-adic Lie group $H$, the super-rigidity asks if a homomorphism $\Gamma \to H$ can be extended to a group homomorphism $G \to H$.

Suppose that both $G$ and $H$ are real semisimple Lie groups with trivial center and no compact factors, and $\Gamma$ is an irreducible lattice, and $\varphi : \Gamma \to H$ is a homomorphism such that the image $\varphi(\Gamma)$ is a lattice in $H$. The Mostow strong rigidity implies that $\varphi$ extends to a homomorphism $\varphi : G \to H$.

In the super-rigidity theory, $G$ and $H$ are assumed to be linear algebraic Lie groups, and the condition that $\varphi(\Gamma)$ is a lattice is replaced by that it is Zariski dense in $H$. Then the super-rigidity theorem of Margulis roughly says that either the homomorphism $\varphi : \Gamma \to H$ extends to $\varphi : G \to H$, or the image $\varphi(\Gamma)$ is bounded in the usual topology.

The extension of a homomorphism $\varphi : \Gamma \to H$ to a homomorphism $\varphi : G \to H$ is natural in certain sense. This is related to the sizes of $\Gamma$ in $G$ in a certain sense. For example, it $\Gamma$ were dense in $G$ in the usual topology, then it is natural to extend the homomorphism of $\Gamma$ by completion. But as we emphasized before, $\Gamma$ is a discrete subgroup of $G$ in the usual topology. If $G$ is the real locus of a linear algebraic group $\mathbf{G}$ defined over $\mathbb{Q}$, then $\mathbf{G} = \mathbf{G}(\mathbb{C})$ has a Zariski topology, and $G$ also has an induced Zariski topology, in which $\Gamma$ is a dense subgroup. Unfortunately, the homomorphism $\varphi : \Gamma \to H$ is not necessarily algebraic to start with, and we could not take the closure in the Zariski topology to get an extension.

The super-rigidity implies the strong rigidity and arithmeticity of lattices of semisimple Lie groups. It is one of the major achievements of Margulis. See [Ji9] for a summary of the work of Margulis on rigidity of lattices and related results. Indeed, the idea and methods from rigidity have been generalized and also applied to many other situations. Besides the books [Zi1] [Marg1], there are also some papers [Cor1] [GrS]. See [Morr1] for some explanations of these rigidity results.

## 11.5. Quasi-isometry rigidities of lattices

The notion of quasi-isometries is important in the proof of the Mostow rigidity and is also natural from many points of view. Recall that an isometry between two metric spaces $(M_1, d_1)$ and $(M_2, d_2)$ is a bijective map $\varphi : M_1 \to M_2$ such that for every pair of points $p, q \in M_1$,

$$d_2(\varphi(q), \varphi(q)) = d_1(p, q).$$

A quasi-isometry between $M_1$ and $M_2$ is a map $\psi : M_1 \to M_2$ such that for some constants $k \geq 1$ and $a > 0$, the following two conditions are satisfied:

(1) for every pair of points $p, q \in M_1$,

$$k^{-1} d_1(p, q) - a \leq d_2(\varphi(q), \varphi(q)) \leq k d_1(p, q) + a.$$

(2) $M_2$ lies in an $a$-neighborhood of the image $\psi(M_1)$, i.e., for every point $r \in M_2$, there exists $p \in M_1$ such that $d_2(r, \varphi(p)) \leq a$.

A simple and instructive example is that the lattice $\mathbb{Z}^n$ of $\mathbb{R}^n$ with the induced metric is quasi-isometric to $\mathbb{R}^n$.

We can easily obtain quasi-isometry invariants of discrete groups. For example, the word metric of a discrete group depends on the choice of the generating set, but different choices lead to quasi-isometric ones. They reflect the large scale geometry of the group. Indeed, quasi-isometry invariants of the word metrics give large scale geometric (or asymptotic) invariants of the discrete group.

In this category, a Lie group $G$ with an invariant metric is quasi-isometric to any cocompact discrete subgroup $\Gamma$. In particular, all co-compact lattices of one common Lie group are all quasi-isometric. More generally, if a discrete group $\Gamma$ acts isometrically and properly on a metric space $M$ with a compact quotient, then $M$ and $\Gamma$ are quasi-isometric.

A natural problem is to classify lattices of semisimple Lie groups up to quasi-isometry. The quasi-rigidity of uniform lattices says any finitely generated discrete group $\Gamma'$ quasi-isometric to a uniform lattice $\Gamma$ in a semisimple Lie group $G$ is, up to a finite group, isomorphic to a uniform lattice in $G$.

It is more difficult to deal with non-uniform lattices $\Gamma$, but there are stronger quasi-rigidity properties. In this case, when $\Gamma$ is considered as an abstract discrete group, then the image of $\Gamma$ in $G$ under any homomorphism $\Gamma \to G$ is commensurable with the discrete subgroup $\Gamma$ under the canonical embedding $\Gamma \subset G$. See the papers [Far] [Esk1] [EskF1-2] [FarbS] [Schw1-2] [KlL1-3] [KapKL] [Pan2-3] for detail.

A closely related problem is to classify symmetric spaces up to quasi-isometry. First, the Euclidean space $\mathbb{R}^n$ is not quasi-isometric to the hyperbolic space $\mathbb{H}^n$. It can be seen easily by comparing the volume growth rates of their metric balls. Therefore, lattices acting compactly on them are not quasi-isometric. It is natural to expect that different symmetric spaces are not quasi-isometric. Indeed, it was conjectured by Margulis and proved by Kleiner-Leeb [KlL2] and Eskin-Farb [EskF2] that if two symmetric spaces of noncompact type are quasi-isometric, then they are isometric up to suitable scaling on factors of the symmetric spaces.

For a characterization of irreducible symmetric spaces and Euclidean buildings of higher rank by their asymptotic geometry, see [Leeb2].

Another important result to be pointed out is that for an irreducible lattice $\Gamma$ acting on a higher rank symmetric space, the word metric on $\Gamma$ is comparable with the induced metric from a $\Gamma$-orbit in $X$ [LubMR].

Besides lattices of Lie groups, hyperbolic groups and CAT(0) groups, another important class of groups in geometric group theory and combinatorial group theory is the class of Baumslag-Solitar groups. For their quasi-isometry classification, see [FarbM1-2] [Wh].

For other results on quasi-isometries, see [PapW] [BouP1-2] [LafP] [Mai1-2] [Xi1-3] [Pap1] [Pap2].

## 11.6. Rank rigidity of locally symmetric spaces

In the study of geometry and harmonic analysis of symmetric spaces and locally symmetric spaces, flat totally geodesic submanifolds play an important role. The maximal dimension of such flat subspaces is called the rank of the symmetric spaces. Higher rank spaces often exhibit rigidity properties due to the presence of many flats and the resulting connection with nontrivial spherical Tits buildings.

For general Riemannian manifolds, we can replace the existence of flat subspaces by the existence of parallel Jacobi fields, and also define the notion of rank as the minimum dimension of vector spaces of parallel Jacobi fields along geodesics. Since the tangent vector field of a geodesic is a parallel Jacobi field, the rank is at least 1. This rank coincides with the earlier one when the manifold is locally symmetric.

The rank rigidity says roughly that if a compact Riemannian manifold whose sectional curvature is non-positive and is not irreducible in the sense that no finite covering splits as an isometric product, then it is a locally symmetric space if its rank is at least 2. Both the higher rank assumption and the irreducible condition are important. See [Bal2-3] [BalBE] [BalBS] [BalGS] [BuS1-2] [EbHS] for this rigidity, geometry of nonpositively curved manifolds and related results on geodesic flow. For some generalizations, see [Ham2] [ShanSW] [Con1] and references there.

## 11.7. Entropy rigidity of locally symmetric spaces and simplicial volume

If $\Gamma\backslash X$ is a symmetric space of rank one, then its sectional curvature is strictly negative. A natural question is how to characterize such locally symmetric spaces among all Riemannian metrics of negative sectional curvature, and more generally among all Riemannian metrics on them. The entropy rigidity gives such a rigidity in terms of the volume growth entropy.

Given a compact Riemannian manifold $M$ of negative sectional curvature, the volume of metric balls in the universal covering $\tilde{M}$ grows exponentially when the radius goes to infinity. The exponential growth rate is called the volume growth entropy of the manifold.

The entropy rigidity says basically that among all negatively curved Riemannian metrics on $\Gamma\backslash X$ of the fixed volume, the invariant metric strictly minimizes this entropy.

A more general form of the entropy rigidity implies the Mostow rigidity of the rank one case. The original proof appears in [BesCG1]. A survey together with a more elementary and geometric proof of a slightly weaker result is given in [BesCG2]. See also the survey [Cro2] for related results.

It is natural to ask if the compactness assumption in the entropy rigidity can be replaced by finite volume, and higher rank locally symmetric spaces of finite volume also satisfy the same type of entropy rigidity. Some results in this direction have been obtained in [ConF1-3] [BolCS]. For related results on minimal volume entropy of graphs, in particular, quotients of Bruhat-Tits buildings of rank 1, see [Lim1].

The barycenter method developed in [BesCG1-2] can be used to show that the simplicial volume of every closed locally symmetric space is positive. Briefly, the notion of simplicial volume of topological spaces was introduced by Gromov in [Gro8] in order to bound from below minimal volumes of smooth manifolds, where the minimal volume of a differential manifold $M$ is defined to be the infimum of volumes of all complete Riemannian metrics on $M$ whose sectional curvature are bounded everywhere in absolute value by 1. The positivity of the simplicial volume of closed hyperbolic manifolds was proved by Thurston, and the positivity of the simplicial volume of general closed locally symmetric spaces of noncompact type was conjectured by Gromov and is proved in [LafS].

## 11.8. Rigidity of Hermitian locally symmetric spaces

Hermitian symmetric spaces are symmetric spaces admitting invariant complex structures. They are special Kähler manifolds. A natural problem is to prove rigidity results among Kähler manifolds.

The projective space $\mathbb{C}P^n$ is a Hermitian symmetric space of compact type, and its first rigid properties are given in [Yau1] as corollaries of the solution of the Calabi conjecture. For example, one result says that the only Kähler metric metric on $\mathbb{C}P^n$ is the standard one, i.e., the invariant metric. Another result says that every complex surface homotopic to $\mathbb{C}P^2$ is biholomorphic to $\mathbb{C}P^2$.

Another major consequence of the proof of the Calabi conjecture is that [Yau1] gave a characterization of compact quotients of the unit ball in $\mathbb{C}^n$ as algebraic manifolds in terms of Chern numbers, i.e., if a compact algebraic manifold over $\mathbb{C}$ satisfies the condition that its canonical bundle is ample and its Chern numbers satisfying the equality in the famous Miyaoka-Yau inequality, $c_1^2 = 3c_2$, i.e., the proportionality principle of Hirzebruch [Hirz3] (see also [Mum3]), then the manifold is quotient of the ball. (This gives a beautiful characterization of this important class of Shimura varieties in terms of characteristic numbers. Ball quotients appear prominently in algebraic geometry as moduli spaces. See [Hol2] [HuKW] [Hun] [DolGK] [DolK] [AllCT1] [Yo1-2] for example. See also [DeM].) [Yau1] then used the Mostow strong rigidity to deduce that there is only one complex structure on the compact ball quotients given by the standard one.

The results in [Yau1] motivated a lot of works in this direction. It also motivated problems of finding and classifying fake projective planes (see §18.4 for more references.) In fact, after spectacular success in [Yau1],

Yau raised the famous question whether a compact Kähler manifold with negative sectional curvature has a unique complex structure. He proposed to use harmonic map to solve such a problem. This was carried in [Siu1-2] which dealt with the case of quotients of irreducible Hermitian symmetric spaces of complex dimension greater than or equal to two.

More precisely, for Hermitian locally symmetric spaces of noncompact type, or equivalently quotients of bounded symmetric domains, the strong rigidity result of [Siu1-2] states that if $M = \Gamma\backslash X$ is a compact quotient of an irreducible bounded symmetric domain of complex dimension at least two, and $N$ is a compact Kähler manifold which is homotopic to $M$, then $N$ is either biholomorphic or antibiholomorphic to $M$. If $N$ is also taken to be a quotient of a bounded symmetric domain, then it is similar to the Mostow strong rigidity for locally symmetric spaces (which states that if locally symmetric spaces of noncompact type not equal to $\Gamma\backslash\mathbb{H}$ or products of such spaces have isomorphic fundamental groups, then they are isomorphic). It is closer to the generalization of the Mostow strong rigidity by Gromov in [BalGS], where only one is assumed to be a locally symmetric space.

Such an approach using harmonic maps was also used in [SiuY2] to prove the Frankel conjecture: Every compact Kähler manifold $M$ of positive bisectional curvature is biholomorphic to the complex projective space. (An independent proof was given in [Mori].)

Another important problem is to characterize invariant metrics of Hermitian locally symmetric spaces in terms of curvature conditions. For example, a special case of the results in [Mok3] [To] says that if $\Gamma\backslash X$ is a compact Hermitian locally symmetric space of noncompact type, then this invariant metric on $\Gamma\backslash X$ is the unique, up to a scaler multiple, Hermitian metric with seminegative curvature, in particular, the unique Kähler metric with seminegative holomorphic bisectional curvature.

It is known that a compact Hermitian locally symmetric space $\Gamma\backslash X$ is a projective variety by the Kodaira embedding theorem [MorrK], and a non-compact Hermitian locally symmetric space of finite volume is also a quasi-projective variety by [BaiB]. A natural problem is to give algebraic geometric characterizations of Hermitian locally symmetric spaces. Such a characterization is given in [Yau2]. As an application, [Yau2] recovers a known result of Kazhdan [Kaz2] that a Galois conjugate of a Shimura variety is also a Shimura variety. Recall that a Shimura variety is a projective variety defined over a number field whose underlying complex space is isomorphic to a compact Hermitian locally symmetric space or the Baily-Borel compactification of a Hermitian locally symmetric space of finite volume. By a result of Shimura (see [Shi3] for example, where he proved that a complex projective non-singular variety $V$ that has no deformation admits an algebraic number field as field of definition, i.e., there is a variety $W$ defined over a number field such that $W$ is isomorphic to $V$), all these Hermitian locally symmetric spaces do admit models defined over number fields. On the other hand, finding good and explicit models of Shimura varieties over specific number fields or even over rings of integers is more difficult (see [Miln2] and its references and [Shi2). See [NoR] for a different proof of this result of Kazhdan. See also [MokY] for related results.

There has been a lot of work on these questions and related questions, and the method of using harmonic maps to prove rigidity has been very fruitful. See [JoY1-4] [JosL] [Mok1-2] [MokSY] [YaZ1-2] and the references there.

The above discussions mainly concentrated on Hermitian locally symmetric spaces of noncompact type. For rigidity of Hermitian symmetric spaces of compact type, see [HwM1-2] [Lan2] and the references there.

Another type of rigidity concerns asymptotically hyperbolic manifolds. Briefly, it says that if the sectional curvature of a noncompact complete Riemannian manifold is asymptotically equal to $-1$, then under suitable further conditions on the manifold, it is isometric to the hyperbolic space $\mathbb{H}^n$. See [ShT] and the references there for various results. It can be interpreted as a sort of Mostow type rigidity in the sense that the asymptotic geometry at infinity determines the geometry in the interior. (In the proof of the Mostow strong rigidity [Most1], the key point is to show that a homotopy equivalence between locally symmetric spaces induces an isomorphism between spherical Tits buildings at infinity of the associated symmetric spaces.) It can also be interpreted as a metric rigidity or characterization of the hyperbolic metric.

See also [Zh] for some results on transformation group theory in complex analysis.

## 11.9. Rigidity of pseudo-Riemannian locally symmetric spaces

Generalizations of (geometric) rigidity theorems for symmetric and locally symmetric spaces so far have concerned with Riemannian manifods, or more specially, with Hermitian manifolds. On the other hand, we can consider another generalization of deformation (or local) rigidity by replacing Riemannian structure with pseudo-Riemannian structure (e.g. Lorentz structure). Then, rigidity tends to fail, or in other words, the deformation space tends to become large.

An intuitive explanation is as follows. Suppose $X = G/K$ is a Riemannian symmetric space of non-compact type and $\Gamma \subset G$ is a cocompact lattice. We note that the Killing form induces a Riemannian metric on $X = G/K$, and a pseudo-Riemannian metric on $G$. Then, both $X$ and $G$ become symmetric spaces. Whereas $G$ is essentially the group of isometries on $G/K$, the direct product group $G \times G$ is that of a pseudo-Riemannian manifold $G$ induced from the Killing form. Then the local rigidity can be formalized via the embeddings of a cocompact lattice $\Gamma$ of $G$ into $G$ up to the conjugation by an element of $G$ in the former case, and into the direct product group $G \times G$ in the latter case. The crucial condition in the latter case is to find when the embedded $\Gamma$ acts properly on $G \simeq (G \times G)/\Delta G$ because the isotropy group $\Delta G$ is non-compact.

For $G = SL(2, \mathbb{R})$, the quotient space by such $\Gamma$ leads us to a three dimensional manifold called a *non-standard Lorentz space form*, see [Goldw3]. For a simple Lie group $G$, local rigidity holds for left actions if and only if $G$ is not locally isomorphic to $SL(2, \mathbb{R})$ (Selberg–Weil rigidity), whereas local rigidity holds for left and right actions if and only if $G$ is not locally isomorphic to $SO(n, 1)$ or $SU(n, 1)$ for some $n$. Thus, as $n$ increases, one can construct a continuous, non-isometric family of locally pseudo-Riemannian symmetric spaces of arbitrarily high dimesnion. For $SO(n, 1)$ $(n = 2, 3)$, the deformation space was studied in [Goldw3] [Gh] and [Sa].

The local rigidity is the beginning step of the many rigidity properties of Riemannian locally symmetric spaces discussed above. In the pseudo-Riemanniaan case, the excluded locally symmetric spaces for the local rigidity above are associated with some rank one semisimple Lie groups, and it is natural to ask if some types of strong or supe-rigidity properties hold for non-Riemannian locally symmetric spaces in other cases, in particular higher rank cases.

## 11.10. Rigidity of non-linear actions of lattices: Zimmer program

The Mostow strong rigidity and the Margulis super rigidity concern actions of lattices of semisimple Lie groups on symmetric spaces and representations of lattices of semisimple Lie groups in other semisimple Lie groups, which can also be interpreted as actions of lattices on vector spaces.

A natural extension is to replace the target semisimple Lie groups by infinite dimensional Lie groups. There is no well-developed theory of infinite dimensional Lie groups similar to that of finite dimensional semisimple Lie groups. One important class of infinite dimensional Lie groups consists of loop groups of Lie groups, and another class consists of diffeomorphism groups of smooth manifolds and special diffeomorphism subgroups of volume preserving diffeomorphisms of the manifolds, or subgroups preserving other geometric structures such as affine connections on the manifolds. (For discussions of geometric structures, see [Gro9] [Fer3]. To compare with algebraic subgroups of $GL(n, \mathbb{C})$, it might be suggestive to point out that algebraic groups are often obtained as subgroups preserving some algebraic conditions, for example, preserving a quadratic form on $\mathbb{C}^n$ or a volume form etc.) These groups of diffeomorphisms of manifolds are important for applications in mechanics and dynamical systems. See §5.11 for more discussions, applications and some references about diffeomorphism groups.

Then representations of a lattice $\Gamma$ of a semisimple Lie group $G$ in the differmorphism group of a manifold $M$ corresponds to smooth actions on $M$, which are certainly fundamental objects in the theory of dynamical systems. In this sense, they give rise to non-linear representations of $\Gamma$.

It is natural to consider analogous rigidity problems for such non-linear representations. A rigorous formulation is the Zimmer program. It was formally formulated in [Zi2]. An important result which partially motivated this program is the strong orbit equivalence rigidity for actions of irreducible lattices of higher rank semisimple Lie groups [Zi4].

The Zimmer program is a substantial generalization of the Mostow strong rigidity theorem and the Margulis super rigidity theorem and basically (or broadly) consists of two related parts:

(1) Classification of ergodic actions of $\Gamma$ preserving volume, affine connections or other geometric structures on compact manifolds in terms of some known algebraic actions, where $\Gamma$ are irreducible lattices of higher rank semisimple Lie groups $G$.
(2) Global rigidity, local rigidity and deformation rigidity of actions of lattices $\Gamma$ which preserve volume, connections or other geometric structures, and rigidity of ergodic actions of $\Gamma$ with respect to orbit equivalence relations.

Roughly, the first part asks what kind of actions can occur, and a general conjecture says that they essentially come from algebraic actions, for example, actions of lattices on homogeneous spaces. Besides volume and connections, there are other so-called rigid geometric structures related to the Zimmer program [Gro9] [Fer3]. The second part asks if the actions are rigid in various senses such as global, infinitesimal and local rigidity, and if they can be recovered from the orbit structures. Of course, these two parts can not be really separated.

There are many variants of the vague conditions and expected results as stated above. To give a glimpse of results on these problems, we describe some early work of Zimmer in this program. One type of results in part (1) is to characterize those actions that arise from lattices acting on natural homogeneous spaces, i.e., the so-called algebraic actions. One result of Zimmer [Zi4, pp. 1250-1251] states that if $G = SL(n, \mathbb{R})$, $n \geq 3$, and $\Gamma = SL(n, \mathbb{Z})$, then the standard action of $\Gamma$ on the flat torus $\mathbb{Z}^n \backslash \mathbb{R}^n$ is of minimal dimension among all actions by lattices of $SL(n, \mathbb{R})$ on compact manifolds $M$ which preserve both the volume and a connection of $M$. Furthermore, for every lattice $\Gamma \subset SL(n, \mathbb{R})$, $n \geq 3$, if $M$ is a compact Riemannian manifold of dimension $n$, and $\Gamma$ acts on $M$ preserving the volume and the connection, then $M$ is flat and $\Gamma$ is commensurable with $SL(n, \mathbb{Z})$. For related results on classification of connection preserving actions on compact manifolds, see [Fer2] [Goe].

There have been many recent works on themes related to part (1) of the Zimmer program, and it is impossible for us to mention all the major papers. An important class of ergodic actions is given by Anosov actions. For classifications of Anosov actions of semisimple Lie groups of higher rank and their lattices, see [GoeS1] [GoeS2]. For classifications of Anosov Cartan actions of abelian groups $\mathbb{R}^n$ and lattices $\mathbb{Z}^n$, see [KatoS1] [KatoS2] [KalS]. See also the survey articles [Spa1] [Spa2] [FerK] and references there.

Results on the global, local, deformation and infinitesimal rigidity of lattices actions are closely related to many rigidity results for locally symmetric spaces and lattices discussed in §11. For a survey on such rigidity results on actions of lattices, see the survey articles [FerK] [Fis1], the paper [KatoL] and many references there.

There are several notions related to orbit equivalence relations. A natural broader context to study volume preserving actions on manifolds is to consider measure preserving actions on measure spaces, which are basic objects in the ergodic theory. Let $G$ and $G'$ be two locally compact groups acting ergodically on standard probability spaces $(X, \mu)$ and $(X', \mu')$ and preserving the probability measures $\mu$ and $\mu'$ respectively. The two actions of $G$ and $G'$ are said to be *orbit-equivalent* if there exists a measure space isomorphism between $(X, \mu)$ and $(X', \mu')$ which takes each $G$-orbit in $X$ onto a $G'$-orbit in $X'$. The two actions of $G$ and $G'$ are said to be *conjugate* if there is an isomorphism between $G$ and $G'$ and a measure space isomoprhism between $(X, \mu)$ and $(X', \mu')$ that is equivariant with respect to the actions of $G$ and $G'$. The group $G$ can either taken to be either a Lie group or a lattice of a Lie group.

Clearly, if two actions of $G$ and $G'$ are conjugate, then they are orbit equivalent; but the converse is not true. In fact, it is known that if $G$ and $G'$ are amenable groups, then the orbit equivalence automatically holds for any two free ergodic actions of $G$ and $G'$ acting on probability measure spaces $(X, \mu)$ and $(X', \mu')$ preserving the probability measures $\mu$ and $\mu'$. On the other hand, actions of non-isomorphic semisimple Lie groups of higher rank and their lattices are not orbit equivalent. More specifically, the rigidity for orbit equivalence in [Zi4, Theorem 4.3] states that under the conditions that $G$ and $G'$ are semisimple Lie groups with trivial center and no compact factors, and the rank of $G$ is at least 2, and that the actions on $(X, \mu)$ and $(X, \mu')$ are irreducible in the sense that every nontrivial normal subgroup of $G$ and $G'$ acts ergodically, then the actions of $G$ and $G'$ are conjugate if they are orbit equivalent.

A corollary of a slight extension of this result is the following rigidity of orbit equivalence for actions of lattices: Let $G, G'$ be connected semisimple Lie groups of rank at least 2, and $\Gamma \subset G$ and $\Gamma' \subset G'$ are irreducible lattices. Suppose that $(X, \mu)$ and $(X', \mu')$ are free ergodic probability spaces of $\Gamma$ and $\Gamma'$ with invariant probability measures $\mu$ and $\mu'$ respectively. If the actions of $\Gamma$ and $\Gamma'$ are orbit equivalent, then $G$ and $G'$ are locally isomorphic.

On consequence of this rigidity result is that the standard action of $SL(n, \mathbb{Z})$ on the torus $\mathbb{Z}^n \backslash \mathbb{R}^n$ are not orbit equivalent for different values of $n$.

Assume that $\Gamma$ is an irreducible lattice of a connected noncompact simple Lie group $G$ with trivial center and of rank at least 2, but $\Gamma'$ is an arbitrary countable group, and they act on probability measure spaces $(X, \mu)$ and $(X', \mu')$ as above. A stronger rigidity of orbit equivalence (also called superrigidity of orbit equivalence) in [Fur3] states roughly that an orbit equivalence between the actions on $(X, \mu)$ and $(X', \mu')$ implies that $\Gamma$ and $\Gamma'$ are virtually isomorphic and the actions of $\Gamma$ and $\Gamma'$ are also virtually isomorphic. The paper [Fur2] also proves a measure equivalence rigidity result for such pairs of groups, which was used in proving the above result in [Fur3]. See [MonS1-2] and the discussion in the next subsection of [Pop4] for related results.

In the proof of these statements and more generally other results in the Zimmer program, a fundamental role is played by the Zimmer cycle superrigidity [Zi1, Theorem 5.2.5], which was motivated by the Margulis superrigidity (see [Marg1, Chap. VII] [Zi1, Theorem 5.1.2]). The Zimmer cocycle superrigidity says roughly if $G$ is a connected semisimple algebraic Lie group of rank at least 2 and $(X, \mu)$ is an irreducible $G$-ergodic probability measure space whose measure $\mu$ is invariant under $G$, and $H$ is also a connected algebraic group, then any cocycle $\alpha : G \times X \to H$ is cohomologous to either a homomorphism $G \to H$ or a cocycle taking values in a compact subgroup of $H$. See [FreL] for another cocycle superrigidity. For a related cocycle superrigidity for negatively curved metric spaces, see [MonS1].

There have been a lot of developments in the Zimmer program since the paper [Zi2]. Besides the papers mentioned above, see also [LubZ2] [LubZ3] [KatoLZ] [FishM1-2] [Fish2] [FerL] [MargQ] [QiY] for other results in this program. For more recent surveys on some results in the Zimmer program, see [Fis1] [Shal9]. See [BurI1-2] [Mon1-2] also for related questions on rigidity phenomenon. See [Mon3] [BurM1-2] for discussions of bounded cohomology and applications to non-linear rigidity of lattices.

Another extension of the Mostow and Margulis type rigidity properties is to consider groups acting on metric spaces which are not Riemannian manifolds. There have been a lot of recent works related to this general theme. See [Mon1-2] [Shal1-8] [MonS1-2] [FisZ] [FisW] and their references.

## 11.11. Rigidity in von Neumann algebras

Many results in the rigidity theory for lattices of semisimple Lie groups have been generalized to operator theory, in particular, von Neumann algebras. And the methods from operator theory and ergodic theory can also be used to prove powerful and surprising results in the Zimmer program.

One such generalization is an analogue of the Mostow strong rigidity, a conjecture made by Connes [Jon]. Briefly, for every countable group $\Gamma$ with discrete topology, there is associated a von Neumann algebra $U(\Gamma)$, which is the weak closure of the group algebra $\mathbb{C}\Gamma$ in the Banach space $\mathcal{B}(\ell^2(\Gamma))$ of all bounded operators on the Hilbert space $\ell^2(\Gamma)$ (note that $\mathbb{C}\Gamma$ acts on $\ell^2(\Gamma)$ by the left regular representation, i.e., the action induced from the left multiplication of $\Gamma$ on itself. We also note that in the Baum-Connes conjecture for $\Gamma$, a $C^*$-algebra $C^*(\Gamma)$ associated with the group algebra $\mathbb{C}\Gamma$ is used, which is defined as the norm closure of $\mathbb{C}\Gamma$ in the Banach space $\mathcal{B}(\ell^2(\Gamma))$. See §11.12).

Then the conjecture of Connes stated in [Jon] says that if $\Gamma_1$ and $\Gamma_2$ are discrete subgroups of semisimple Lie groups with property (T) and the von Neumann algebras $U(\Gamma_1)$ and $U(\Gamma_2)$ are isomorphic, then $\Gamma_1$ is isomorphic to $\Gamma_2$. A more precise version [Jon, Problem 8] says that up to inner automorphisms, any isomorphism between the von Neumann algebras $U(\Gamma_1)$ and $U(\Gamma_2)$ comes from a group isomorphism between $\Gamma_1$ and $\Gamma_2$ (see also [Pop1]).

The same strong rigidity conjecture for $\Gamma_1$ and $\Gamma_2$ equal to free groups $F_{n_1}$ and $F_{n_2}$ (i.e., if $n_1 \neq n_2$, then $U(F_{n_1})$ and $U(F_{n_2})$ are not isomorphic) is still open [Jon, Problem 7], though the more precise version on realizing automorphisms of the vou Neumann algebra $U(F_n)$ by automorphisms of the group $F_n$ up to inner automorphisms fails for free groups.

There are also related rigidity results in von Neumann algebras similar to the Zimmer cocycle superrigidity, and the second part in the Zimmer program on the rigidity of orbit equivalences. Many striking and unexpected results on these rigidity problems and applications to von Neumann algebras have been obtained by Popa and his collaborators (see [Pop1] [Pop2] [Pop3] [Pop4] [GabP] [V] and references there).

One deep result of Popa [Pop2] [Pop3] is the following von Neumann strong rigidity (a special case): Let $\Gamma$ be a group with infinite conjugacy classes and $\Gamma$ acts on $(X, \mu) = \prod_{\gamma\in\Gamma}(X_0, \mu_0)$ by the Bernoulli action, where $(X_0, \mu_0)$ is the standard non-atomic probability measure space. Let $\Lambda$ is an infinite group with property (T) that acts freely and ergodically on a probability measure space $(Y, \eta)$ leaving the measure $\eta$ invariant. If there is a *-isomorphism between the crossed product von Neumann algebras:

$$L^\infty(Y) \rtimes \Lambda \cong p(L^\infty(X) \rtimes \Gamma)p$$

for some projection $p \in L^\infty(X) \rtimes \Gamma$, then $p = 1$, and the groups $\Gamma$ and $\Lambda$ are isomorphic and the actions of $\Lambda$ and $\Gamma$ are conjugate through this isomorphism.

Such a rigidity result has surprising applications to the fundamental group $\mathcal{F}(M)$ of $\mathrm{II}_1$-factors $M$, which are subgroups of $\mathbb{R}_+^\times$. In fact, Popa [Pop2] showed every arbitrary countable subgroup of $\mathbb{R}_+^\times$ can be realized as the fundamental group $\mathcal{F}(R \rtimes \mathrm{SL}_2(\mathbb{Z}))$, where $R$ is a hyperfinite $\mathrm{II}_1$-factor with an action of $\mathrm{SL}_2(\mathbb{Z})$. This brings new light to a long standing open problem of Murray and von Neumann on the nature of the fundamental group of $\mathrm{II}_1$ factors in the theory of von Neumann algebras.

A major result in [GabP] says that each free group $F_n$, $2 \leq n \leq \infty$, admits an uncountable family of non-stably orbit equivalent free ergodic and probability measure preserving actions.

Another fundamental result of Popa is a cocycle superrigidity theorem for Bernoulli actions [Pop4]. We state one sample version as follows: Let $\Gamma$ be a discrete group with property (T), let $\Gamma_0 \subset \Gamma$ be a subgroup of infinite index, $(X_0, \mu_0)$ an arbitrary probability measure space, and let $\Gamma$ act on the induced measure space $(X, \mu) = \prod_{\gamma\in\Gamma/\Gamma_0}(X_0, \mu_0)$ by the generalized Bernouli action.

Then for any discrete countable group $\Lambda$ and any measurable cocyle $\alpha : \Gamma \times X \to \Lambda$, there exists a homomorphism $\rho : \Gamma \to \Lambda$ and a measurable map $\phi : X \to \Lambda$ such that

$$\alpha(g, x) = \phi(gx)\rho(g)\phi(x)^{-1}.$$

The Popa cocycle superrigidity has many applications. For example, it implies the superrigidity of orbit equivalence (or orbit equivalence superrigidity) [Pop4], a special case of which can roughly be stated as follows: Let $\Gamma$ be a countable group acting on a probability measure space $(X, \mu)$ by the Bernoulli action and preserving the measure $\mu$. Suppose that $\Gamma$ does not have any finite nontrivial subgroups. Let $Y$ be a measurable subset of $X$. If the restriction to $Y$ of the equivalence relation given by the $\Gamma$-action is given by the orbits of a countable group $\Lambda$ acting freely and ergodically on $Y$, then up to subsets of measure 0, $Y = X$ and the actions of $\Gamma$ and $\Lambda$ are conjugate through a group isomorphism $\Gamma \cong \Lambda$.

For related results on superrigidity orbit equivalence, see [Fur2-3] and [MonS1-2]. For more details about this cocycle superrigidity theorem and applications, see [V] [Pop1] and [Fur4], and for comparison with the Zimmer cocycle superrigidity theorem, see [Fur4]. See also the original paper [Pop4].

For other related rigidity results in the theory of von Neumann algebras, see [GabP] [Gabo] [Pop1] [Jon] and references there.

## 11.12. Topological rigidity and the Borel conjecture

Recall that an aspherical manifold $M$ is a manifold satisfying the conditions $\pi_i(M) = \{1\}$ if $i \geq 2$. This is equivalent to that its universal covering space $\tilde{M}$ is contractible. Then by a theorem of Whitehead, the homotopy type of $M$ is determined by its fundamental group $\pi_1(M)$.

Locally symmetric spaces of noncompact type provide an important class of aspherical manifolds. More generally, for any noncompact connected Lie group $H$, not necessarily semisimple, and a maximal compact subgroup $K \subset H$, the homogeneous manifold $H/K$ is diffeomorphic to $\mathbb{R}^n$, where $n = \dim H/K$. For any discrete torsion-free subgroup $\Gamma \subset G$, then $\Gamma\backslash H/K$ is an aspherical manifold.

The Borel conjecture asserts that any two closed aspherical manifolds are homeomorphic if they are homotopic, or equivalently if their fundamental groups are isomorphic. Therefore, this conjecture only depends on the fundamental group under discussion.

It follows from the Mostow strong rigidity that if two finite volume locally symmetric spaces of noncompact type are homotopic, then they are diffeomorphic. In fact, if they are not hyperbolic surfaces or products of them up to finite covers, then they are isometric up to suitable scaling. For compact hyperbolic surfaces, their diffeomorphism type is determined by their genus. For finite volume hyperbolic surfaces, they are determined by their genus and the Euler characteristic, in particular by their fundamental groups.

For general aspherical manifolds, we cannot replace the homeomorphism by the diffeomorphism in the Borel conjecture. This follows from counter examples constructed by connected sums with exotic spheres, i.e., differential manifolds which are homeomorphic to but not diffeomorphic to $S^n$.

The Borel conjecture is basically equivalent to the Farrell-Jones isomorphism conjecture and implies the integral Novikov conjecture, which asserts that the assembly map in the surgery theory from a generalized homology group to the surgery group is injective.

There is a weaker stable version of the Borel conjecture asserting that if two closed aspherical manifolds $M$ and $N$ are homotopic, then $M \times \mathbb{R}^3$ and $N \times \mathbb{R}^3$ are homeomorphic. This is called the stable Borel conjecture.

This stable Borel conjecture follows from the integral Novikov conjecture in surgery theory for the group $\pi_1(M)$ [Ji7]. (The usual Novikov conjecture on homotopy invariance of higher signatures is equivalent to that the assembly map in surgery theory is rationally injective, i.e., when tensored with $\mathbb{Q}$, the assembly map becomes injective. See [Ran1-2] [Dav] [Wein3].)

The Borel conjecture has generated a lot of work in geometric and algebraic topology. For precise definitions of the Novikov conjectures and related conjectures such as the Farrell-Jones conjecture, Baum-Conne conjectures, results about them and methods to prove these conjectures, see §14.6 below, and [FeRR] [BloW] [Farr] [FarrJ1-3] [FarrJO] [Wein2] [KrLu] [ChaWe] [CarP] [Ji7] [Ros1-2] [Ran1-2] [Rost] [Gold] [GueHW] [Yu1-2] [FarrL] and references there.

## 11.13. Methods to prove the rigidities

There are several methods to prove rigidity results of locally symmetric spaces or lattices of semisimple Lie groups. One approach uses compactifications of symmetric spaces, or rather the boundary at infinity contained in the compactifications. The basic idea is that once things are pushed to infinity, some finite ambiguities are washed away, and the essential things are left. Then the rigidity of the structure at infinity implies the desired rigidity result.

For example, in the proof of the Mostow strong rigidity when the rank of the symmetric spaces is at least 2, the maximal Furstenberg boundary is used crucially. In fact, a homotopy equivalence between two locally symmetric spaces $\Gamma_1\backslash X_1$ and $\Gamma_2\backslash X_2$ induces a homeomorphism of the maximal boundaries of $X_1$ and $X_2$ and an isomorphism of the associated Tits buildings of $X_1$ and $X_2$. Then the rigidity of the Tits buildings implies an isomorphism of the associated Lie groups $G_1$ and $G_2$ (where $X_1 = G_1/K_1$ and $X_2 = G_2/K_2$) and then an isometry of the locally symmetric spaces after suitable scaling. See [Most1] for the original proof. See [Ji5] for a summary and explanation of the argument. As pointed out in [Ji5], the topological buildings introduced in [BuS2] are actually used in the proof.

The Furstenberg boundaries are also used crucially in the super-rigidity of lattices. See [Marg1] [Zi1].

Another approach to prove the Mostow strong rigidity is to deform a homotopy to a more canonical harmonic map between locally symmetric spaces or Bruhat-Tits buildings. Then Bocher type inequalities imply the desired rigidities for the harmonic maps. This method has also been used successfully in other situations, for example, in rigidity of Kahler manifolds [Siu1-2]. This motives a lot of work of harmonic maps into metric spaces. See [Cor1] [JosL] [JoY1-3] [MokSY]. See [Ji5] for a brief summary and more references.

## 11.14. Dynamics, flows on locally symmetric spaces and number theory

As mentioned above, geodesics and flat subspaces of locally symmetric spaces play an important role in the study of geometry and rigidity properties. They are closely related to the geodesic flow and higher dimensional abelian flows (or Cartan flows.) There are also rigidity type questions and classifications for higher dimensional flows such as the higher rank rigidity of locally symmetric spaces (see [KatoS1-2] [Tom2-3] [Dan1-9]) and applications to quantum chaos in §12.6 (see [Lin1-2] [SilbV] and the references there). See also [Min] for a different kind of flows on metric spaces, which is used in [BarLR1-3] to prove the Farrell-Jones conjecture in algebraic $K$-theory for hyperbolic groups.

Besides abelian subgroups, semisimple and reductive Lie groups contain other important classes of subgroups, for example, unipotent subgroups. The flows associated with unipotent subgroups are unipotent flows. Their dynamics have had a lot of striking applications through Ratner's theorems to number theory such as counting of integral points (see §15.7), Diophantine approximation, and the Oppenheim conjecture etc. See [KleM1] [KleSS] [Marg2] [Spa1-2] [ClU1] [ClOU] [Dan1-9] [Furr5] and references there. See the book [Morr2] for a systematic exposition of Ratner's theorems. The Oppenheim conjecture roughly states that if a quadratic form in more than three variables is indefinite and irrational (i.e., not a multiple of a quadratic form with rational coefficients), then its values at the integral points form a dense subset of $\mathbb{R}^*$. For the history of the Oppenheim conjecture and its connection to structures of the closure of orbits of group actions on homogeneous spaces, see [Ji9].

See [ClU2] for applications of homogeneous flow to distributions of subvarieties. See also [ConzG1-2].

CHAPTER 12

# Automorphic forms and automorphic representations for general arithmetic groups

For automorphic forms on $\mathbb{H}$ with respect to arithmetic subgroups of $SL(2,\mathbb{R})$, see §3.4.11 and §3.4.12. The books [Bum1-2] [GelGP] [HariC] [Gold] give some general and comprehensive introductions to automorphic forms on general semisimple Lie groups with respect to arithmetic subgroups.

When the locally symmetric spaces are three dimensional hyperbolic spaces, a detailed exposition of spectral theory of automorphic forms and the Selberg trace formula, etc, is given in [ElsGM]. For Hilbert modular forms, see [Fre] [Gat1]. For automorphic forms on classical domains, see [PiaS1], and arithmetic theory of automorphic forms [Shi1].

## 12.1. Automorphic forms

For a general Riemannian manifold $M$, the most important differential operator is the Laplace operator. This operator is functorial, for example in the sense that it commutes with any isometric action. For a symmetric space $X = G/K$ of noncompact type, the Laplace operator is invariant under $G$, i.e., it commutes with the group action of $G$. If the rank of $X$ is greater than 1, there are many other differential operators on $X$ which commute with $G$. If fact, if the rank of $X$ is equal to $r$, then there are $r$ basic differential operators which generate the whole ring of all invariant differential operators on $X$. A probably oversimplified or misleading example is to consider symmetric spaces of products of $r$ rank 1 symmetric spaces.

The group $G$ acts on the homogeneous space $\Gamma\backslash G$ and hence also on $L^2(\Gamma\backslash G)$, called the regular representation of $G$. The space of $K$-invariant functions can be identified with $L^2(\Gamma\backslash X)$. In fact, $L^2(\Gamma\backslash G)$ can be decomposed into a sum of $L^2(\Gamma\backslash X, E_\sigma)$, where $E_\sigma$ is a homogeneous bundle associated with a finite dimensional representation $\sigma$ of $K$, $E_\sigma = G \otimes_K V$, where $V$ is the representation vector space of $\sigma$.

A fundamental problem in representation theory is to decompose this regular representation of $G$ into its irreducible summands. The invariant differential differential operators on $X$ induce unbounded operators on $L^2(\Gamma\backslash X, E_\sigma)$. One important step towards decomposing the regular representation is to obtain the spectral decomposition of all these invariant differential operators on $L^2(\Gamma\backslash X, E_\sigma)$. In fact, each joint eigenspace is invariant under $G$ and gives a subrepresentation of $G$. (If we use only the

Laplace operator, its spectral decomposition is rougher than the joint spectral decomposition. So to get a better approximation, we need all invariant differential operators.) Eigenfunctions in such decompositions are basically automorphic forms.

In general, automorphic forms do not have to be square integrable, but they need to satisfy some growth bounds. These bounds are important in proving basic results such that the space of automorphic forms of a fixed type is finite dimensional. One reason why we need non-square integrable eigenfunctions is that if $\Gamma\backslash X$ is noncompact, it has a nonempty continuous spectrum, and their corresponding generalized eigenfunctions are not square integrable. Another reason is that some naturally occurring automorphic forms are simply not square integrable.

The structure of infinity of $\Gamma\backslash X$ provided by the reduction theory is crucial in understanding the behaviors at infinity of automorphic forms. For example, one important result says roughly that the asymptotic bounds on Siegel sets of an automorphic form is controlled by its constant terms with respect to $\mathbb{Q}$-parabolic subgroups, and this result is one instance of the so-called theory of constant terms of automorphic forms. A very useful reference is [HariC]. Other references include [MoeW] [GelGP] [Wal1].

As in the case of $G = SL(2, \mathbb{R})$, automorphic forms can be studied in terms of automorphic representations over adeles.

## 12.2. Boundary values of eigenfunctions and automorphic forms

For a joint eigenfunction of the invariant differential operators on $X$, a natural problem is to understand its behaviors at infinity. This is an essential part of the Helgason conjecture which states that every such eigenfunction on $X$ is the Poisson transformation of a hyperfunction on the maximal Furstenberg boundary of $X$.

The theory of regular singularities can be used for such a purpose, and good compactifications of $X$ such as the Oshima compactification are needed for this theory. Briefly, the Oshima compactification is a closed real analytic manifold and contains the symmetric space as an open, but not dense submanifold. It is basically obtained by gluing together finitely many maximal Satake compactifications, and hence the Oshima compactification contains the maximal Furstenberg boundary. For example, if $X$ is of rank one, then two copies of the maximal Satake compactification of $X$ are glued along the boundary together into a closed manifold. In this case, since the Oshima compactification of $X$ is a closed real analytic manifold, each maximal Satake compactification of $X$ , which is equal to the geodesic compactification $X \cup X(\infty)$ of $X$ obtained by adding the sphere at infinity, is a real analytic manifold with boundary, as mentioned before in §10.3.

From this gluing construction, the existence of the real analytic structure on the Oshima compactification is not obvious. One way to solve this is to complexify the symmetric space $X$ into a symmetric variety $X_{\mathbb{C}}$, which is automatically a quasi-projective variety, and to make use of the real locus of a smooth compactification of $X_{\mathbb{C}}$. (Since $X_{\mathbb{C}}$ is a quasi-projective variety, it is easier to imagine that there are compactifications given by smooth projective varieties.) An important fact is that the invariant differential operators on symmetric spaces $X$ have regular singularities along the maximal Furstenberg boundary of $X$. Boundary values of eigenfunctions are also important for representation theory of Lie groups. (see [Schl] [Osh] [Kor1] [Furr1-3] [BoreJ] [Lew] and the references there.)

Without any growth bounds on the eigenfunction on $X$, its boundary is usually a hyperfunction on the maximal Furstenberg boundary. On the other hand, if it grows moderately as in the came automorphic forms, the boundary value is a distribution.

For an automorphic form on $\Gamma\backslash X$, we can consider its lift to $X$ and its boundary value is a distribution [MilW1-2].

## 12.3. Spectral decomposition

As in the case of $\Gamma\backslash\mathbb{H}$, the spectral theory of automorphic forms on general locally symmetric spaces $\Gamma\backslash X$ is important for many applications. Assume that $\Gamma\backslash X$ is noncompact. Then as in the above case, the Laplace operator has both the continuous spectrum whose generalized eigenfunctions are given by the Eisenstein series, and the discrete spectrum, whose eigenfunctions are square integrable.

The space $\Gamma\backslash X$ has compactifications whose boundary components are locally symmetric spaces of lower dimension, and the discrete spectrum of these spaces enter into the continuous spectrum of $\Gamma\backslash X$ via Eisenstein series associated with the square integrable eigenfunctions on the boundary locally symmetric spaces. In general, the continuous spectrum consists of countably infinitely many half intervals of the form $[a, +\infty)$, where $a > 0$ and depends on the discrete spectrum of the boundary locally symmetric spaces. (In the case of $\Gamma\backslash\mathbb{H}$, each of the boundary symmetric spaces consists of one point, and hence $a$ is equal to $\frac{1}{4}$ as mentioned earlier. If $G = SL(n, \mathbb{R})$, $X = SL(n, \mathbb{R})/SO(n)$ and $\Gamma \subset SL(n, \mathbb{Z})$, then the boundary symmetric spaces of $\Gamma\backslash X$ are arithmetic quotients of $SL(k, \mathbb{R})/SO(k)$, where $k \leq n-1$.)

As in the case of $\Gamma\backslash\mathbb{H}$, it is a very important but difficult problem to understand the discrete spectrum. On the other hand, some properties of the continuous spectrum of $\Gamma\backslash X$ and their generalized eigenfunctions, given by Eisenstein series, can be described. For example, the desired meromorphic continuation of Eisenstein series can be proved, which gives a spectral decomposition of $\Gamma\backslash X$ and shows that the continuous spectrum of $\Gamma\backslash X$ is absolutely continuous.

The fundamental book on the spectral theory of locally symmetric spaces is [Lang3], and an interpretation/exposition is given in [MoeW]. Another exposition is given in [OsbW1].

An important spectral invariant of $\Gamma\backslash X$ is the resolvent kernel, which is basically the Green function of $\Gamma\backslash X$. See [MiaW1] [Te1-2] [Wal2].

For a detailed exposition of the spectral theory through examples of hyperbolic 3-dimensional manifolds, see [ElsGM].

## 12.4. Weyl law

As mentioned earlier, the Selberg trace formula was motivated by the problem of proving the existence of infinitely many eigenvalues.

Unfortunately, the Arthur-Selberg trace formula for higher rank arithmetic groups is much more complicated and it is difficult to derive a Weyl law from it. One reason is that some parts of the discrete spectrum is mixed up with (or rather embedded into) the continuous spectrum. Therefore, both the existence of the discrete spectrum and the Weyl asymptotic formula for the counting function of the discrete spectrum are difficult to prove.

People often have to use some special tricks to handle the trace formula, i.e., to pick out some particular terms, or use the idea of the trace formula but avoid the actual trace formula. Some references include [BruMP] [Gan2] [LinV] [Mul1-4] [LaM] [Wal3-4]. For some results on counting cuspidal eigenvalues and Diophantine approximation, see [Hof].

If the rank of $X$ is greater than 1, then there are other invariant differential operators besides the Laplace operator. In fact, the ring of invariant differential operators is generated by $r$ operators, where $r$ is equal to the rank of $X$. It is natural to count the distribution of their joint eigenvalues of these $r$ differential operators. When $\Gamma\backslash X$ is compact, this was done in [DuKV]. For some non-compact $\Gamma\backslash X$, for example, the Hilbert modular varieties, see [BruMP].

## 12.5. Counting of eigenvalues for a tower of spaces

In the Weyl law, the space is fixed and the upper bound $\lambda$ on the eigenvalues in the counting function $N(\lambda)$ goes to infinity. In this problem about a tower of covering spaces, the volumes of the spaces go to infinity, but the upper bound on the eigenvalues stays fixed. More specifically, let $X = G/K$ be a symmetric space, and let $\Gamma_i$, $i = 1, 2, \cdots$, be a sequence of shrinking lattices in $G$ with $\cap\Gamma_i = \{e\}$. Then for every two $\Gamma_i, \Gamma_j$, $i < j$, there is a covering map $\Gamma_j\backslash X \to \Gamma_i\backslash X$, and $\mathrm{vol}(\Gamma_i\backslash X)$ goes to infinity as $i \to +\infty$. The problem is to understand how the eigenvalues of $\Gamma_i\backslash X$, for example, their multiplicities and large scale distributions, change when the volumes of the spaces go to infinity. In certain sense, this is about uniform asymptotics in the Weyl law.

Such results can be used to show that some irreducible representations of $G$ can be realized in $L^2(\Gamma\backslash G)$. See [DegW] [DH1-2] [RohSp4] [DeitH2-3] [Sav] and the references there.

For similar results on growth of Betti numbers of a tower of finite coverings of a compact CW-complex, see [Luc4] [ClaW].

## 12.6. Quantum chaos

One purpose of the broad theory of quantum chaos is to interpret classical chaotic behaviors of classical systems in terms of the corresponding quantum systems.

If $\Gamma\backslash\mathbb{H}$ is a compact hyperbolic surface, then its geodesic flow is a chaotic dynamical system. (In fact, if $M$ is a compact Riemannian manifold with negative sectional curvature, then its geodesic flow is ergodic.) The corresponding quantum dynamical system is described by the eigenfunctions of $\Gamma\backslash\mathbb{H}$. The passage from the quantum system to the classical system corresponds to the process that the eigenvalues go to infinity. For some purposes, a simpler and more concrete model of a chaotic dynamical system is the Bunimovich stadium whose billiard dynamics is ergodic. An interesting phenomenon is the so-called quantum scarring on some distinguished proper subregions where the maximum values of some sequences of eigenfunctions concentrate and some (unstable) paths of the classical system take.

Therefore, an important aspect of quantum chaos is to understand behaviors of eigenfunctions of such Riemannian manifolds as the eigenvalues go to infinity. A basic problem concerns behaviors of the probability measures defined by these eigenfunctions. A known question here is the quantum unique ergodicity (QUE), which says that when the classical system, i.e., the geodesic flow, is ergodic, then there is only one limiting measure given by the normalized volume element. If the quantum unique ergodicity holds, then there is no scarring.

More specifically, as in [Ze3], assume that $\Gamma\backslash\mathbb{H} = \Gamma\backslash PSL(2,\mathbb{R})/PSO(2)$ is a compact hyperbolic surface, and $\varphi_k$ an orthonormal eigen-basis of $L^2(\Gamma\backslash\mathbb{H})$:

$$\Delta\varphi_k = \lambda_k\varphi_k.$$

Then the microlocal lift of each $\varphi_k$ defines a distribution, denoted by $dU_k$, on the unit cosphere bundle $S^*\Gamma\backslash\mathbb{H}$, which can be identified with $\Gamma\backslash PSL(2,\mathbb{R})$:

$$\langle Op(a)\varphi_k, \varphi_k\rangle = \int_{\Gamma\backslash PSL(2,\mathbb{R})} a\, dU_k,$$

where $Op : C^\infty(S^*\Gamma\backslash\mathbb{H}) \to \mathcal{B}(L^2(\Gamma\backslash\mathbb{H}))$ is a map from smooth zeroth order symbols $a$ to bounded pseudo-differential operators $Op(a)$. The limits of these microlocal lifts $dU_k$ of the eigenfunctions $\varphi_k$ always tend to measures on $\Gamma\backslash PSL(2,\mathbb{R})$ which are invariant under the geodesic flow. The scarring and QUE problem is to determine the limits.

In [Ze3], Zelditch proved that that almost all $dU_k$ tend to the Liouville measure on $\Gamma\backslash PSL(2,\mathbb{R})$. In particular, almost all eigenfunctions $\varphi_k$ become equi-distributed as $\lambda_k \to +\infty$. This paper motivated a lot of work. For example, Colin de Verdiére proved in [Coli] that an analogous result holds for a compact Riemannian manifold whose geodesic flow is ergodic. An important and natural remaining question is whether all distributions $dU_k$, i.e, without any exception, converge to the Liouville measure.

It was conjectured by Rudnick and Sarnak in [RuS] that any compact Riemannian manifold $M$ with negative sectional curvature is quantum uniquely ergodic, i.e., the Liouville measure is the only weak limit of the measures $dU_k$ (or the sequence of probability measures $|\varphi_k|^2 d\text{vol}$ on $M$ converges to the normalized unit Riemannian measure $d\text{vol}$ of $M$). (See also [AnaN] [Don] for related results.) For arithmetic hyperbolic surfaces $\Gamma\backslash\mathbb{H}^2$ associated with congruence subgroups, there is a modified version, called *arithmetic* QUE, where the eigenfunctions $\varphi_k$ are restricted to joint eigenfunctions of the Laplace operator and the Hecke operators. Then this arithmetic QUE conjecture is proved for such arithmetic surfaces in [Lin1].

Another problem studies sharp bounds on the $L^\infty$-norm of orthonormal eigenfunctions in terms of eigenvalues. In fact, it is believed that for general Riemannian manifolds, in particular locally symmetric spaces, these norms of eigenfunctions go to infinity when the eigenvalues go to infinity. It is indeed true for compact arithmetic hyperbolic surfaces $\Gamma\backslash\mathbb{H}$, and such sharp bounds are useful for many purposes [IwS].

Closed geodesics corresponding to closed orbits of the geodesic flow. A problem closely related to the above one about the distribution of the eigenfunctions is to study distributions of simple closed geodesics when their lengths go to infinity. Since the geodesic flow of compact manifolds of constant negative sectional curvature is ergodic, it is natural to conjecture that they become uniformly distributed. See [Bowe] [Ze1-2] [Marg5] for positive results on this problem. (See also [ClO] for some results of related nature on uniform distribution on modular correspondences.)

The above problems are in general very difficult. Besides arithmetic hyperbolic surfaces, for general locally symmetric spaces associated with arithmetic groups, some results are known about these questions. For the higher rank spaces $\Gamma\backslash X$, see [SilbV] [Lin2] [BuO6]. Some other references are [LuSa1-3] [BernR] [Jak1-2] [AnaN] [Ana].

## 12.7. Arthur-Selberg trace formula

The idea of the Arthur-Selberg trace formula is the same as the original Selberg trace formula, i.e., it is an equality between the spectral and geometric invariants of a noncompact locally symmetric space $\Gamma\backslash X$ associated with an arithmetic subgroup $\Gamma$. Hence, it can be used to study both types of invariants.

The Arthur-Selberg trace formula for general semisimple Lie groups $G$ and arithmetic subgroups $\Gamma$ is very complicated. It is different from the original Selberg trace formula in which all terms involving the discrete spectrum are separated from terms involving the continuous spectrum. Instead, some discrete spectrum terms are combined with some continuous spectrum terms in order to ensure convergence of the sums. (As pointed out earlier, the continuous spectrum consists of countably infinitely many half lines

$[a, +\infty)$ in general.) Hence, it is difficult to use the formula to understand the distribution of the discrete spectrum, i.e., eigenvalues of square integrable eigenfunctions. On the geometric side, the terms are not expressed via lengths of closed geodesics, or volumes of tori or other higher dimensional flat submanifolds. Hence, it is also difficult to use the formula to understand distributions of these basic geometric objects.

On the other hand, it is very important and powerful for proving some cases of the functoriality. The reason is that one can compare the terms of the trace formula for related groups, and one can use such a comparison to understand how automorphic forms on different groups are related. (We do not need to understand completely these terms individually.)

The original paper by Selberg is [Sele1]. Some references on the Selberg trace formula and the Author-Selberg trace formula are [Art1] [Lang1] [Gelb2] [Wal5] [KnL1]. For other applications of the Selberg trace formula, see [Has1].

For some problems in analytic number theory, there is a variant of the Selberg trace formula, called the Kuznetsov formula, which is obtained by computing the Fourier coefficients of the integral kernel (an important feature is that Kloosterman sums appear in one side, and this Kuznetsov formula can be used to get sharp bounds on them). There are also other related trace formulas, for example, the relative trace formula, Petterson formula, Eichler-Selberg trace formula. See [Iw1] [Jac] [LiY] [KnL1-2] [MiaW2-3] [Goo] [Brug1] [BruM] [Gold] and the references there.

## 12.8. Selberg zeta function

As mentined before, for a hyperbolic surface $\Gamma\backslash\mathbb{H}$, its Selberg zeta function $\zeta_\Gamma(s)$ is important for the purpose to understand both the structure of the lengths of closed geodesics of $\Gamma\backslash\mathbb{H}$, for example, its counting function, and the spectrum of the Laplace operator.

If the rank of $X$ is equal to 1, the sectional curvature of $\Gamma\backslash X$ is strictly negative. Then closed geodesics of $\Gamma\backslash X$ are isolated (in fact, there exists a unique geodesic in every homotopy class of closed loops), and their lengths form an increasing sequence with finite multiplicity and going to infinity. The Selberg zeta function can be similarly defined and can be used for counting the length spectrum.

If the rank of $X$ is at least 2, the geodesics are not isolated anymore but come in positive dimensional continuous (or rather smooth) families. The reason is similar to that geodesics in the flat torus $\mathbb{Z}^2\backslash\mathbb{R}^2$ are not isolated. In particular, the lengths of closed geodesics do not have finite multiplicity. On the other hand, if we count each connected family of closed geodesics of the same length as multiplicity 1, then we still have an increasing sequence of finite multiplicity and going to infinity, and their counting function is finite and goes to infinity. The counting problem is to determine its asymptotic behaviors.

In the higher rank case, even when $\Gamma\backslash X$ is compact, the geometric side of the Selberg trace formula is more complicated, and its dependence on the lengths of closed geodesics is not so clear.

A geometric Selberg zeta function for $\Gamma\backslash X$ of higher rank is constructed in [Dei1] after the initial success on related functions in [MoS]. It could also be used to count lengths of closed geodesics. For more results and references, see [Dei3-4] [DeiP] [Pav].

See [Dei1-2] [Gan1-3] [GanW] [May] [Fri1] [MoS] [Pe1-4] for various points of view on the Selberg zeta function. For zeta functions rising from dynamics, see [Ju] [Fri2] and [ParP1].

## 12.9. Counting of lengths of geodesics and volumes of tori

Counting primes is a basic question in number theory. Simple and closed geodesics of locally symmetric spaces $\Gamma\backslash X$ of rank 1 (or more generally Riemannian manifolds of strictly negative curvature) are analogues of prime numbers, and counting the lengths of closed geodesics is similar to counting the logarithms of prime numbers. For more discussions of this topic, see §3.4.7. Some references on counting geodesics and related topics such as class number asymptotics are [BruMW] [Deg] [Gan3] [Dei3-4] [Pav] [DeitH1] [Fri1] [Sar2] [SarW] [Sun].

If the rank of $X$ is equal to 1, the sectional curvature of $X$ is strictly negative, and the counting of lengths of closed geodesics of $\Gamma\backslash X$ is relatively easy, as pointed out in the previous subsection. Assume that locally symmetric spaces $\Gamma\backslash X$ have rank at least two. Besides closed geodesics, it also contains compact quotients of $\mathbb{R}^n$, $n \geq 2$. They are flat submanifolds and are images of flat totally geodesic submanifolds of $X$. They are called tori in $\Gamma\backslash X$.

A natural and important problem is to understand structures of these tori in $\Gamma\backslash X$. For example, one question asks whether volumes of these tori in each fixed dimension satisfy some asymptotic laws. Very little is known about them. It is also natural to ask if we can construct a Selberg type zeta function out of volumes of these higher dimensional flat submanifolds.

Another question asks about bounds and meanings of the minimum volumes of these higher dimensional flat submanifolds, similar to systoles, which is the minimum length of closed geodesics. See [Bav2-3].

There are various other types of counting problem related to number theory and geometry of numbers. For counting of extensions of number fields, see also [EllV1-2] and references there.

## 12.10. $L$-functions of automorphic representations

As mentioned earlier, for each modular form or Mass form of $SL(2,\mathbb{Z})$ or a subgroup of a finite index, there is an associated $L$-function via Mellin transform. We also mentioned that in order to get the Euler product for the $L$-function, it is important to identify the automorphic forms with automorphic representation over adeles.

For a general automorphic representation of a semisimple algebraic group $\mathbf{G}$ over a number field, we can also define a $L$-function.

These $L$-functions enjoy some good properties, for example, meromorphic continuation and functional equations, which follow from the spectral theory of automorphic forms, specifically, the meromorphic continuation of Eisenstein series. They also have Euler products. An influential and classical reference is [BoreC1], which contains many papers and references. For other introductions, see [Art1-4] [ArtG] [Cog1-2] [Gelb4].

## 12.11. Meromorphic continuation of Eisenstein series

As mentioned above, when $\Gamma\backslash X$ is noncompact, it contains both discrete spectrum and continuous spectrum, and the generalized eigenfunctions are given by Eisenstein series. The Eisenstein series is first defined by an infinite sum which converges absolutely for large spectral parameters. Briefly, for the cusp at infinity $i\infty$ of $\Gamma\backslash\mathbb{H}$, its Eisenstein series $E(z,s)$ is a sum of translates of $\mathrm{Im}(z)^s$ over $\Gamma/\Gamma_\infty$, where $\Gamma_\infty$ is equal to the stabilizer of $i\infty$ and also equal to the intersection of $\Gamma$ with the parabolic subgroup $\mathbf{P}_\infty$ of $SL(2)$ corresponding to $i\infty$. For general $\Gamma\backslash X$, they are defined similarly, and there is one Eisenstein series for every $\Gamma$-conjugacy class of $\mathbb{Q}$-parabolic subgroup $\mathbf{P}$ and an square integrable eigenfunction $\varphi$ on the locally symmetric space associated with $\mathbf{P}$, and the main difference is that $\mathrm{Im}(z)^s$ is replaced by the product of a similar function (in terms of the $A_{\mathbf{P}}$-component of the horospherical decomposition of $X$ with respect to $\mathbf{P}$) with the eigenfunction $\varphi$ on the boundary. (Note that for $\Gamma\backslash\mathbb{H}$, the boundary locally symmetric spaces consist of points, and hence there is only one Eisenstein series for every $\Gamma$-conjugacy class of $\mathbb{Q}$-parabolic subgroups of $SL(2)$.)

But the generalized eigenfunctions correspond to spectral paramters which lie outside the region of absolute convergence. An important and difficult problem is to meromorphically continue the Eisenstein series to all the spectral parameters. The original proof is contained in [Lang3], and the meromorphic continuation of all Eisenstein series is proved only after the whole spectral decomposition of $L^2(\Gamma\backslash G)$ is achieved (or rather at the same time).

The meromorphic continuation of Ensenstein series associated with cuspidal forms (i.e., those automorphic forms whose constant terms vanish along all $\mathbb{Q}$-parabolic subgroups, or equivalently they decay rapidly at infinity) is relatively easy. The difficulty is to handle other non-cuspidal Eisenstein series. An important problem is to give an independent proof of the meromorphic continuation of non-cuspidal Eisenstein series without using the spectral decomposition. The idea of a proof of J.Bernstein of such nature is explained in [Bore9] through a simple example.

## 12.12. Constant term of Eisenstein series

For the modular group $SL(2, \mathbb{Z})$, there is one Eisenstein series associated with the cusp point $\{i\infty\}$. Its constant term can be expressed in terms of the Riemann zeta function. The meromorphic continuation of the Eisenstein series implies the meromorphic continuation of the constant terms, and hence the meromorphic continuation of the Riemann zeta function. On the other hand, the meromorphic continuation of the constant terms can be used to prove the meromorphic continuation of the Eisenstein series.

For general arithmetic groups of semisimple algebraic groups, the constant terms along $\mathbb{Q}$-parabolic subgroups of Eisenstein series can also be expressed in terms of $L$-functions of the automorphic forms, which define the Eisenstein series.

Then the meromorphic continuation of the Eisenstein series implies that these $L$-functions can also be meromorphically continued. $L$-functions occur naturally in number theory and their meromorphic continuation and functional equations are often difficult but sought after. If these $L$-functions can be identified with the $L$-functions of automorphic forms, then they automatically have meromorphic continuation etc. This is an important part of the Langlands program. See [GelbS] [Shah2] and the references there.

## 12.13. Langlands program

One of the original papers is [Lang4]. For a friendly introduction to the Langlands program, see [Gelb4]. For some other general introductions to the Langlands program, see [BaeN] [BernG] and [Art1-4] [ArtG]. For a thorough discussion of the Langlands program through the example of Picard modular surfaces, see [LanR]. See [AsG1-2] for an accessible exposition to part of the Langlands program with an emphasis on a conjectural non-abelian reciprocity between algebraic varieties over $\mathbb{Q}$ and Hecke eigenvectors in the homology of locally symmetric spaces $\Gamma\backslash SL(n, \mathbb{R})/SO(n)$, where $\Gamma$ are congruence subgroups. The book [AsG1] is elementary and self-contained.

There are several different aspects of the Langlands program. One aspect is to relate automorphic forms or representations of different groups, in particular, relate general semisimple (or rather reductive) algebraic groups to the general linear group $GL(n)$.

Another is to show that naturally occurring $L$-functions such as those from the Galois representations on the cohomology of Shimura varieties are actually $L$-functions of automorphic forms.

One important tool to prove the Langlands functorial principle is the converse theorem, a vast generalization of the Weil converse theorem. See [CogSP]. Another important method is to use the Arthur-Selberg trace formula, i.e., by comparing terms in the trace formula for different groups. See §12.7 for references about the Arthur-Selberg trace formula.

See [Shah1-2] for the Langlands-Shahidi methods and applications of the funatoriality to the Ramanujan conjecture, which is a general version of the

Selberg $\frac{1}{4}$-conjecture in §3.4.3. See also [Lang2] [Shah3] for an update on the functoriality conjecture.

## 12.14. Spectral theory over function fields: an example

In the above subsections, we have discussed spectral theory of automorphic forms over number fields, i.e., spectral theory of arithmetic locally symmetric spaces. There is a corresponding theory for automorphic forms over function fields. We discuss this through an example of rank 1 group $\mathbf{G} = PGL(2)$.

Let $X$ be the Bruhat-Tits tree of $G = PGL(2, K)$, where $K = \mathbb{F}_q(C)$ is a function field of a curve over a finite field $\mathbb{F}_q$, and let $\Gamma$ be a non-uniform lattice subgroup of $G$. Let $T$ denote the adjacency operator operating on functions on the set $V(X)$ of vertices of $X$.

Specifically, a $\Gamma$-automorphic function $f : V(\Gamma\backslash X) \longrightarrow \mathbb{C}$ on $\Gamma\backslash X$ is a complex-valued function $f$ on the vertices of $\Gamma\backslash X$. Such a function commutes with the action of $\Gamma$ on $X$. Define an inner product

$$\langle f, g\rangle \quad = \sum_{v\in V(\Gamma\backslash X)} f(v)\overline{g(v)}\mu(v) \quad = \quad \int_{V(\Gamma\backslash X)} f\overline{g}d\mu.$$

where $\mu(v) = 1 \ / \ |\Gamma_v|$. Then

$$L^2(\Gamma\backslash X) \ = \ \{f \ | \ \langle f, f\rangle \ < \ \infty\}.$$

Let

$$L_0^2(\Gamma\backslash X) \ = \ \{f \ | \ \langle f, 1\rangle \ = \ 0\}$$

be the subspace of functions orthogonal to the constant functions. Define the *adjacency operator* $T$ by

$$(Tf)(x) \quad = \quad \sum_{\partial_0 e = x} i(e)\ f(\partial_1 e),$$

where $x \in V(\Gamma\backslash X)$ and $\sum_{\partial_0 e = x} i(e) = q+1$. Then $T : L^2(\Gamma\backslash X) \longrightarrow L^2(\Gamma\backslash X)$. The Laplace operator is defined by

$$\Delta \quad = \quad (q+1)\, I \ - \ T,$$

and $\Delta : L^2(\Gamma\backslash X) \longrightarrow L^2(\Gamma\backslash X)$. The set of all eigenvalues of $T$ is denoted $\mathrm{Spec}(T)$ and consists of all $\lambda \in \mathbb{C}$ such that $T - \lambda I$ is not invertible, and the discrete spectrum $\mathrm{Spec}_{dis}(T)$ is defined by that $\lambda \in \mathrm{Spec}_{dis}(T)$ if there exists an $f \in L^2(\Gamma\backslash X)$ such that $Tf = \lambda f$.

The operator $T$ is bounded and self-adjoint (i.e., $\langle Tf, g\rangle = \langle f, Tg\rangle$), and the Laplace operator $\Delta$ is positive, bounded and self-adjoint. Consequently, $\mathrm{Spec}(T) \subseteq \mathbb{R}$. It is known that

$$\inf(\mathrm{Spec}(\Delta\,|_{L_0^2(\Gamma\backslash X)})) \leq (q+1) - 2\sqrt{q}.$$

Efrat showed [Ef] that for $\Gamma = PGL(2, \mathbb{F}_q[t])$, the discrete spectrum of the adjacency operator $T$ on $\Gamma\backslash X$ consists only of $\pm(q+1)$. Thus two one dimensional eigenspaces exist, namely the constant functions with eigenvalue $(q+1)$ and the alternating functions with eigenvalue $-(q+1)$.

Cusp forms are discrete eigenfunctions of the adjacency operator which sum to zero along quotients of horocycles by stabilizers of ends. Cusp forms arise from the non-constant part of the discrete spectrum of the adjacency operator. For $\Gamma = PGL(2, \mathbb{F}_q[t])$, there are no cusp forms. Furthermore, Effrat showed that there are arithmetic non-congruence lattices in $G = PGL(2, K)$ of arbitrarily large covolume that admit no cusp forms. (This gives support to the Phillips-Sarnak conjecture in §3.4.6). In contrast, for Hecke congruence subgroups of $G$, Harder, Li and Weisinger [HarLW] proved that the dimension of the space of cusp forms grows asymptotically to the covolume.

There are similarly continuous spectra described explicity by Eisenstein series $E(g, s)$ and parametrized by a parameter $s$ in the interval $[-2\sqrt{q}, 2\sqrt{q}]$. The constant and alternating eigenfunctions correspond to poles of Eisenstein series $E(g, s)$ at $s = 1, 1-\pi i/\log(q)$ respectively. [Ef] then gave a decomposition of $L^2(\Gamma\backslash X)$ into the $T$-invariant subspaces generated by non-cuspidal $L^2$-eigenfunctions and Eisenstein series, that is $L^2(\Gamma\backslash X) = R \oplus E$, where $R$ is generated by the constant and alternating functions, and $E$ is generated by Eisenstein series.

In the case where $\Gamma$ is a principal congruence subgroup of $PGL(2, \mathbb{F}_q[t])$, Nagoshi [Nag] showed that there are both discrete and continuous spectra, and that the continuous part can be described by Eisenstein series. These Eisenstein series are invariant under $\Gamma$ and hence can be expanded as a Fourier series at each cusp. Winnie Li computed the Fourier series at each cusp and showed that the constant term satisfies a functional equation ([LiWi2]). A Selberg zeta function can also be defined [Nag] [Iha] [StarT].

CHAPTER 13

# Cohomology of arithmetic groups

The starting point of the cohomology of groups is the fact that for any discrete group $\Gamma$, for any choice of the classifying space $B\Gamma$, the cohomology of $H^*(B\Gamma, \mathbb{Z})$ is independent of the choice of $B\Gamma$, and hence the cohomology groups $H^*(\Gamma, \mathbb{Z})$ can be defined to be equal to $H^*(B\Gamma, \mathbb{Z})$. (Note that since $B\Gamma$ is unique up to homotopy equivalence, this independence is clear.)

Though there are abstract definitions of $H^*(\Gamma, \mathbb{Z})$ purely in terms of the group $\Gamma$, i.e., in terms of projective resolutions of the $\mathbb{Z}\Gamma$-module $\mathbb{Z}$, the identification with $H^*(B\Gamma, \mathbb{Z})$ is often the best way to understand $H^*(\Gamma, \mathbb{Z})$ when a good model of $B\Gamma$ can be found.

For an arithmetic group $\Gamma$, when $\Gamma$ is torsion free, the locally symmetric space $\Gamma\backslash X$ is a $B\Gamma$-space. Even when $\Gamma$ contains torsion elemets, we still have the identification $H^*(\Gamma, \mathbb{Q}) = H^*(\Gamma\backslash X, \mathbb{Q})$.

If $\Gamma\backslash X$ is noncompact, there are several difficulties, and there are other cohomology theories besides $H^*(\Gamma\backslash X, \mathbb{Z})$, $H^*(\Gamma\backslash X, \mathbb{Q})$ for different needs.

Besides being fundamental objects and of intrinsic interests, cohomology groups of arithmetic subgroups have many applications, for example, in computing the algebraic $K$-groups of rings of integers [Borel2] [Borel9], in the Langlands program.

A short and accessible introduction to cohomology of arithmetic subgroups is [Borel8]. For a book on cohomology of arithmetic subgroups, in particular, cocompact ones, see [BoreW]. For a recent survey on noncompact cases, see [LiS2]. See also survey articles [Borel4-15]. For a brief summary of some results of cohomology groups of arithmetic groups from the point of view of general groups, see [Bro2].

For numerical computations of cohomology groups, see [AsGM] and its references. For various results on cohomology of arithmetic groups not covered below, see [Har1-6] [LabS1-2] [LeeS1-2] [LeeW1-4] [Roh1-3] [RohSp1-4] [RohSc1-2] [HarrT]. For cohomology of arithmetic subgroups with infinite dimensional coefficients, see [DeitH].

For some striking and unusual applications to spaces of metrics of Riemannian manifolds, see [NabW] [Wein1].

## 13.1. Cohomology groups

For a discrete group $\Gamma$, a basic question is if the groups $H^i(\Gamma, \mathbb{Z})$ are finitely generated in every degree $i$, and $\Gamma$ satisfies other cohomological finiteness conditions such as being of type $FP_\infty$ and $FL$, and a stronger condition of the existence of a finite $B\Gamma$.

Another question is to determine cohomological dimension of $\Gamma$, i.e., the minimal degree above which all cohomology groups vanish. If $\Gamma$ contains torsion elements, there are arbitrary high degrees in which the cohomology does not vanish, and hence the cohomological dimension is equal to infinite. By passing to torsion free subgroups of finite index, we can define the virtual cohomological dimension of $\Gamma$. A systematic exposition of the general theory of cohomolgy of groups is given in [Bro2]. Another is [Bier1].

By using the Borel-Serre compactification $\overline{\Gamma\backslash X}^{BS}$ of $\Gamma\backslash X$, the answer to the above questions on cohomological finiteness properties is positive for arithmetic subgroups $\Gamma$. If $\Gamma$ is torsion-free, then it also admits a finite $B\Gamma$-space given by $\overline{\Gamma\backslash X}^{BS}$.

The virtual cohomological dimension of arithmetic groups can also be computed explicitly and is equal to $\dim \Gamma\backslash X - r$, where $r$ is the $\mathbb{Q}$-rank of $\Gamma\backslash X$. For this purpose, the topology of the boundary of the Borel-Serre partial compactification of $\overline{X}^{BS}$ plays an important role. In fact, it is homotopic to the spherical Tits building of the algebraic group $\mathbf{G}$, whose simplexes are parametrized by proper $\mathbb{Q}$-parabolic subgroups of $\mathbf{G}$. This fact can be used to show that arithmetic subgroups are duality groups of dimension $\dim \Gamma\backslash X - r$. (Note that if $\Gamma\backslash X$ is compact and $\Gamma$ is torsion-free, then $H^*(\Gamma, \mathbb{Z})$ satisfies the Poincaré duality. The notion of duality group is a generalization of the Poincaré duality.) The determination of the cohomological dimension follows from this (see [BoreS1] and also [Ji10]).

A nice and short exposition of cohomology of groups together with connections to compactifications of locally symmetric spaces is given in [Borel8].

## 13.2. $L^2$- and $L^p$-cohomology of arithmetic groups

If $M$ is a differentiable manifold, $H^*(M, \mathbb{R})$ can be identified with the De Rham cohomology $H^*_{DR}(M, \mathbb{R})$ defined in terms of differential forms. If $M$ is a compact Riemannian manifold, the Hodge theorem says that in each cohomology class of $H^*_{DR}(M, \mathbb{R})$, there is a unique harmonic representative. Such a representative is convenient for many purposes.

If $M$ is noncompact, there is no such Hodge theorem for the De Rham cohomology. To do this, the notion of $L^2$-cohomology is introduced, which is basically defined in terms of $L^2$ differential forms. When $M$ is a complete Riemannian manifold, then the Hodge theorem holds if the $L^2$-cohomology groups are finite dimensional.

Assume $\Gamma\backslash X$ is noncompact, finite dimensionality of the $L^2$-cohomology groups are determined in [BoreC2]. Since the $L^2$-cohomology groups are defined in terms of differential forms, a natural and important problem is to understand how it depends on the topology of the manifold, for example, to identify $L^2$-cohomology groups of $\Gamma\backslash X$ with some other natural cohomology groups which are topological in suitable sense. This is related to the Zucker conjecture below.

An important fact about such $L^2$-cohomology groups is that they have some stability properties when $\Gamma$ ranges in a family of arithmetic subgroups of a common type. See [Bore7]. They are important for applications to algebraic $K$-groups of integers [Bore12, 19].

For introductions to these topics, see [Sap3] [SapZ] [Gore1]. Other references include [SapS] [Ste] [Zu1] [Zu3-4]. For $L^2$-cohomology of convex cocompact discrete subgroups of rank 1 semisimple Lie groups, see [Ol3]. For $L^p$-cohomology of discrete groups, see [BouMV] and [Bou1-3].

## 13.3. Intersection cohomology

If $M$ is a compact smooth manifold, its cohomology (or homology) groups satisfy the Poincaré duality. It can be shown easily by examples that the Poincaré duality fails for singular spaces.

The intersection homology (or cohomology) is a homology theory for singular spaces which satisfies the Poincaré duality. It is not canonical but depends on the choice of perversity, which controls how cycles in the homology intersect the singular strata. But for complex singular varieties, there are canonical choices of perversity, the so-called middle perversities. See the original papers [GorM1-2] for definitions and basic properties of intersection homology and cohomology, and [Kl] for a historical survey.

When $\Gamma\backslash X$ is a Hermitian locally symmetric space, its Baily-Borel compactification is a normal projective variety. Hence it admits a canonical intersection cohomology group.

The Zucker conjecture says that this intersection cohomology of the Baily-Borel compactification of $\Gamma\backslash X$ is canonicallyy isomorphic to the $L^2$-cohomology of $\Gamma\backslash X$.

This identification has far reaching consequences. See [Gore1] for a survey for various related results and applications. See also [Gore2] [SapS] [Loo4] [Sap3] [Zu4] [Lau1-3] [Pin2] [More] [HarrT] and the references there.

The Zucker conjecture is only for the Baily-Borel compactification of Hermitian locally symmetric spaces. There is a related conjecture of Rapoport-Goresky-MacPherson asserting that the middle perversity intersection cohomology groups of the reductive Borel-Serre compactification of $\Gamma\backslash X$ are equal to the intersection cohomology groups of the Baily-Borel compactification. There are also related results for $\Gamma\backslash X$ which are not necessarily Hermitian. See [Sap3-5] [Rapp].

## 13.4. Weighted cohomology

As mentioned earlier, automorphic forms are not necessarily square integrable. But they satisfy some growth property.

A natural generalization of the $L^2$-cohomology group is the weighted cohomology groups, where the integration is against suitable weights. If the weights are small, it will allow more automorphic forms into the weighted $L^2$-space. See [Fra].

There is another related notion of weighted cohomology, where the weights refer to the weights of some torus actions, and the weighted cohomology is defined by truncating according to weights of these torus actions. For a survey, see [Gore1]. See also [GorHM] [GorHMN] [LooR] and the references there.

Relations between these two types of weighted cohomology groups are established in [Nai1-2]. There are also analogues of the Zucker conjecture and Rapoport-Goresky-MacPherson conjecture involving the weighted cohomology groups. See [Sap3-5].

## 13.5. Continuous cohomology

Even when $\Gamma\backslash X$ is compact, computing cohomology groups $H^*(\Gamma, \mathbb{C})$ is not easy. The De Rham cohomology and representation theory allow one to reduce it to the following two problems:

(1) compute the multiplicity of the irreducible sub-representations of the regular representation of $G$ in $L^2(\Gamma\backslash G)$.
(2) compute continuous cohomology group with coefficients coming from the irreducible sub-representations.

By the continuous cohomology, we mean the $H^*(\mathfrak{g}, K, \pi)$, where $K$ is a maximal compact subgroup of a reductive Lie group $G$ with Lie algebra $\mathfrak{g}$, and $\pi$ is a unitary irreducible representation of $G$. An important point is that these continuous cohomology groups can be computed relatively easily using results from representation theory of $G$. In particular, some cohomology vanishing results can be obtained this way. See the book [BoreW], the foundational paper [VogaZ], a recent survey [Spe], and the references there.

## 13.6. Applications of automorphic forms to cohomology

Both the $L^2$-cohomology groups and weighted cohomology groups of $\Gamma\backslash X$ can be decomposed using automorphic representations.

Understanding these summands often require detailed information about the automorphic forms which appear in the spectral decomposition of $\Gamma\backslash G$. Cuspidal automorphic forms are always square integrable, and they play an important role.

Once the cohomology groups of arithmetic subgroups are computed in terms of automorphic forms, non-vanishing results on the cohomology groups can be used to prove existence of some automorphic forms. Other properties of automorphic forms, for example, congruence property, can also be studied using cohomology of arithmetic groups.

See the survey [LiS2] on cohomology of locally symmetric spaces $\Gamma\backslash X$, which concentrates on noncompact spaces and is a sequel to [Schwe6] and [BoreW]. For other references, see [AsGM] [AsM] [Fra] [Har1-5] [LabS1-2]

[LeeW3-4] [LiS1-3] [Mah2] [Ra6-7] [RajV] [Ro1-2] [RohSc1-2] [RohSp1-2] [Schwe1-5].

## 13.7. Construction of cycles and relations to automorphic forms

Since the space $\Gamma\backslash X$ is defined in terms of Lie groups and arithmetic subgroups, it is natural to expect that some cycles can be constructed by suitable sub-Lie groups and their discrete subgroups. When $\Gamma\backslash X$ is compact and of rank 1, more results have been obtained. One difficulty is to find a suitable Lie subgroup such that $\Gamma$ induces a cocompact discrete subgroup. Some cycles are related to theta functions and modular forms. See [KuM1-2] [KudRY] [Mill1-2] [LeeS1] [MilR] [Ramk] [Sou6] [Ve1-4] [Scz] [Od1-2]. For the three dimensional hyperbolic manifolds, see the survey [Schwe7]. For related modular embeddings, see [Has2].

Another problem is to show that such cycles represent nontrivial cohomology classes. This was motivated by a conjecture of Thurston that every compact real hyperbolic manifold admits a finite cover with non-zero first Betti number, which implies that the cover is large in the sense being Haken. See [Li] [LiM] [Mill1-2] [RajV] [Ve2-4] [Lub3] [Berge1-2] [BergeC] [Raj] [DuT] [FuM].

If $\Gamma\backslash X$ is non-compact, there are natural supplies of Lie subgroups to construct the cycles. In fact, for every $\mathbb{Q}$-parabolic subgroup $\mathbf{P}$ of $\mathbf{G}$, the factor $M_{\mathbf{P}}$ in its Langlands decomposition of $P = N_P A_{\mathbf{P}} M_{\mathbf{P}}$ is a reductive subgroup, and can be used for this purpose. These cycles are called modular symbols. Their relations to Eisenstein series are not clear. See [AsB] [Gun1-3] [AsR] [Mah3] [SpV] for more about these modular symbols.

## 13.8. Hecke trace formula on the cohomology groups

The space $\Gamma\backslash G$ is a homogeneous space with a left $G$-action, but $\Gamma\backslash X$ is not a homogeneous space. The lack of the $G$-action on $\Gamma\backslash X$ is made up by many Hecke correspondences on $\Gamma\backslash X$. Hecke correspondences also induce actions on the cohomology groups. It is important to compute their traces. They are related to the action of Frobenius endormorphism on the associated Shimura varieties, and hence play a fundamental role in studying the Hasse-Weil zeta function of such Shimura varieties, which is a basic problem in the Langlands program. For a thorough introduction in the case of $\mathbf{G} = SL(2)$, see [KnL1]. See [Fre] for related results on Hilbert modular groups. For general $\Gamma\backslash X$, see the survey article [Gore1], and other papers [GorM3-4] [GorHM] [GorHMN] [Pin2] [Lau3] [HarrT] and the references there.

## 13.9. Euler characteristics, Gauss-Bonnet formula

A basic formula in Riemannian geometry is the Gauss-Bonnet formula relating the curvature properties and topology of a compact Riemannian manifold $M$. For example, if $M$ is a closed surface and $K_p$ denotes its curvature at $p \in M$, then

$$\int_M K_p dp = 2\pi\chi(M),$$

where $\chi(M)$ is the Euler characteristic of $M$. Since it relates geometry and topology of $M$, this formula is both interesting and useful. For example, it implies that the sphere $S^2$ does not admit any Riemannian metric of nonpositive curvature. There is also a similar formula for compact surfaces with boundary. For noncompact surfaces, see [Hub].

In higher dimension, there is also such a formula for every compact Riemannian manifold $M$ with possibly non-empty boundary. If $M$ is closed, i.e., compact without boundary, then the Gauss-Bonnet formula states:

$$\int_M \omega_M = \chi(M),$$

where $\omega_M$ is the Euler-Poincaré form of $M$ and can be expressed in terms of polynomials of the curvature of $M$.

The compactness of $M$ certainly implies that the integral on the left is convergent. A natural question is whether it holds for noncompact complete Riemannian manifolds with bounded curvature and finite total volume.

The first testing class consists of locally symmetric spaces $\Gamma\backslash X$ of finite volume. The Gauss-Bonnet formula for $\Gamma\backslash X$ was first proved by Harder [Har3] for arithmetic quotients $\Gamma\backslash X$, i.e., when $\Gamma$ is arithmetic. Following Raghunathan [Ra6], he explicitly constructed a smooth exhaustion function $h$ on $\Gamma\backslash X$ which has no critical points outside a compact set. Then he applied the Gauss-Bonnet formula for the sub-level sets of $h$, which are compact submanifolds with boundary. The crucial point is to show that the contribution in the Gauss-Bonnet formula from the boundary goes to 0 as the level goes to infinity. A certain defect of the function $h$, however, was the quite complicated geometry of its sublevel sets. See also [Serr1-2] for an exposition and discussion of some related results.

In several articles Cheeger and Gromov investigated the Gauss-Bonnet theorem for *open* complete Riemannian manifolds with bounded sectional curvature and finite volume (see e.g. [CheG1].) Locally symmetric spaces $\Gamma\backslash X$ provide important examples for this class of manifolds.

In [Leu5] a new, more geometric proof of the Gauss-Bonnet theorem for locally symmetric spaces is given, which avoids the technically complicated estimates of [Har3] and also provides an explicit (and independent) illustration of general results of Cheeger and Gromov. The approach is based on an exhaustion by compact submanifolds with corners (see [Leu4].) The essential new feature of this exhaustion is that the boundary consists of subpolyhedra which are projections of pieces of horospheres. As a consequence their second fundamental forms are uniformly bounded. This property together with the generalized Gauss-Bonnet formula for Riemannian polyhedra of

Allendoerfer-Weil and Chern leads to a considerably simplified new proof of the Gauss-Bonnet theorem for locally symmetric spaces. See also [Hor].

## 13.10. Cohomology of $S$-arithmetic subgroups

As mentioned above, $S$-arithmetic subgroups are natural generalizations of arithmetic subgroups. Cohomology of arithmetic groups has been intensively and fruitfully studied and is useful for many purposes, and it is natural to study the cohomology of $S$-arithmetic subgroups. Some work has been done in this direction. For example, see [Bore20] and [Har6].

For arithmetic subgroups of reductive algebraic groups defined over number fields, we can use automorphic forms on the symmetric spaces to study the cohomology groups and such study can feed back to automorphic forms. For $S$-arithmetic subgroups of reductive algebraic groups over number fields, their cohomology groups can also be expressed in terms of automorphic forms. In fact, their structure is simpler in a certain sense. See [BlFG].

For $S$-arithmetic subgroups over function fields, the Bruhat-Tits buildings have canonical simplicial structures which can be used effectively to study the cohomology groups. See [Har6] [Stu2-3].

On the other hand, over number fields, $S$-arithmetic subgroups act on products of Riemannian symmetric spaces and Bruhat-Tits buildings. It is natural to expect that suitable combination of these two approaches can also be used to study cohomology of $S$-arithmetic subgroups in a more explicit and combinatorial way.

Some references on cohomology of $S$-arithmetic groups over number fields also include [BoreLS] [Kuh].

## 13.11. Boundary cohomology

Compactifications of locally symmetric spaces are important in the study of cohomology of arithmetic groups. One such particular instance is the notion of boundary cohomology. There are several aspects. First, it is important to understand the local topology and homology of the boundary points (or rather their neighborhoods.) Second, it is important to understand contributions of the boundary cohomology to the cohomology of arithmetic groups and various trace formulas on the cohomology groups. See [HarrZ1-2] [GorHMN] [Fra] [MG4] [LeeS2] [Mah1] [Nai1-2] [RohSp3].

CHAPTER 14

# $K$-groups of rings of integers and $K$-groups of group rings

Algebraic $K$-groups are generalized cohomology groups. For the ring $\mathbb{Z}$ or more generally the ring of integers of a number field, they encode a lot of information about the arithmetic of the number field. The well-developed theory of cohomology of arithmetic groups, in particular $SL(n, \mathbb{Z})$, can be used for understanding $K$-groups of these rings.

## 14.1. Definitions of algebraic $K$-groups

For a ring $R$, $K^0(R)$, the $K$-group in the zeroth degree, is abelian group induced from the monoid of finitely generated projective modules over $R$, which is similar to the topological $K$-group defined in terms of vector bundles.

The degree 1 group $K^1(R)$ is the stablized form of the quotient group $GL(n, R)/[GL(n, R), GL(n, R)]$ when $n \to +\infty$, or equivalently is equal to the group $GL(R)/[GL(R), GL(R)]$, where $GL(R) = GL(\infty, R)$ is the direct limit of $GL(n, R)$ as $n \to +\infty$ under the standard inclusion $GL(n, R) \hookrightarrow GL(n + 1, R)$ in the upper left corners.

Higher $K$-groups are more difficulty to define. Briefly, let $BGL(R)$ be the classifying space of the group $GL(R)$. Since $[GL(R), GL(R)]$ is a normal subgroup, Quillen's plus construction gives a space $BGL(R)^+$ which has the same homology groups as $BGL(R)$ but whose fundamental group is equal to the abelian group $GL(R)/[GL(R), GL(R)]$. Then for $n \geq 1$, the group $K^i(R)$ is defined to be $\pi_i(BGL(R)^+)$.

$K$-groups form a generalized cohomology theory, but there are some differences. For example, unlike the usual cohomology groups, there are also $K$-groups in negative degrees.

A comprehensive introduction starting from the basic definitions is given in [FrW]. An introduction and together with an overview is given in [LLGS]. For more comprehensive discussions, see [Lod] [Kar].

## 14.2. Finite generation of $K^i(\mathbb{Z})$

The simplest ring is $\mathbb{Z}$, the ring of integers. A basic question is to prove that $K^i(\mathbb{Z})$ is finitely generated.

Since the homology groups $H_m(GL(n,\mathbb{Z}))$ stabilize when $n \gg 1$, and homotopy groups and homology groups of topological spaces are related, it is not surprising that finite generation of the homology groups of the arithmetic groups $GL(n,\mathbb{Z})$ (or $SL(n,\mathbb{Z})$) can be used to prove finite generation of $K^i(\mathbb{Z})$. See [Sou3] [Sou6] for more details.

The next class of rings is the class of ring $\mathcal{O}_k$ of integers of number fields $k$. More generally, we can also consider rings of $S$-integers of both number fields and function fields.

The $K$-groups of all these rings are finitely generated. See [Qu] for the ring of integers in number fields, and [Gra3] for function fields.

In these proofs, both symmetric spaces and buildings (Tits and Bruhat-Tits) and reduction theory of arithmetic subgroups and $S$-arithmetic subgroups are used and have played an important role. One version of the reduction theory in terms of stability [Gra1-2] was motivated by the proofs of these results.

## 14.3. Relations between $K^i(\mathbb{Z})$ and cohomology of the arithmetic groups $SL(n,\mathbb{Z})$

Since the abelian groups $K^i(\mathbb{Z})$ are finitely generated, it is natural to determine their ranks and torsion elements.

The stable ranks of cohomology groups $H^i(SL(n,\mathbb{Z}),\mathbb{C})$ can be used to determine the ranks of $K^i(\mathbb{Z})$. In fact, for any number field $k$, the ranks of $K^i(\mathcal{O}_k)$ can also be determined by the notion of Borel regulator. See [Bore12] [Bore7] [Sou3]. Some torsion parts can also be determined by using the Borel-Serre compactification of locally symmetric spaces and cohomology groups of arithmetic groups [LeeZ1-3].

## 14.4. Torsion elements of $K^i(\mathbb{Z})$

The torsion parts of the groups $K^i(\mathbb{Z})$ are important and have deep arithmetic meaning. In fact, there is a famous conjecture of Lichtenbaum relating the orders of torsion elements in $K^i(\mathbb{Z})$ to special values of the Riemann zeta function at integers. See [Sou3].

Some upper bounds on the orders of the torsion parts of $K$-groups can be obtained via the reduction theory of arithmetic groups, see [Sou1].

The precise determination of the torsion parts of the $K$-groups is much more difficult. The torsion of $K^i(\mathbb{Z})$ is now (almost) completely determined. See [We] [RogW] [Sou3-4] and the references there. Cohomology of arithmetic groups also played an important role in computing some torsion parts [LeeZ1-3].

On the other hand, for the $S$-ring of integers of function fields, their $K$-groups are torsion [Har6], but their orders are not known. Using the action of $S$-arithmetic subgroups on the Bruhat-Tits buildings, it might be possible to give some upper bounds on them by following the method in [Sou1].

## 14.5. Applications of $K_i(\mathbb{Z}[\Gamma])$ in topology

In topology, the most important class of rings consists of group rings $\mathbb{Z}\Gamma$ of the fundamental group $\Gamma$ of manifolds $M$.

The lower $K$-groups are often related to obstructions in topology. For example, the obstruction for a homotopy equivalence to $M$ to be a *simple* homopoty equivalence is the Whitehead torsion, which belongs to a natural quotient of $K_1(\mathbb{Z}[\Gamma])$, i.e., the Whitehead group

$$Wh(\mathbb{Z}\Gamma) = K_1(\mathbb{Z}[\Gamma])/\{\pm\Gamma\}.$$

(Roughly, a simple homotopy is a homotopy equivalence obtained by collapsing and expanding cells with null-homotopic attaching maps.) The group $Wh(\mathbb{Z}\Gamma)$ also contains an obstruction for an h-corbordism of the manifold $M$ to be trivial.

The Wall finiteness obstruction for a finitely dominated CW-complex to be homotopic to a finite CW-complex lies in the quotient $\tilde{K}_0(\mathbb{Z}\Gamma)$ of $K_0(\mathbb{Z}\Gamma)$ by the canonical subgroup $\mathbb{Z}$ generated by the free $\mathbb{Z}\Gamma$-module.

The higher $K$-groups are related to some global properties of the topology of the manifolds, for example, the homotopy types of diffeomorphism group. See [Ros1-2] [Wag1-2] [Hatc] and the references there for these applications.

There is also a Hermitian version of algebraic $K$-theory, which arises in surgery theory. They are called the surgery groups or L-groups. They will appear in the next subsection. See also [LucR] [BarLR] [Sta1-2] [Ran1-2] [Dav].

## 14.6. Farrell-Jones conjecture, Borel conjecture and Novikov conjecture

As pointed out in §14.4, it is difficult to compute the groups $K^*(\mathbb{Z})$, and it is even more difficult to compute $K^*(\mathbb{Z}\Gamma)$.

For general groups $\Gamma$, even the finite generation of $K^*(\mathbb{Z}\Gamma)$ is not known. (When $\Gamma$ is the trivial group, $\mathbb{Z}\Gamma = \mathbb{Z}$, and this finite generation property is known by §14.2.) One possible way to settle this problem and compute $K_*(\mathbb{Z}\Gamma)$ is to use the Farrell-Jones conjecture, which says that these groups are equal to some generalized homology groups of suitable classifying spaces with coefficient in $K_*(\mathbb{Z})$. In fact, they are related by a so-called assembly map. If the classifying spaces are compact, then this conjecture implies the finite generation of $K_*(\mathbb{Z}\Gamma)$. The Farrell-Jones conjecture is closely related to the Borel conjecture (see §11.12.)

A weaker conjecture that the assembly map is injective is called the integral Novikov conjecture, whose version in surgery theory implies a stable version of the Borel conjecture, i.e., two homotopic aspherical manifolds become homeomorphic if multiplied by $\mathbb{R}^3$ [Ji7]. This is called the stable Borel conjecture.

There is also an even weaker rational Novikov conjecture which states that the assembly map tensored with $\mathbb{Q}$ is injective. For the assembly map in surgery theory, this rational injectivity is equivalent to homotopy invariance of the higher singatures, which is the original Novikov conjecture.

A comprehensive survey is given in [LucR] [BarLR]. Other references include [Ran1-2] [Dav] [FarrJ1-3] [Sta1-2] [BloW] [BoHM] [FeRR] [Gold] [Wein2-3] [Yu1-3].

CHAPTER 15

# Locally homogeneous manifolds and period domains

If $G$ is noncompact semisimple Lie group, and $K$ a maximal compact subgroup, then $G/K$ with an invariant metric is a Riemannian symmetric space of noncompact type.

On the other hand, if $H$ is a non-maximal compact subgroup, i.e., $H \subsetneq K$, then $G/H$ with an invariant metric is not a symmetric space.

If $\Gamma$ is a discrete subgroup of $G$, then we get a space $\Gamma\backslash G/H$. This is a natural generalization of locally symmetric spaces $\Gamma\backslash X$.

An important point is that period domains in the theory of variation of Hodge structures are often of this form $\Gamma\backslash G/H$.

Even the case when $H$ is trivial is interesting, since the regular representation of $G$ on $L^2(\Gamma\backslash G)$ is a fundamental object in automorphic representation theory.

## 15.1. Homogeneous manifolds as special Riemannian manifolds

In dimension three, Thurston's geometrization conjecture says that every manifold can be cut into pieces and every one of them is a homogeneous manifold. For a detailed description of these eight homogeneous manifolds, see the survey article [Scot], and also the book [Kap].

Though it is naturally expected, it is not obvious from definition that symmetric spaces of compact type are compact. (Recall that the types of symmetric spaces are defined in terms of the Lie algebras of the Lie groups of isometry.) They are important examples of compact homogeneous manifolds and admit Riemannian metrics of positive sectional curvature. Compact homogeneous manifolds which admit Riemannian metrics of positive curvature have been extensively studied and completely classified. See [Wal7] [Wal8] [Bera] [Shan] [Wi1] [ShanSW] and the references there. For other related results on homogeneous spaces such as homogeneous Einstein metrics, see [WanZ2], and for strongly isotropy irreducible homogeneous spaces, see [WanZ1].

Lie groups $G$, and their quotients $\Gamma\backslash G$ by discrete subgroups $\Gamma$, and quotients $H\backslash G$ by closed non-discrete subgroups $H$ are very important class of homogeneous manifolds, and the $G$-action on them induces the regular representations on $L^2(G)$, $L^2(\Gamma\backslash G)$ and $L^2(H\backslash G)$. A fundamental problem in representation theory is to understand irreducible constituents of $L^2(G)$, and the basic problem in automorphic representations is to understand the decomposition of $L^2(\Gamma\backslash G)$ into irreducible sub-representations. It is natural to expect that the decomposition of $L^2(H\backslash G)$ into irreducible sub-representations is important and useful as well.

If $K$ is a compact subgroup of $G$, then $L^2(G)$ is decomposed into sums of $L^2(G/K, E_\sigma)$, where $E_\sigma$ is a homogeneous bundle on $G/K$ associated with representations $\sigma$ of $K$. A similar decomposition holds for $L^2(\Gamma\backslash G)$ and $L^2(H\backslash G)$. For analysis on homogeneous spaces and homogeneous bundles, see [Hel1-2] [Schl] [Wal6] [Fle]. For a concise exposition and further developments in this direction, see [Kob8]. See also many papers in [Gin] for related results.

For locally symmetric spaces, the most important Riemannian metric is the invariant metric induced from the symmetric spaces. But it is also important to consider other Riemannian metrics on them. Since the topology of locally symmetric spaces is relatively easy to understand, they allow us to understand better relations between topology and geometry of manifolds. For some applications of arithmetic groups to spaces of metrics and algorithms in computer science, see [Wein1] [NabW].

## 15.2. Non-symmetric, but homogeneous spaces

As mentioned earlier, bounded symmetric domains in $\mathbb{C}^n$ are Hermitian symmetric spaces. On the other hand, there are many homogeneous domains in $\mathbb{C}^n$ which are homogeneous but not symmetric. With respect to the Bergman metric, they provide a natural class of homogeneous but non-symmetric manifolds. See the classical book [PiaS1] and the survey article [PiaS2]. There are also many other natural complex homogeneous manifolds. See the book [Ak1] and a survey article [Ak2] for detailed discussion and references. See also [Zh] for group actions on complex manifolds. See [Kan] for results on rigidity of Grauert tubes over homogeneous Riemannian manifolds.

The Killing form of the Lie algebra of $G$ induces a bi-invariant pseudo-Riemannian metric on $G$, and hence the homogeneous space $\Gamma\backslash G$ admits a natural $G$-invariant pseudo-Riemannian metric, but has no $G$-invariant Riemannian metric if $Ad(\Gamma)$ is noncompact. In this case, it is a non-symmetric space. For example, this happens when $G$ is a noncompact semisimple Lie group and $\Gamma$ is a lattice subgroup.

On the other hand, it is natural to consider homogeneous spaces $\Gamma\backslash G$. For example, the quotient $SL(n,\mathbb{Z})\backslash SL(n,\mathbb{R})$ is the moduli space of unimodular lattices in $\mathbb{R}^n$, or equivalently the moduli space of all flat metrics on the torus $\mathbb{Z}^n\backslash\mathbb{R}^n$ of total volume 1.

Considering the homogeneous spaces $\Gamma\backslash G$ instead of the locally symmetric spaces $\Gamma\backslash X$ is important for the trace class conjecture in the theory of Selberg trace formula. See [Ji2]. Applications of compactifications of $\Gamma\backslash G$

to compactifications of $\Gamma\backslash X$ and extensions of homogeneous bundles over $\Gamma\backslash X$ are discussed in [BoreJ].

## 15.3. Hodge structures, period domains and period maps

For every Kähler manifold $M$, there is Hodge decomposition of

$$H^n(M,\mathbb{C}) = \oplus_{p+q=n} H^{p,q}(M,\mathbb{C}).$$

This Hodge decomposition gives a Hodge structure on the complex vector space $H^n(M,\mathbb{C})$. It is equivalent to a filtration by linear complex subspaces, or a flag of subspaces. The type of a flag of subspaces is determined by the dimensions of the subspaces. All flags of subspaces of a fixed type form a flag variety, a generalization of a usual Grassmann variety, which is the case when there is only one subspace in the flag. (The flag varieties are closely related to the Furstenberg boundaries. In fact, the latter is contained in the real locus of the former.)

The Kähler metric defines a polarization on this Hodge structure, and the points corresponding to flags satisfying the polarization condition form an open domain in the flag variety and is called the period domain. This is similar to the more classical picture that a noncompact Hermitian symmetric space can be embedded into its compact dual, a compact Hermitian symmetric space, as an open domain.

Suppose $M_t$, $t \in T$, is a family of complex manifolds such that they are all diffeomorphic but not biholomorphic. Then as complex vector spaces, for every degree $n$, all $H^n(M_t,\mathbb{C})$ are isomorphic, and the Hodge decompositions $H^n(M_t,\mathbb{C}) = \oplus_{p+q=n} H^{p,q}(M_t,\mathbb{C})$ are of the same type but depend on the complex structure on $M_t$, i.e., depend on $t$. Therefore, every complex manifold $M_t$ determines a point in the period domain. This is the period map from $T$ to the period domain. If $T$ is not simply connected, the identification between the complex vector spaces $H^n(M_t,\mathbb{C})$ is well-defined up to monodromy. Hence the period map has a well-defined map into a quotient of the period domain by a discrete subgroup, which is often an arithmetic subgroup.

The theory of variation of Hodge structures basically studies how changes in the Hodge decomposition reflect changes in the complex structures. Equivariantly, it tries to understand the period map, for example, its injectivity and surjectivity properties, which are related to the so-called Torelli theorem.

More generally, given a complex vector space, all possible Hodge decompositions of it form a period domain, which is of the form $G/H$, where $G$ is a noncompact Lie group and $H$ is a compact, but not necessarily maximal, subgroup.

An important question for a degenerating family of complex manifolds $M_t$, $t \in \mathbb{C}$, $|t| < 1$, where $M_t$ is non-singular if $t \neq 0$, asks about the behaviors of the period map near the singular parameter $t = 0$. One question asks if the period map extends across the singular point to a suitable compactification of the period domain. The Borel extension theorem in [Bore6] and related results in [Ji4] [KiK1-2] [KobO] were motivated by this problem. See [Us1] [Us2] [KatU1-4] for general results on extension of period maps. See also [GreG1-2] for some possible applications of extension properties (or boundary properties of compactifications) to the Hodge conjecture, which asserts some cohomological classes are represented by linear combinations of sub-varieties.

Several recent books on period maps and period domains are [Voi1-2] [CarMP]. See also the paper [GriS]. For general overviews of Hodge theory, see the long survey article [BryZ] and a more recent survey [Hai]. For other papers, see [Cat1-3] [CatK] [CatKS1-3] [Gri] [Kau1-3] [Ra2] [Sche1-2] [Schm1] [O12].

For relations between Hodge conjecture, Hodge theory and Kuga-Satake varieties, and related topics, see [vanG] [KugS] [Voi3] [Moro].

## 15.4. Homogeneous, non-Riemannian manifolds

In the above subsections, we have considered homogeneous spaces of the form $G/H$, where $H$ is a compact subgroup of $G$.

It is natural and important to consider homogeneous spaces when $H$ is non-compact. For example, when $\Gamma$ is an arithmetic subgroup of $G$, it is non-finite and hence noncompact if $G$ is noncompact. The quotient $G/\Gamma$ (or equivalently $\Gamma\backslash G$) has been considered before and is fundamental in automorphic representation theory.

There are some other noncompact Lie subgroups $H$ for which the space $G/H$ arise naturally. In such a case, $G/H$ in general does not admit Riemannian metric, but may admit some other interesting geometric structures. For example, when $p, q > 0$, $H = SO(p, q)$ is noncompact, and $SL(n, \mathbb{R})/SO(p, q)$, where $p + q = n$, is an example of semisimple symmetric spaces.

For example, consider the homogeneous variety $SL(n, \mathbb{C})/SO(n, \mathbb{C})$. It is defined over $\mathbb{R}$. Its real locus certainly contains the symmetric space $SL(n, \mathbb{R})/SO(n)$. But it also contains non-Riemannian homogeneous spaces $SL(n, \mathbb{R})/SO(p, q)$, where $p + q = n$, $p, q > 0$, and $SO(p, q)$ is a noncompact reductive Lie group. In fact, $SL(n, \mathbb{R})/SO(p, q)$ is the example of semisimple symmetric spaces mentioned above.

More generally, if $G$ is a semisimple Lie group and $H$ is a closed subgroup such that $H$ is reductive in $G$, then the homogeneous space $G/H$ carries a $G$-invariant pseudo-Riemannian metric induced from the Killing form. Whereas a Riemannian metric is given by a positive definite 2-form at every point of a manifold, a *pseudo-Riemannian metric* is the generalization obtained by replacing the positive difinite condition on the form by the non-degenerate condition. The case of *Lorentz manifold* corresponds to the non-degenerate 2-form having signature $(n-1,1)$. Semisimple symmetric spaces are important examples of pseudo-Riemannian homogeneous manifolds.

A topic closely related to homogeneous spaces is the theory of transformation groups. See [AdeD] [AmG] for surveys of topics related to geometric topology. See also [AdR, Remark 5.8] for $K$-theory of orbifolds, where the Borel-Serre partial compactification $\overline{X}^{BS}$ as a cofinite $\underline{E}\Gamma$-space, i.e., a universal space for proper actions of $\Gamma$, is used.

## 15.5. Clifford–Klein forms of homogeneous spaces

Let $G/H$ be a pseudo-Riemannian symmetric space or more generally a pseudo-Riemannian homogeneous space, and $\Gamma$ is a discrete subgroup of $G$ such that it acts properly on $G/H$. Quotients $\Gamma\backslash G/H$ are called Clifford-Klein forms. This is related to the usual space forms in §4.10, which are complete Riemannian manifolds of constant sectional curvature. See more discussions about space forms in §4.10 and the next subsection.

A fundamental problem here is to understand how local homogeneous geometric structure given by $G/H$ affects the global nature of the manifold $\Gamma\backslash G/H$.

If $X = G/K$ is a Riemannian symmetric space, a theorem of Borel [Borel6] says that there always exist discrete subgroups $\Gamma$ acting isometrically and cocompactly on $X$. This is equivalent to that $G$ always admits cocompact discrete subgroups. The compactness criterion of §4.11 is used in the construction.

Suppose that $H$ is a noncompact closed subgroup of $G$ and $\Gamma$ is a cocompact discrete subgroup of $G$. Unlike the previous case that a proper action of $\Gamma$ on $G$ induces a proper action on $G/K$, the group $\Gamma$ in general does not act properly on $G/H$. In fact, for a cocompact discrete subgroup $\Gamma$ of $G$, $\Gamma$ does not act properly on $G/H$ if and only if $H$ is non-compact. The intuitive reason is clear: since $\Gamma$ is cocompact in $G$, it is of the same size as (or rather quasi-isometric to) $G$. If $\Gamma$ still acts properly on $G/H$, its quotient is certainly compact. This will imply that $\Gamma$ is also of the same size as $G/H$, which implies that $G$ and $G/H$ are of the same size. Since $H$ is noncompact, $G$ and $G/H$ are certainly not of the same size. This implies that $\Gamma$ never acts properly on $G/H$.

Some homogeneous spaces such as the de Sitter space $SO(n,1)/SO(n-1,1)$ do not admit infinite discrete subgroups $\Gamma$ that act properly on them. Such a phenomenon is called the *Calabi–Markus phenomenon* named after the paper [CalaM]. It is a natural problem to classify homogeneous spaces $G/H$ for which the Calabi–Markus phenomenon occurs. For pseudo-Riemannian homogeneous manifolds $G/H$ such that $G$ and $H$ are reductive Lie groups, this problem was completely solved by Kobayashi [Kob4]: $\mathrm{rank}_{\mathbb{R}}\, G > \mathrm{rank}_{\mathbb{R}}\, H$ is the necessary and sufficient condition. The key idea of the proof is to find a criterion of proper actions of subgroups of $G$ on $G/H$. Such a criterion has been obtained for reductive groups $G$ in terms of the Cartan decomposition $G = KAK$, see [Beno3] [Kob4] [Kob9].

For solvmanifolds (i.e. homogeneous spaces of solvable Lie groups), the Calabi–Markus phenomenon does not occur in general. That is, if $G$ is a simply connected solvable Lie group and $H$ is a (proper) connected subgroup of $G$, then there always exists an infinite discrete subgroup of $G$ that acts properly on $G/H$, see [Kob7]. However, the problem of finding a useful criterion of proper actions for general Lie groups is not completely solved. See [Milno4] [Lip] [BakK] [Yos] for solvable cases or affine transformation groups.

For the existence problem of compact Clifford–Klein forms, earlier references are [Kob4] [Kob6] [KobO]. After those papers, the existence problem of compact Clifford–Klein forms of the homogeneous space $SL(n)/SL(m)$ ($n > m$) have been extensively studied by various approaches including cohomogoly of discrete groups [Kob5], ergodic theory [Zi3] [LabZ], and unitary representations [Shal2]. Recently, Margulis [Marg6] introduced a notion in representation theory which is akin to proper actions, and then applied it to the existence problem of compact Clifford–Klein forms for pseudo-Riemannian homogeneous manifolds, see also [Oh3]. See [Kob2] [Kob3] for the survey on the existence problems of compact Clifford–Klein forms. See also [IoW] [Kob1] [Labo] [Marg6] [OhW1] for some other perspectives on related topics.

A natural problem about actions of discrete groups on pseudo-Riemannian homogeneous spaces is to relax the compactness condition of the quotient but require that the quotients have finite volume. The reason is that among many locally (Riemannian) symmetric spaces, for example, those coming from arithmetic quotients, typical ones are not compact but have finite volume. For non-Riemannian locally symmetric spaces $\Gamma\backslash G/H$ of finite volume, the reduction theory, automorphic forms and spectral theory have not been developed yet.

## 15.6. Space forms: non-Riemannian case

As defined in §4.10, a complete Riemannian manifold of constant sectional curvature is called a *space form.*

More generally, for pseudo-Riemannian manifolds, sectional curvature is also defined for 2-planes in a general position at each tangent space. Thus, we can generalize the notion of space forms: a pseudo-Riemannian manifold is called a *space form* if the sectional curvature is constant.

For example, for signature $(n-1, 1)$ (Lorentz manifold), de Sitter space is a space form of positive curvature, Minkowski space is a space form of zero curvature, and anti-de Sitter space is a space form of negative curvature. The semisimple symmetric space $O(p, q+1)/O(p, q)$ $(q \geq 2)$ is a simply connected space form of a pseudo-Riemannian metric of signature $(p, q)$ with negative curvature. In general, a space form is an example of locally symmetric spaces.

It is natural to ask whether or not there exists a compact space form. For the Lorentz case, a complete answer is known: there exists an $n$-dimensional compact Lorentz manifold of constant curvature $\kappa$ if and only if $\kappa = 0$, or $\kappa < 0$ and $n$ is odd. The last condition on parity of $n$ can be explained by using the Euler class, see [KobO].

For pseudo-Riemannian manifolds of general signature $(p, q)$ $(p, q > 0)$ there exists a compact space form of negative sectional curvature if $q = 1$ and $p$ is even, $q = 3$ and $p$ is a multiple of 4, or $(p, q) = (8, 7)$. The *space form conjecture* asks if the converse statement is true. See [KobY] for the current status of the space form conjecture.

Another interesting example of Clifford–Klein forms is an affinely flat manifold. A complete affinely flat manifold is expressed as a Clifford–Klein form of the homogeneous space $(GL(n, \mathbb{R}) \ltimes \mathbb{R}^n)/GL(n, \mathbb{R})$.

The *Auslander conjecture* concerns with the structure of the fundamental group of complete flat affine manifolds: *if $\Gamma$ is a discrete subgroup of $G$ such that $\Gamma$ acts properly on $G/H$ ($\simeq \mathbb{R}^n$) and $\Gamma\backslash G/H$ is compact, then $\Gamma$ contains a solvable subgroup of finite index.*

Without the compactness assumption of $\Gamma\backslash G/H$, Margulis proved that an analogous statement fails (a counterexample to Milnor's problem [Milno4]).

The Auslander conjecture is known to be true for Riemannian cases by Bieberbach [Bieb1-2] (see [Milno1] for a summary), for Lorentz cases [GoldK] and [Tom1], and for lower dimensional cases. See [AbeMS4] and the references therein.

For other related actions and references, see [Labo2] and [Dr].

## 15.7. Counting lattice points on homogeneous varieties

As mentioned in §3.2, counting lattice points in $\mathbb{R}^n$ is a basic but still not completely solved problem in geometry of numbers.

$\mathbb{R}^n$ is the simplest kind of homogeneous spaces. There have been a lot of recent work on counting lattice (or integral) points on homogeneous varieties, by using flows on homogeneous manifolds, in particular Ratner's theorems.

More specifically, let $W$ be a real finite dimensional vector space with a $\mathbb{Q}$-structure, and $V \subset W$ be a Zariski closed real subvariety defined over $\mathbb{Q}$. Let $G$ be a real linear group defined over $\mathbb{Q}$ and act on $W$ through a representation on $W$ defined over $\mathbb{Q}$. Suppose $G$ acts transitively on $V$. Let $B_R$ be the ball in $W$ of radius $R$ with center at the origin with respect to an Euclidean norm. Let $\Lambda$ be a lattice in $W$ such that $\Lambda \otimes \mathbb{Q} = W(\mathbb{Q})$. Define $N(R, V)$ to be the number of points in $V \cap B_R \cap \Lambda$. The counting problem is to understand the asymptotic behaviors of $N(R, V)$ when $R \to +\infty$.

If $W = \mathbb{R}^n$, $V = W$, $G = GL(n, \mathbb{R})$, and $\Lambda = \mathbb{Z}^n$, then this is reduced to the problem of counting integral points in §3.2. See [EsMS] [EsMc] [Oh1-2] and the references there.

The theory of unipotent flows, in particular Ratner's theorems, are important for this problem. See [Marg2-3] [Esk2] [EsMaM] [EskO] [KleSS] [KleM2] [Kle] [GuiS] [Morr2] [Dan1-9] and also [Ji9] for related results.

CHAPTER 16

# Non-cofinite discrete groups, geometrically finite groups

By definition, a lattice $\Gamma$ of a Lie group $G$ is a discrete subgroup such that the volume of $\Gamma\backslash G$ is finite. This means that $\Gamma$ is not too small in $G$.

But there are other discrete subgroups which are not cofinite. For example, let $M$ be a compact surface with nonempty boundary. Endow the interior of $M$ with a complete hyperbolic metric of infinite volume and no cusp end. Near each boundary circle, there is a funnel end instead of a cusp end. By the uniformization theorem, $M$ corresponds to a non-cofinite discrete subgroup of $SL(2,\mathbb{R})$.

There are similar such discrete subgroups acting on the real hyperbolic space $\mathbb{H}^n$ of dimension $n$, $n \geq 3$. For example, convex cocompact (non-uniform) discrete subgroups have the above property. There are several other ways to construct such non-cofinite discrete subgroups, for example, Schottky groups [Mas].

## 16.1. Geometrically finiteness conditions

If a discrete subgroup $\Gamma$ is cofinite in a semisimple Lie group $G$, finiteness of the volume of $\Gamma\backslash G$ imposes structure of $\Gamma\backslash G/K$ at infinity and implies finiteness properties of $\Gamma$.

For a non-cofinite discrete subgroup $\Gamma$ of $G$, it is important to require some finiteness conditions. One such condition is called the geometrically finite. There are various ways to define it. They are related to, but different from, the condition that there exists a Dirichlet fundamental domain with finitely many sides. One consequence of this condition says that the locally symmetric space $\Gamma\backslash X$ is the interior of a topological compact manifold with boundary. Such a structure makes many questions more accessible. See the next subsection for more results about dimensional 3 real hyperbolic manifolds.

For precise definitions of geometrically finiteness conditions and other related results, see [Bow1-2] [Ap1-2] [ApX1-2].

## 16.2. Applications in low dimensional topology

When $X = G/K$ is the real hyperbolic space $\mathbb{H}^n$, i.e., the symmetric space of constant negative curvarture, then discrete subgroups of $G$ are called Kleinian groups.

When $n = 3$, they correspond to discrete subgroups of $PSL(2, \mathbb{C})$. This is the original class of Kleinian groups studied by Poincaré and Klein and is probably the most interesting case in view of its connections with three dimensional hyperbolic manifolds. See [CanMT] [Can] [Kap] [Mas1] [Kr] [BeneP] [Th1-2] [Mard] and the references there.

In this dimension, Kleinian groups often enjoy various finiteness. For example, the positive solution of Marden's tameness conjecture says every finitely generated Kleinian group $\Gamma$ is tame in the sense that $\Gamma\backslash\mathbb{H}^3$ is the interior of a compact topological manifold. Several recent important consequences of the solution of the Marden conjecture include the following:

(1) Proof of the Bers-Thurston density conjecture that every Kleinian subgroup of $PSL(2, \mathbb{C})$ is an algebraic limit of geometrically finite Kleinian groups.
(2) Proof of the end lamination conjecture that every tame Kleinian subgroup of $PSL(2, \mathbb{C})$ is determined up to isometry by its end invariants, a generalization of the Mostow strong rigidity to hyperbolic three dimensional manifolds of infinite volume. (Recall that an important step of the proof of the Mostow strong rigidity is to push an equivariant quasi-isometry of symmetric spaces to infinity, and to recover an isometry from an isomorphism at infinity.)
(3) Proof of the Ahlfors conjecture that the limit set of a finitely generated Kleinian subgroup $PSL(2, \mathbb{C})$ is either equal to the entire 2-sphere at infinity $\mathbb{H}^3(\infty)$ or has Lebesgue measure 0.

See [Ag2] [CaleG] [Mins] and the references there for details.

The theory of Kleinian group theory allows us to construct hyperbolic manifolds and also to understand the structure of these manifolds. Conversely, hyperbolic geometry and manifold theory can be used to understand Kleinian groups.

Arithmetic Kleinian groups are particularly important, for example, for the purpose of finding three dimensional hyperbolic manifolds of minimal volume.

For the general theory of Kleinian groups, see the book [Mas1]. For a thorough discussion of arithmetic Kleinian groups, see the book [MaclR]. For a general introduction to discrete groups with emphasis on hyperbolic groups and Kleinian groups, see the book [Ohs]. Two other books with detailed information about fundamental domains in the three dimensional hyperbolic space $\mathbb{H}^3$ are [Bea] [BeneP]. For detailed discussion of spectral theory and relations to geometry of hyperbolic manifolds, see [ElsGM]. For geometry and topology of three dimensional manifolds, see [Th1] and [Rat]. For hyperbolic manifolds of general dimension and conformal discrete groups, see the books [Ap1-2]. For connections between homogeneous manifolds and all three dimensional manifolds, see the survey [Scot]. For other related results on hyperbolic manifolds and the geometry at infinity, see [BouP1-2].

The real hyperbolic spaces $\mathbb{H}^n$ form one special subclass of rank one symmetric spaces of noncompact type. Among the remaining three types of rank one symmetric spaces, the complex hyperbolic spaces share more properties with the real hyperbolic spaces. For complex hyperbolic spaces and groups acting on them, see [Goldw1-2] [Bele2].

## 16.3. Spectral theory of geometrically finite groups

When $\Gamma$ is an arithmetic subgroup, the spectral theory of the local symmetric space $\Gamma\backslash X$ is essentially the spectral theory of automorphic forms. Such a study is foundational to the Langlands program.

It is natural to develop the spectral theory of locally symmetric spaces $\Gamma\backslash X$ when $\Gamma$ is not a lattice. They also serve as models of spectral theory of Riemannian manifolds of infinite volume but of finite topology.

It turns out that for this purpose, it is important to assume that $\Gamma$ is geometrically finite. A particularly important class consists of discrete subgroups in semisimple Lie groups $G$ of rank 1. They include the classical Kleinian groups and are closely related to them. There are systematic methods to construct them, for example, various combination theorems, groups of Schottky type groups, [Mas1] [Kr]. When the rank of $G$ is greater than or equal to 2, there are also some methods to construct them [Link3] [Beno2].

Assume for the rest of this subsection that $G$ has rank 1. Then for a geometrically finite $\Gamma$, $\Gamma\backslash X$ admits a natural compactification. The continuous spectrum of $\Gamma\backslash X$ is described by Eisenstein series, which are parametrized by the boundary points; in particular, there are uncountably infinitely many Eisenstein series (in fact, positive dimensional families of them.) (Recall that when $\Gamma\backslash X$ is a finite volume noncompact locally symmetric space and the rank of $G$ is equal to 1, $\Gamma\backslash X$ has finitely many Eisenstein series. If the rank of $G$ is greater than 2, there are countably infinitely many Eisenstein series.) There are also similar issues of meromorphic continuation of Eisenstein series and scattering theory. See [BuO1-5] [Pe1-4] for detail.

Some natural invariants associated with the continuous spectrum are the resonances. For resonances of hyperbolic manifolds of infinite volume, see [GuiZ1-2] [Zwo1-2] [Pe1-4].

# CHAPTER 17

# Large scale geometry of discrete groups

The basic point here is to study discrete groups as metric spaces when endowed with word metrics. Since discrete groups are not intrinsically metric spaces and there are no canonical word metrics on them (though they are equivalent on large scale), only their large scale geometry of the word metrics gives intrinsic invariants of the groups.

Another point is that if $M$ is a compact Riemannian manifold with an infinite fundamental group $\Gamma$, then the large scale geometry of $\Gamma$ as a metric space is the same as the large scale geometry of $\tilde{M}$.

## 17.1. Word metric on discrete groups and growth of groups

As discussed in §11.5, quasi-isometric properties of symmetric spaces and lattices acting on them are important for various rigidity results.

Quasi-isometry has also been very important to understand the group theoretic properties of abstract groups. For example, for a group $\Gamma$ endowed with a word metric $d_S$, let $N(R)$ be the number of elements $\gamma$ of $\Gamma$ with $d_S(\gamma, e) \leq R$. Though the explicit value of $N(R)$ depends on the word metric $d_S$ and hence the choice of the set $S$ of generators, the nature of its growth, for example whether it is of exponential, subexponential or polynomial growth, does not depend on $S$.

Clearly when $\Gamma = \mathbb{Z}^n$, $N(R)$ grows like a polynomial as $R \to +\infty$. The same is true if $\Gamma$ is nilpotent. On the other hand, if $\Gamma$ is a free group on more than one generators, then it is of exponential growth. If $\Gamma$ is a uniform lattice of a semisimple Lie group, then it is also of exponential growth. A famous result of Gromov [Gro5] says that $\Gamma$ is virtually nilpotent if and only if $N(R)$ grows like a polynomial.

## 17.2. Geometric group theory and property (T)

In order to prove that for an irreducible lattice $\Gamma$ of a higher rank semisimple Lie group $G$, its cohomology group $H^1(\Gamma, \mathbb{Z})$ is finite, the property (T) was introduced by Kazhdan in [Kaz1] (see also [Wan1].) He proved that under this rank assumption, $G$ has property (T) and hence $\Gamma$ has property (T) as well, which together with finite generatedness of $\Gamma$ implies that $H^1(\Gamma, \mathbb{Z})$ is finite.

Basically, a group $\Gamma$ satisfies property (T) if the trivial representation is isolated in the space of all unitary representations of $\Gamma$. This means that if $\Gamma$ acts on a Hilbert space unitarily with an almost fixed point, then it has a fixed point. See [Va2] for a cohomological characterization of property (T). For relations between the Matsushima formulas and property (T) of discrete groups acting on symmetric spaces and Bruhat-Tits buildings, see [Pan4]. For a survey for recent results and relations between geometric and algebraic approaches, see [Shal8].

There is also a relative notion of property (T). Specifically, a group $\Gamma$ is said to have relative property (T) with respect to a closed subgroup $A$ if every unitary representation of $\Gamma$ with almost invariant vectors has $A$-invariant vectors. See [Fern] for references and examples of such groups. For applications of this notion of relative property (T) in the theory of operators, see [Pop1].

The property (T) can be made quantitative in terms of a Kazhdan constant. For explicit computation of this constant, see [Shal3-4] [Shal8] [Zuk] [GelaZ].

The property (T) has many applications, for example, to construct expander graphs and construct new von Neumann algebras (i.e., type $II_1$ factors) (see [Pop1]). For such applications, it is closely related to bounding from below the first positive eigenvalue of the Laplace operator of suitable Riemannian manifolds, in particular to the Selberg $\frac{1}{4}$-conjecture. See [HoLW] [Lub1] [DaSV] [HarpV] [Shal8] and also [Ji9].

For relations to the Hausdorff dimension of limit sets, the critical exponent of the Poincaré series, see [Cor2] [Leu3]. For some constructions of groups with property (T), see [DyJ].

By definition, the property (T) of a group is defined by considering its action on Hilbert spaces. For actions of discrete groups with property (T) on Banach spaces and related concepts together with rigidity properties of lattices of semisimple Lie groups, see [BadFGM] and references there. See also [Fis2] for related spectral gap results.

For some generalizations and variants of property (T), see [LubZ1]. A property opposite to property (T) is the Haagerup property. See [CheJV] [RamRS].

Property (T) and Haagerup property are important notions of geometric group theory. Books on geometric group theory include [Harp] [LyS] [MaKs] [FiR]. Other references are [Hum1] [JaS1].

## 17.3. Ends of groups

For noncompact topological spaces and infinite groups, an important invariant of their structure at infinity is the notion of ends.

Since manifolds can be triangulated and hence have structures of locally finite simplicial complexes, and other spaces considered in this book also have the structure of locally finite simplicial complexes, we discuss only ends of locally finite simplicial complexes, following the basic reference of this subsection [ScoW].

Let $X$ be a connected locally finite simplicial complex. For every finite subcomplex $K$ of $X$, the complement $X - K$ contains only finitely many connected components. Denote the number of unbounded connected components by $n(K)$. Clearly $n(K)$ is increasing (or rather non-decreasing) with respect to $K$. The number of ends of $X$, denoted by $e(X)$, is defined to be

$$e(X) = \sup n(K),$$

where $K$ ranges over all finite subcomplexes of $X$. The same definition works if $X$ is a connected manifold and $K$ a compact submanifold. If $e(X)$ is finite, then it is realized by some $K$ and stabilizes for compact subsets containing $K$. Then every unbounded component of $X - K$ is called an end of $X$.

It is clear that $e(X) = 0$ if and only if $X$ is compact. It is also easy to see that $e(\mathbb{R}) = 2$ and $e(\mathbb{R}^n) = 1$ if $n \geq 2$. On the other hand, if $X$ is the infinite regular tree, for example, the binary tree, then $e(X) = \infty$.

If $\Gamma$ is a finitely generated group, then for every finite set $S$ of generators, the associated Cayley graph of $\Gamma$ is a connected 1-complex. Its number of ends is defined to be the number of ends of $\Gamma$ and denoted by $e(\Gamma)$. In fact, it can be shown that this number is independent of the choice of the generating set $S$ and can be identified with a number defined purely in terms of the group $\Gamma$ [ScoW, p. 173]. For this purpose, the assumption that $X$ is a simplicial complex is convenient.

A basic result is that for every finitely generated group $\Gamma$, $e(\Gamma)$ is equal to either $0, 1, 2$, or $\infty$ [ScoW, p. 176]. This restriction on the number of ends of groups does not hold for spaces $X$, since we can easily constructed noncompact spaces with any given finite number of ends.

A basic formula relating the numbers of ends of groups and spaces is the following: if a group $\Gamma$ acts freely on a connected simplicial complex $X$ with a compact (i.e., finite) quotient, then

$$e(\Gamma) = e(X).$$

From this it follows immediately that $e(\mathbb{Z}) = 2$, and $e(\mathbb{Z}^n) = 1$ for $n \geq 2$. It also follows that if $X$ is a symmetric space of non-positive sectional curvature, in particular of noncompact type, and $\Gamma$ is a torsion free co-compact discrete subgroup acting on $X$, then $e(\Gamma) = 1$, by observing that $X$ is diffeomorphic to $\mathbb{R}^{\dim X}$. In fact, the torsion free condition can be removed by using the following fact: if $\Gamma' \subset \Gamma$ is a subgroup of finite index, then $e(\Gamma) = e(\Gamma')$ [ScoW].

If the quotient $\Gamma \backslash X$ is non-compact, then relations between $e(X)$ and $e(\Gamma)$ are not so clear. For example, assume that $X = G/K$ is a symmetric space of noncompact type, $G = \mathbf{G}(\mathbb{R})$ the real locus of a linear algebraic group $\mathbf{G}$ defined over $\mathbb{Q}$, and $\Gamma \subset \mathbf{G}(\mathbb{Q})$. Suppose that $\Gamma \backslash X$ is noncompact, i.e., the $\mathbb{Q}$-rank $r$ of $\mathbf{G}$ is positive. Then there are examples of $\Gamma$ with $e(\Gamma) = 1$ and $e(\Gamma) = \infty$.

In fact, when $\Gamma = SL(2, \mathbb{Z})$, the spine of the tesselation of $\mathbb{H}$ by the fundamental domain $\Omega$ discussed in §3.3 is a regular tree which is $SL(2, \mathbb{Z})$ invariant with a compact quotient and a subgroup of $SL(2, \mathbb{Z})$ of finite index acts on it freely. This implies that the number of ends of $SL(2, \mathbb{Z})$ is equal to the number of ends of the tree, which is equal to infinity.

On the other hand, if $\Gamma$ has Kazhdan's property (T), for example, when $\Gamma$ is irreducible and the rank of $G$ is at least 2, then $e(\Gamma) = 1$. To see this, we need the famous result in [St1] which says a finitely presented torsion-free group $\Gamma$ with infinitely many ends is a nontrivial free product, which in turn, by the Bass-Serre theory (see [ScoW]), implies that $\Gamma$ acts on a tree without any fixed point. Note that the tree is a countable simplicial tree. But by [Wat] (see also [AdaS]), if $\Gamma$ has property (T), then $\Gamma$ must fix a point on the tree. This gives a contradiction. Different proofs are also given in [Marg7] [Alp].

If $G$ is the isometry group of the real or complex hyperbolic spaces, then $G$ is of rank 1 and does not have property (T), and hence their lattice subgroups $\Gamma \subset G$ do not have property (T) either. On the other hand, the other rank 1 simple Lie groups have property (T). A natural conjecture is that if $G$ is of rank one and does not have property (T), then $G$ admits lattices $\Gamma$ with $e(\Gamma) = \infty$. Some hyperbolic groups have more than one ends.

It is also a natural question to determine the number of ends of other groups such as the mapping class groups and outer automorphism groups of free groups discussed later.

For other papers related to ends of groups and spaces on which they act, see [Coh] [Dun] [ Nib] [St2] [St3] [Bri] [Pesc] [Mih] [ConMT] [MihR]. Closely related to the ends of groups is the so-called JSJ decomposition of groups, motivated by the JSJ decomposition of three dimensional manifolds. See [Sel1-2] [RiS] [DunS] [ScoS]. For quasi-isometry invariance of splittings of groups and JSJ decompositions, see [Pap1] [Pap2].

## 17.4. Ends of locally symmetric spaces and bottom of the spectrum

If a locally symmetric space $\Gamma\backslash X$ is not compact, then it is natural to consider the number of its ends and more detailed structures of the ends. They are quite different from the number of ends of the group $\Gamma$.

Assume as above that $X = G/K$ is a symmetric space of noncompact type, $G = \mathbf{G}(\mathbb{R})$ the real locus of a linear algebraic group $\mathbf{G}$ defined over $\mathbb{Q}$, and $\Gamma \subset \mathbf{G}(\mathbb{Q})$ is an arithmetic subgroup. Suppose that $\Gamma\backslash X$ is noncompact, i.e., the $\mathbb{Q}$-rank $r$ of $\mathbf{G}$ is positive.

If $r \geq 2$, then it follows from the connectivity of the spherical Tits building $\Delta_{\mathbb{Q}}(\mathbf{G})$ of $\mathbb{Q}$-parabolic subgroups of $\mathbf{G}$ that $\Gamma\backslash X$ has only one end, i.e., $\Gamma\backslash X$ is connected at infinity. Furthermore, all fundamental groups $\pi_i$, $i \leq r-2$, of the infinity of $\Gamma\backslash X$ is equal to 0, while the group $\pi_{r-1}$ of the infinity is non-zero if $\Gamma$ is sufficiently small. This is related to the fact the boundary of the Borel-Serre partial compactification of $X$ is homotopic to $\Delta_{\mathbb{Q}}(\mathbf{G})$, which is homotopic to a bouquet of spheres of dimension $r-1$ by the Solomon-Tits theorem. This fact is used to show that $\Gamma$ is a virtual duality group of dimension $\dim X - r$, where $r$ is the $\mathbb{Q}$-rank of $\mathbf{G}$.

To show that $\Gamma\backslash X$ has only one end when the rank $r \geq 2$, we note that the Borel-Serre partial compactification $\overline{X}^{BS}$ of $X$ is a manifold with corners, and a tubular neighborhood of the Borel-Serre boundary deformation retracts to it, and its image in $\overline{\Gamma\backslash X}^{BS}$ (or rather its intersection with $\Gamma\backslash X$) can be taken as a neighborhood of the infinity of $\Gamma\backslash X$. This implies that the complement of all suitable large compact subsets of $\Gamma\backslash X$ in $\Gamma\backslash X$ is homotopic to their complement in the Borel-Serre compactification $\overline{\Gamma\backslash X}^{BS}$. Since the latter can be deformation retracted to its boundary $\partial\overline{\Gamma\backslash X}^{BS}$ when $\Gamma$ is torsion-free, which is homotopic to a quotient of the Tits building $\Delta_{\mathbb{Q}}(\mathbf{G})$, and hence is connected, it follows that $\Gamma\backslash X$ has only one end when $\Gamma$ is torsion-free. If $\Gamma$ is not torsion-free, we can pass to a torsion-free subgroup $\Gamma'$ of finite index. Since $\Gamma'\backslash X$ has one end and covers $\Gamma\backslash X$, $\Gamma\backslash X$ certainly also has only one end.

On the other hand, if $r = 1$, $\Gamma\backslash X$ has finitely many ends which correspond obijectively to the set of $\Gamma$-conjugacy classes of $\mathbb{Q}$-parabolic subgroups of $\mathbf{G}$.

The reduction theory of arithmetic subgroups also gives fairly good description of the structures of these ends, which are important in applications to spectral theory and geometry of $\Gamma\backslash X$.

For infinite volume locally symmetric spaces $\Gamma\backslash X$, the number of ends and their structures are not completely understood. For convex-compact infinite covolume discrete subgroups $\Gamma$ acting on the complex hyperbolic spaces, or the quarternionic hyperbolic spaces, or the Cayley plane, i.e., a rank one symmetric space which is not a real hyperbolic space, then $\Gamma\backslash X$ has only one end [Cor2]. (Note that this is false for the real hyperbolic manifolds.) By [KlL4] [Quin2], no such convex-compact infinite covolume discrete groups exist in higher rank semisimple Lie groups $G$. See [Ol3] for various results about convex compact discrete subgroups of semisimple Lie groups of rank 1.

For other infinite volume locally symmetric spaces $\Gamma\backslash X$, the number of ends is not clear. Their geometry and topology at infinity are not understood either.

Besides the number of ends of $\Gamma\backslash X$, the sizes of ends are also important. In fact, they have been studied intensively for general noncompact complete Rielammian manifolds with applications to existence and bounds on dimension of spaces of harmonic functions and spectral theory on them.

Briefly, assume that $M$ has finitely many ends, i.e., $e(M) < +\infty$. Then there exists a compact submanifold $K \subset M$ such that $M - K$ has $e(M)$-unbounded components. Then for every compact submanifold $K'$ containing

$K$, $M - K'$ also has $e(M)$-unbounded components. Fix such a compact submanifold $K$ and call each of the unbounded components of $M - K$ an end of $M$.

The first obvious way to measure sizes of an end is to ask if the end has finite or infinite volume. If $\Gamma\backslash X$ is an arithmetic locally symmetric space, then each end has finite volume. On the other hand, if $\Gamma$ is convex-compact and $\Gamma\backslash X$ has infinite volume, then every end of $\Gamma\backslash X$ has infinite volume. For an end of infinite volume, we could also measure the rate of growth of volume of balls in terms of the radius and define large and small ends. They determine the dimension of the space of bounded nontrivial harmonic functions [LiT3]. A simple example of a manifold with both one small end and a large end is the quotient of $\mathbb{H}$ by the cyclic group generated by a hyperbolic element of $SL(2, \mathbb{R})$.

Also for applications to harmonic functions, another way to measure sizes of an end depends on whether the end has a positive Green function. If it does, then the end is called a parabolic end; otherwise, it is called a non-parabolic end. This classification of ends is needed to describe structure of harmonic functions on $M$ satisfying various growth conditions, for example, bounded, or of linear growth, or of polynomial growth. See papers [LiT1-2] [ColM1-4] [LiW1-2] [SunTW] [CheCM] for various types of results and a preprint of a book [LiP] for a basically self-contained summary on harmonic functions on noncompact Riemannian manifolds. The foundational paper of this subject is [Yau3].

If $\Gamma\backslash X$ is a locally symmetric space of finite volume, then the bottom of spectrum $\lambda_0(\Gamma\backslash X)$ of $\Gamma\backslash X$ is equal to 0, since the constant function is a square integrable eigenfunction of $\Gamma\backslash X$ with eigenvalue 0. On the other hand, if $\Gamma$ is trivial or a cyclic subgroup generated by a semisimple element of $G$, then $\lambda_0(\Gamma\backslash X) > 0$. In fact, in this case, $\lambda_0(\Gamma\backslash X) = \lambda_0(X)$, where the latter is known to be positive by the spherical Fourier transformation on the symmetric space [He1l-2]. More generally, it is shown in [JiLW] that if $\Gamma$ is amenable, then $\lambda_0(\Gamma\backslash X) = \lambda_0(X)$ also holds.

For general locally symmetric spaces $\Gamma\backslash X$, is can be shown easily that

$$\lambda_0(\Gamma\backslash X) \leq \lambda_0(X).$$

As pointed out earlier, the strict inequality holds when $\Gamma\backslash X$ has finite volume. A natural question is to describe those locally symmetric spaces $\Gamma\backslash X$, or rather discrete subgroups $\Gamma$, satisfying the equality

$$\lambda_0(\Gamma\backslash X) = \lambda_0(X).$$

It turns out that this question is closely related to the study of harmonic functions and types of ends. A lot of work has been done in [LiW3-6] and they give a different type of rigidity property of non-finite volume $\Gamma\backslash X$ in terms of the bottom of the spectrum $\lambda_0(\Gamma\backslash X)$, as special cases of their extensive study of noncompact complete Riemannian manifolds $M$ satisfying a similar extremal condition on $\lambda_0(M)$. In [JiLW], it is proved that if $X$ is any symmetric space of noncompact type, and $\lambda_0(\Gamma\backslash X) = \lambda_0(X)$, then $\Gamma\backslash X$ either has *one* end, which is necessarily of infinite volume; or has *two* ends, one of finite volume and another of infinite volume, and is diffeomorphic to a cylinder $\mathbb{R} \times N$, where $N$ is a compact manifold. This reminds one of the restriction on number of ends of a group discussed in the previous subsection. A different characterization of $\Gamma\backslash X$ for $X$ of rank 1 similar to the case (2) is given in [BeleK].

## 17.5. Asymptotic invariants

Given an infinite finitely generated group, there are several asymptotic invariants. They are defined in terms of word metrics of the group. For example, two important asymptotic invariants of noncompact metric spaces are:

(1) the tangent cone at infinity,
(2) the asymptotic dimension.

The foundational paper on asymptotic invarisnts of geometric groups is [Gro1]. For an earlier survey, see [Gro5] and also [Gro2].

A famous result in [Gro5] states that a finitely generated group $\Gamma$ is virtually nilpotent, i.e., $\Gamma$ contains a nilpotent subgroup of finite index, if and only if $\Gamma$ is of polynomial growth, i.e., with respect to any word metric $d_S$ on $\Gamma$, the number of elements in the ball $B(R, e)$ of radius $R$ with center the identity element is bounded by a polynomial in $R$ as $R \to \infty$. In this proof, the tangent cone at infinity is used crucially. A continuous (or non-discrete) metric space is produced from the discrete metric space $(\Gamma, d_S)$ under successive scaling (shrinking), and continuous (Lie) groups are obtained from discrete groups through a limiting process.

Though the word metric $d_S$ and the number of elements in the ball $B(R, e)$ depend on the choice of the generating set $S$, but its growth order (exponential or not) does not. Many important properties of $\Gamma$ depend on asymptotic behaviors of $\Gamma$. For example, the Novikov conjecture on the homotopy invariance of higher signatures and various stronger integral versions hold for groups $\Gamma$ which have finite asymptotic dimension together with other topological conditions, see [Yu1] for the first result of this type (and the references of [Ji1] and [Ji7] for other generalizations.)

Hyperbolic groups have finite asymptotic dimension [Roe] (a stronger equivariant version of finite asymptotic dimension is proved for hyperbolic groups in [BarLR3], which is used to prove the Farrell-Jones conjecture in algebraic $K$-theory for hyperbolic groups in [BarLR2].) Finiteness of asymptotic dimension of arithmetic subgroups was proved in [Ji1] and that of $S$-arithmetic subgroups and general linear groups over global fields are proved in [Mat] and [Ji7].

The tangent cone at infinity of noncompact locally symmetric spaces $\Gamma\backslash X$ are determined in [JiM] [Hatt2-3] [Leu1] and are isometric to metric cones over finite simplicial complexes, which are quotients by $\Gamma$ of the Tits buildings $\Delta_{\mathbb{Q}}(\mathbf{G})$ of the algebraic group $\mathbf{G}$.

The tangent cone at infinity of a symmetric space is an $\mathbb{R}$-building [KlL2]. For the real hyperbolic space, it is an $\mathbb{R}$-tree (see [Best2].) See [Ji5] for a summary of these results and more references.

For characterization of tangent cone at infinity of hyperbolic groups and relative hyperbolic groups, and applications to quasi-isometric rigidity, see [DruS] and references there.

The quasi-isometry rigidity of discrete groups and lattices of Lie groups is discussed in §11.5 and is closely related to quasi-isometry of Riemannian manifolds and other metric spaces. For large scale geometry questions about mapping class groups, see [BehM] and [Ham2], and for graph manifold groups, see [BehN]. See also [OlsS].

For a lattice subgroup $\Gamma$ of a semisimple Lie group $G$, its action on the symmetric space $X = G/K$ is important for many purposes. A natural question is if $X$ is of the smallest dimension among all contractible manifolds on which $\Gamma$ acts properly. The answer is positive and given in [BestF2]. It should be pointed out that if $\Gamma$ is a uniform lattice, then $\dim X$ is also equal to the smallest dimension of contractible simplicial complexes on which $\Gamma$ acts properly. But this is not the case if $\Gamma$ is not a uniform lattice. This follows from the determination of the virtual cohomological dimension of $\Gamma$ (see §8.6). If $X$ is a linear symmetric space, i.e., a homothety section of a symmetric cone, then there exists a $\Gamma$-equivariant deformation retraction to a simplicial complex contained in $X$ whose dimension is equal to the virtual cohomological dimension of $\Gamma$. In general, the existence of such a deformation retraction of $X$ is still open. See §10.8.

Another large scale notion of discrete groups concerns uniform embedding into Hilbert spaces, which is very important for applications to Baum-Connes conjecture, Novikov conjecture etc. See [Yu2-3] [HigR].

As it is know, an important invariant of a countable group $\Gamma$ is its virtual cohomological dimension, i.e., the cohomological dimension of a torsion free subgroup $\Gamma'$ of finite index, which is independent of the choice of $\Gamma$ and hence well-defined. For a finitely generated group $\Gamma$, its virtual cohomological dimension is a quasi-isometric invariant of $\Gamma$ [Ger1].

## 17.6. $L^2$-invariants

Another type of large scale invariants comes from analysis. One such example for discrete infinite subgroups is the Novikov invariants describing the spectral density near the bottom of the corresponding covering spaces [GrSh] [Lot]. It should be emphasized that these $L^2$-invariants are not purely large scale invariants, since they also contain information about volume and Euler characteristics of the compact manifolds.

Briefly, given a compact Riemannian manifold $M$ (or more generally of finite volume). Assume that $\pi_1(M)$ is infinite and hence its universal covering space $\tilde{M}$ is non-compact. It is natural to combine analysis of the Laplace operator on $L^2(\tilde{M})$ together with the action of $\pi_1(M)$ to get invariants of $M$. For example, the space of square integrable differential forms on $\tilde{M}$ is invariant under the action of $\Gamma$. Such a space is of infinite dimension, but the dimension theory of von Neumann algebras associated with $\Gamma$ allows one to get real valued dimensions and obtain the $L^2$-Betti numbers of $M$. See [At] [Dod].

Though these $L^2$-Betti numbers of $M$ are not necessarily equal to the usual Betti numbers of $M$, their alternative sums give the same Euler characteristic $\chi(M)$ of $M$. An easy consequence of this equality is the following result [Luc2, Theorem 2]: If $M$ is a closed hyperbolic manifold of even dimension $2n$, then $(-1)^n\chi(M) > 0$. This statement is clearly true when $n = 1$, since $H^0(M, \mathbb{Z}) = H^2(M, \mathbb{Z})$, $H^1(M, \mathbb{Z}) = \mathbb{Z}^{2g}$, where $g \geq 2$ is the genus of $M$. On the other hand, for $n > 1$, it is not so obvious. On the other hand, the universal covering space of $M$ is the real hyperbolic space $\mathbb{H}^{2n}$, and its space of square integrable differential forms vanish except for the middle dimension. This implies immediately that $(-1)^n\chi(M)$ contains only one positive term.

An important application of the $L^2$-index theorem in [At] is to prove the existence of and realize discrete series representations of semisimple Lie groups $G$ via the space of suitable square integrable bundle-valued harmonic differential forms in [AtS].

It should probably be emphasized that the above method is different from the usual trick of studying $M$ in terms of equivariant theory of a finite covering $N$ of $M$. Briefly, if there is a manifold $N$ admitting a finite group $F$ such that the quotient $F\backslash N$ is equal to $M$, then a differential form on $M$ can be identified with a $F$-invariant differential form on $N$, and analysis of $M$ can be treated as a special case of analysis on $N$. On the other hand, if the covering $N \to M$ is infinite, differential forms on $N$ invariant under the covering group are never square integrable forms. Therefore, in the above definition of $L^2$-invariants, forms are not invariant under $\Gamma$. This is reasonable from the point of view of representation theory, since spaces consisting of invariant functions give trivial representations.

A comprehensive book on $L^2$-invariants is [Luc1]. For a quick introduction, see [Luc3]. See also [CheG1-2] [Ol1] [LucS] [LotL] for $L^2$-invariants such as $L^2$-Betti numbers, the Novikov-Shubin invariants and the $L^2$-torsion of locally symmetric spaces and more general complete Riemannian manifolds of finite volume.

## 17.7. Boundaries of discrete groups

In the proofs of Mostow strong rigidity and the Margulis super-rigidity, actions of lattices on the Furstenberg boundaries are very important [Most1]

[Marg1] [Zi1] [Furr1-4]. As mentioned before, the Furstenberg boundaries are closely connected with compactifications of symmetric spaces, and also with harmonic analysis on symmetric spaces, in particular boundary values of harmonic functions.

Briefly, for any group $G$, which could be a Lie group, a discrete subgroup, or a locally compact group, a compact $G$-space $B$ is called a *boundary* of $G$ if for every probability measure $\mu$ on $B$, there exists a sequence $g_i \in G$ such that $g_i^* \mu$ converges in the weak-$\star$ topology to a Dirac measure $\delta_b$ for some point $b \in B$. The set of all probability measures on $B$ is a convex affine subspace of the dual space of the linear space of all continuous functions on $B$, and the action of $G$ on it gives an affine representation. In general, there are many boundaries of every group. Using the theory of affine representations of $G$ and the notion of the barycenter of a probability measure, it can be shown that there exists a unique *maximal boundary* $B_G$ in the sense that all other boundaries are quotients of $B_G$.

For general $G$, this maximal boundary $B_G$ is very large and difficult to be identified. On the other hand, if $G$ is a semisimple Lie group, then $B_G = G/P$, where $P$ is a maximal amenable subgroup of $G$, which turns out to be a minimal real parabolic subgroup of $G$. In this case, the boundary $B_G$ is the *maximal Furstenberg boundary* mentioned above. It turns out that this maximal Furstenberg is related to the maximal Satake compactification of the associated symmetric space $X = G/K$, which is also called the maximal Satake-Furstenberg compactification of $X$. For definitions and more discusssions about applications of Furstenberg boundaries, see [Furr1-3] [Kor1] [Sch] [BoreJ].

This explicit identification of the maximal Furstenberg boundaries can be used to prove rigidity of lattices of some semisimple Lie groups. For example, they were used by Furstenberg [Furr4] to show that lattices in $SL(2, \mathbb{R})$ are not isomorphic to lattices in $SL(3, \mathbb{R})$. Besides the applications mentioned above of the Furstenberg boundaries to the Mostow strong rigidity and the Margulis superrigidity, more general notions of boundaries have also be introduced and used to prove other rigidity properties of more general spaces. See [BadFS] and the references there.

Furstenberg boundaries are closely related to random walks on groups and hence to Poisson boundaries and Martin boundaries (or compactifications) in probability and potential theory. See [KaiW] [Woe] [Kai2] [CartW] [KaiM1-2] [GuiJT]. For another notion of boundary and applications to nonlinear rigidity, see [BurM1-2].

A useful notion related to the actions of discrete groups on the boundary is that of (linear) proximal maps [Furr4]. See [Abe5] for an introduction and a survey of applications. See also [ConzG2] and [Sof].

For groups arising from geometric topology, we can also define other types of boundary connected with asymptotic properties of the groups or the spaces they act on. An important class of groups consists of Gromov hyperbolic groups. These are groups whose Cayley graphs are Gromov hyperbolic spaces, which are characterized by the condition that triangles in them are uniformly thin, i.e., one side of a triangle is contained in a uniform neighborhood of two other sides. For such groups, there is a canonical Gromov boundary, consisting essentially of equivalence classes of rays in the Cayley graph (or the Rips complex). See [Ohs] and [BestM] for more detail and §19 for more references.

For more general groups, one notion of boundaries is introduced in [Best1] motivated by properties of the sphere at infinity, which is defined to be the set of equivalence classes of geodesics, of simply connected and nonpositively curved Riemannian manifolds, and the Gromov boundary of hyperbolic groups.

There are other notions of boundary for discrete groups, and actions on them are important for various purposes. For example, for a discrete group $\Gamma$, the existence of a suitable compactification of a cofinite space $\underline{E}\Gamma$ for proper actions of $\Gamma$ are important for conjectures such the integral Novikov conjectures in surgery theory and algebraic $K$-theory, the Baum-Connes conjecture in $C^*$-algebras. See [HigR] [CarP] [KamY] [Rost] [EmM].

Good models of cofinite spaces $\underline{E}\Gamma$ for arithmetic groups and mapping class groups can be constructed by either compactifications of symmetric spaces and Teichmüller spaces or truncations of them. They could be used for such purposes.

See also [Rob] and references there for $K$-groups of boundary actions.

## 17.8. Asymptotic geometry of locally symmetric spaces $\Gamma\backslash X$

Compactifications can be understood from many points of views. One is from the asymptotic geometric point of view. For example, one natural question is to understand behaviors of geodesics. Unlike the case of symmetric spaces, not all geodesics of locally symmetric spaces go to infinity. Some are unbounded and do not go to infinity; some may go to infinity, but slowly; and some go directly to infinity. There is also a problem, called the Siegel conjecture, on comparison of different metrics on Siegel sets induced from the metrics of $X$ and $\Gamma\backslash X$.

Another question concerns sizes of compactifications of $\Gamma\backslash X$. For example, in the one point compactification, every two sequences of points of $\Gamma\backslash X$ going to infinity converge to the same limit point. This implies that the one point compactification is small, since it does not distinguish these sequences.

In the geodesic compactification of a CAT(0) manifold, suppose two sequences $x_n, x_n'$ of points converge to boundary points $x_\infty$ and $x_\infty'$; if the distance $d(x_n, x_n')$ is bounded, then $x_\infty = x_\infty'$.

It turns out that not all compactifications of locally symmetric spaces satisfy this condition. For example, the Borel-Serre compactification does not. But the Baily-Borel compactification of Hermitian locally symmetric spaces does satisfy this condition. This has an important implication to the theory of period domains, the so-called Borel extension theorem on maps from the punctured disc to suitable compactifications of period spaces. See [Bore6] [KiK1-2] [Ki] [Ji4] [KobO].

Another way to understand the sizes of $\Gamma\backslash X$ is to scale down the metric and find the limit of pointed metric spaces $(\Gamma\backslash X, \varepsilon d_{\Gamma\backslash X})$ when $\varepsilon \to 0$, where $d_{\Gamma\backslash X}$ is the distance function of the invariant metric. This limit, if exists, is called the tangent cone at infinity. For $\Gamma\backslash X$, it turns out to be a simplicial cone over a finite simplex, which is the quotient by $\Gamma$ of the rational spherical Tits building $\Delta_{\mathbb{Q}}(\mathbf{G})$ of $\mathbf{G}$. See [JiM] [Ji4] [Hatt2-3] [Leu1] and the references there.

## 17.9. Isoperimetric profile, Dehn functions of arithmetic subgroups

The isoperimetric inequality for closed curves in the plane is a well-known result: the circle encloses the maximal area among all curves of the same length. In particular, there is a universal quadratic function of the length which bounds areas of regions in $\mathbb{R}^2$ enclosed by curves of the given length.

For a finitely presented group $\Gamma$, we can define a Cayley complex for every pair of sets of finite generators and finite relations. Briefly, the Cayley complex is a two dimensional cell complex such that its 1-dimensional cell sub-complex is the Cayley graph, and each cell corresponds to a conjugate of one of the fixed relations. Assume that each edge (i.e., 1-cell) has length 1, and each 2-cell has area 1. Then the notion of isoperimetric functions can be defined for such Cayley complexes. Specifically, for every $n \in \mathbb{N}$, the Dehn, or filling, function $f(n)$ is the largest area enclosed by a closed loop of chains of length $n$.

Though the Dehn function of $\Gamma$ depends on the presentation used in defining the Cayley complex, its "type" (e.g. polynomial, exponential etc.) is invariant under quasi-isometries. A natural problem is to relate types of Dehn functions to geometric properties of $\Gamma$.

Gromov-hyperbolic groups are characterized by having a linear Dehn-function. A *uniform* lattice in a higher rank semisimple Lie group is quasi-isometric to the corresponding symmetric space. Since Hadamard manifolds have quadratic filling-functions, this implies that Dehn functions of uniform lattices (arithmetic or not) are at most quadratic.

The Dehn functions of *non-uniform* lattices in semi-simple Lie groups of *rank one* are either linear or quadratic or cubic (see [Gro1].) The group $SL(3, \mathbb{Z})$ and more generally any non-uniform irreducible lattice in a semi-simple Lie group of real rank 2 has an exponential Dehn function (see [LeuP].) The Dehn functions of irreducible non-uniform lattices in semi-simple Lie groups of real rank 3 or more, are expected to be quadratic. This is proved for lattices of $\mathbb{Q}$-rank 1 in semi-simple Lie groups of $\mathbb{R}$-rank $\geq 3$ in [Dru].

Thurston conjectured that $SL(n, \mathbb{Z})$, $n \geq 4$ has a quadratic Dehn function. This is still unproven.

The above examples of groups have Dehn functions of integral power growth. There are also groups which have fractional power Dehn functions. For these and some other results on existence of groups with given types of Dehn functions of finitely generated groups, and relations between isoperimetric functions or Dehn functions and relation to complexity of the word problem, see [BirRS-2] [Brid] [Bra]. See also [Now] for connections to $C^*$-algebras.

## 17.10. Trees and applications in topology

Trees are special hyperbolic spaces in the sense of Gromov. In fact, for any triangle in a tree, every side is contained in the union of other two sides. Hence, trees can also be interpreted as spaces with curvature equal to the negative infinity. (The reason is that if the curvature of a manifold is more negative, then triangles in it are thinner.)

There is another way to see that trees have curvature equal to negative infinity. Let $(\mathbb{H}^n, g_0)$ be the hyperbolic space of constant curvature equal to $-1$. For a positive $\varepsilon < 1$, then the curvature of the new Riemannian metric $\varepsilon g_0$ is equal to $-\varepsilon^{-2}$. By definition, the tangent space at infinity of $\mathbb{H}^n$ is equal to the limit of $(\mathbb{H}^n, \varepsilon g_0)$. It turns out that it exists in this case and is equal to a $\mathbb{R}$-tree, whose curvature is equal to $-\infty = \lim_{\varepsilon \to 0} -\varepsilon^{-2}$.

Recall that a general $\mathbb{R}$-tree is a metric space such that

(1) for every pair of points, there is a unique geodesic connecting them, where by a geodesic in a metric space, we mean an isometric embedding of an interval $[a, b]$,
(2) if a connected path is the finite union of finitely many geodesics, then it is also a geodesic.

We emphasize that a usual tree is a simplicial tree, in the sense that it is an one dimensional simplicial complex. The $\mathbb{R}$-trees here that arise as limits of hyperbolic spaces branches everywhere. See [Best2] [Ji5] and the references there.

The $\mathbb{R}$-trees are one kind of $\mathbb{R}$-buildings. In fact, if $X$ is a symmetric space of noncompact type, then its tangent space at infinity is an $\mathbb{R}$-building.

For more discussions about $\mathbb{R}$-trees and $\mathbb{R}$-buildings, see [CuM] [Best2] [DasDW] [Ji5].

CHAPTER 18

# Tree lattices

The basic point of the Erlanger program is that a geometry is described in terms of its symmetry group and for each group there is a corresponding geometry.

If $G$ is a noncompact simple real Lie group (a complex Lie group is treated as a real Lie group), then $G$ is essentially the full isometry group of the associated symmetric space $X = G/K$.

If $G$ is a noncompact simple $p$-adic Lie group, then $G$ acts isometrically on the associated Bruhat-Tits building $\Delta(G)$. If the rank of $G$ is greater than or equal to 2, then $G$ is also essentially the full isometry group of $\Delta(G)$.

When the rank of $G$ is equal to 1, $\Delta(G)$ is a bi-regular tree, in particular there are two kinds of valences for vertices. In this case the full isometry group $Aut(\Delta(G))$ is much larger than $G$. A natural question is to understand this group and its subgroups, in particular lattices.

## 18.1. Structures of tree lattices

Given a locally finite tree $X$, it is also natural to study its automorphism group $G = Aut(X)$ and its discrete subgroups. A subgroup $\Gamma \subset G$ is discrete if for every point $x \in X$, the stabilizer $\Gamma_x$ is a finite group. For every discrete subgroup $\Gamma \subset G$, if there are nontrivial stabilizers $\Gamma_x$, it is natural to keep track of the stabilizers with the quotient $\Gamma\backslash X$. The quotient $\Gamma\backslash X$ is a graph, and the enhanced quotient is a graph of groups, where each vertex $\Gamma \cdot x \in \Gamma\backslash X$ of the graph is assigned the conjugacy class of stabilizers $\{\gamma\Gamma_x\gamma^{-1} \mid \gamma \in \Gamma\}$. Denote this graph of groups by $\Gamma\backslash\backslash X$. Define its volume by

$$vol(\Gamma\backslash\backslash X) := \sum_{x\in V(\Gamma\backslash X)} \frac{1}{|\Gamma_x|}.$$

Then $\Gamma$ is a *lattice* in $G$ if $Vol(\Gamma\backslash\backslash X)$ is finite, and a *uniform lattice subgroup* of $G$ if $\Gamma\backslash X$ is a finite graph ([BasK], [BasL]). (Note that for a nontorsion-free lattice subgroup $\Gamma$ acting on a symmetric space, consideration of nontrivial stabilizers will result in an orbifold but will not affect the volume of the quotient. On the other hand, for lattices acting on trees, this consideration is important.)

The group $G = Aut(X)$ and its lattices have striking similarities with lattices in Lie groups, particularly over non-archimedean local fields. We refer the reader to [Lub2] for a survey of the comparisons between tree lattices and lattices in Lie groups. While $G = Aut(X)$ is not simple, Tits has shown that when $G$ acts minimally on $X$, fixing no end of $X$, then $G$ has a large simple normal subgroup, $G^+$, generated by all edge stabilizers ([Tit4].)

For a general locally finite tree $X$, it was shown in [BasK] that $G = Aut(X)$ contains uniform lattices if and only if $G$ is unimodular and $G\backslash X$ is finite. In analogy with Borel's theorems in the classical case about the co-existence of uniform and non-uniform lattices in connected non-compact semisimple Lie groups ([Bore4], [Bore16]), it was shown in [Car1] that when $G = Aut(X)$ contains uniform lattices, under some natural assumptions, $G$ also contains non-uniform lattices. [Car2] describes the necessary and sufficient conditions for $G$ to contain lattices, both uniform and non-uniform, answering earlier conjectures of Bass and Lubotzky in full.

If $X$ has more than one end and $G = Aut(X)$ contains a non-uniform $X$-lattice $\Gamma$, then it was shown in [CarbR] that $G$ contains an infinite ascending chain

$$\Gamma_1 \subset \Gamma_2 \subset \Gamma_3 \subset \cdots$$

of non-uniform $X$-lattices. Hence $vol(\Gamma_i \backslash\backslash X) \longrightarrow 0$ as $i \longrightarrow \infty$.

The Kazhdan-Margulis property for lattices in Lie groups in [KazM] states that the covolume of a lattice is bounded away from zero. Hence the existence of infinite towers of lattices in $G = Aut(X)$ shows that the Kazhdan-Margulis property is violated for $G$.

The analogue of the spectral theory of automorphic forms has not been fully developed for tree lattices. See §12.14 for some results in this direction.

For a detailed introduction to properties and constructions of tree lattices and other related topics mentioned above, see the book [BasL].

## 18.2. Arithmeticity and density of commensurability groups of tree lattices

There is no natural notion of arithmeticity of lattices in the automorphism group $G = Aut(X)$ of a general locally finite tree $X$. For a non-compact semisimple Lie group $H$ with trivial center and no compact simple factors, Margulis characterization of arithmeticity subgroups [Marg1] [Zi1] states that an irreducible lattice $\Gamma$ in $H$ is arithmetic if and only if its commensurability group

$$C_H(\Gamma) = \{h \in H \mid h\Gamma h^{-1} \cap \Gamma \text{ is of finite index in } \Gamma \text{ and in } h\Gamma h^{-1} \}$$

is not discrete, or equivalently that $C_H(\Gamma)$ is dense in $H$, which is also equivalent to that $C_H(\Gamma)$ contains $\Gamma$ as a subgroup of infinite index. It is not known if these conditions are equivalent for lattices in $G = Aut(X)$. We can however adopt one of these conditions as the definition of arithmeticity for tree lattices, including lattices in rank 2 Kac-Moody groups over finite fields.

By [Liuy], for uniform tree lattices $\Gamma$, $C_G(\Gamma)$ is dense in $G = Aut(X)$, thus in this sense, all uniform lattices are "arithmetic". We also know of examples ([BasL]) of non-uniform $X_{q+1}$-lattices $\Phi$ such that $\Phi < G - H$, where $G = Aut(X_{q+1})$, $H = PSL_2(\mathbb{F}_q(t^{-1}))$ and $C_G(\Gamma)$ is discrete. It remains to investigate the structure of the commensurability group of a general non-uniform tree lattice.

## 18.3. Rigidity of lattices in products of trees and CAT(0) groups

As discussed before, lattices of semisimple Lie groups of higher rank have many rigidity properties.

Trees are related to hyperbolic spaces and hence can be considered as rank 1 spaces. Products of trees are of higher ranks. Irreducible lattices acting on them also enjoy rigidity properties. See [BurMo1-3] [Moz] and the references there.

Products of trees are not hyperbolic spaces, but rather CAT(0)-spaces. Groups acting properly and isometrically on them with compact quotients provide a large class of CAT(0)-groups. For rigidity of such CAT(0)-groups, see [Burge] [Mon1-2] [Moz].

For restriction on actions on trees, see [Leeb1].

## 18.4. Building lattices and applications to fake projective planes

Bruhat-Tits buildings of rank 1, i.e., when the related groups are of rank one, are trees. Certainly, there are many, even regular, trees which do not come from Bruhat-Tits buildings.

Bruhat-Tits buildings are Euclidean buildings. A known result of Tits [Tit1] says that every irreducible Euclidean building of rank at least 3 arises from the Bruhat-Tits building of an reductive algebraic group. Suppose that $k_\nu$ is a locally compact and totally disconnected field such as $\mathbb{Q}_p$ and $\mathbf{G}$ a linear semisimple algebraic group defined over $k_\nu$. Let $X_\nu$ be the Bruhat-Tits building associated with $\mathbf{G}(k_\nu)$. As mentioned above, $\mathbf{G}(k_\nu)$ acts simplicially on $X_\nu$ and the stabilizer of every simplex is a compact and open subgroup, a so-called paraholic subgroup. In particular, they are quite large.

In general, the automorphism group of an Euclidean building in general does not act transitively on the set of vertices. If the building is irreducible and the automorphism group acts transitively on the set of vertices, then the building is of type $\tilde{A}_n$ [CartMS1-2].

A natural and important problem is to find discrete subgroup of the automorphism group of Euclidean buildings which acts *simply transitively* on the set of vertices. Euclidean buildings of dimension 1 are trees, and a classification of groups acting simply transitively on the vertices of a locally finite tree is given in [Tit7].

Groups acting simply transitively on vertices of buildings of type $\tilde{A}_2$ are constructed and classified in [CartMS1-2] [CartMSt] [RobS1-2] [RamRS]. For $n \geq 3$, such groups are constructed in [CartS].

Groups acting simply transitively on buildings of type $\tilde{A}_2$ have striking applications to construction of the so-called fake projective planes. Roughly, a fake projective plane is a projective surface of the general type with the same characteristic numbers as $\mathbb{C}P^2$. See [Kat2] for an introduction and survey of fake projective planes. (Note that the characterization of $\mathbb{C}P^2$ in terms of its homotopy type in [Yau1] motivated the search for fake projective planes.) See [Mum4] [IsK] [Kat1] [KatO] for construction of some of them. For a complete classification of fake projective planes and construction of higher dimensional fake projective spaces, see [PYe1-2].

For a gentle introduction to non-Archimedean uniformization, see [CorK]. For the strong rigidity theorem for non-Archimedean uniformization by co-compact lattices in $PGL(n, K)$, where $K$ is a non-Archimedean local field with finite residue field, see [IsK].

CHAPTER 19

# Hyperbolic groups

The goal was to give a combinatorial group theoretical characterization of the fundamental group of negatively curved compact Riemannian manifold. It is known that a randomly chosen finitely presented group is hyperbolic, In this sense, hyperbolic groups are *generic* discrete groups.

On the other hand, most interesting arithmetic subgroups are not hyperbolic subgroups. More specifically, let $\Gamma$ is a lattice of a semisimple Lie group $G$. If either the rank of $G$ is not equal to 1, or $\Gamma$ is not a co-compact (i.e., uniform) lattice, then $\Gamma$ is not a hyperbolic group. One could argue that these special and non-generic discrete groups are at least equally interesting. They are special and reflect different kinds of properties of discrete subgroups. For example, in the case of numbers in $\mathbb{C}$, transcendental numbers are generic, but the special numbers such as the integers, rational numbers are interesting and need special study.

## 19.1. Basic properties of hyperbolic groups

A crucial notion of Riemannian manifolds is the sectional curvature. One way that the curvature affects the geometry of the manifold is through comparison of triangles. It is known that if $M$ is a simply connected and nonpositively curved Riemannian manifold, then its triangle is thinner than a corresponding triangle of $\mathbb{R}^2$ with the same side lengths. If the negative sectional curvature is uniformly bounded away from zero, then triangles in $M$ are uniformly thin. Motivated by this, Gromov introduced the notion of hyperbolic spaces. They are geodesic metric spaces whose triangles are uniformly thin in the sense that there is a fixed constant $\delta$ such that one side of every triangle is contained in the $\delta$-neighborhood of the other two sides.

Given a finitely generated group $\Gamma$ and a choice of a finite set of generators, we can embed it into a geodesic metric space, its Cayley graph, where every edge has unit length. Then the group is called hyperbolic if the Cayley graph is a Gromov hyperbolic space.

As pointed out above, hyperbolic groups are generic groups in a sense similar to that transcendental numbers are generic. There are many natural groups which are not hyperbolic, for example, an arithmetic group is hyperbolic if and only if it is a cocompact subgroup of a rank one semisimple Lie group.

## 19.2. Rips complex and Gromov boundary

As mentioned and emphasized before, the action of arithmetic subgroups on symmetric spaces of noncompact type is important for understanding the group.

For a general discrete group $\Gamma$, it is not easy to find a good contractible space on which $\Gamma$ acts properly. On the other hand, if $\Gamma$ is hyperbolic, then it admits natural contractible simplicial complexes on which $\Gamma$ acts properly. These simplicial complexes are called Rips complexes and have the same set of vertices as the Cayley graph, but their simplexes, in particular edges, depend on the choice of a set of generators and a positive number $d$. Briefly, let $d_S$ be the word metric on $\Gamma$ associated with a set of generators $S$. A finite set of elements $\gamma_1, \cdots, \gamma_{k+1}$ of $\Gamma$ form the vertices of a $k$-simplex if the diameter of this set with respect to the metric $d_S$ is less than or equal to $d$. Denote this complex by $R_d(\Gamma)$. If $d$ is equal to 1, then $R_d(\Gamma)$ is exactly the Cayley graph of $\Gamma$ associated with $S$. The crucial property is that if $\Gamma$ is hyperbolic, then for $d \gg 1$, the Rips complex $R_d(\Gamma)$ is contractible. See [Oh, §2.7] for detail. In fact, it is a $\underline{E}\Gamma$-space, i.e., a universal space for proper actions of $\Gamma$ [MeiS]. For $d \geq 1$, $R_d(\Gamma)$ is roughly obtained by filling in some higher dimensional simplexes to kill off nontrivial homotopy classes of the Cayley graph.

The spaces $R_d(\Gamma)$ are non-compact, and can be compactified by adding the common Gromov boundary $\partial\Gamma$, which is basically defined to be the set of equivalence classes of points going to infinity (see [Oh, §2.6].) In fact, every $R_d(\Gamma)$ is a hyperbolic space, and for different values of $d$, $R_d(\Gamma)$ have the same set of vertices and are quasi-isometric to each other; hence all $R_d(\Gamma)$ have the same Gromov boundary as hyperbolic spaces.

The $\Gamma$-action extends continuously to the boundary $\partial\Gamma$ and also to the compactification $R_d(\Gamma) \cup \partial\Gamma$, and the action near the boundary is small. This extended action is important for many applications. For example, together with the fact that they also have finite asymptotic dimension [Roe], it can be used to prove the integral Novikov conjecture for hyperbolic subgroups [CarP] and the Farrell-Jones isomorphism in algebraic $K$-theory for hyperbolic groups [BarLR1-3]. (A stronger equivariant version of finite asymptotic dimension of the hyperbolic group or the Rips complex and its compactification is needed in [BarLR2-3]. See [Min] for the flow on metric spaces.) For the integral Novikov conjecture and the Baum-Connes conjecture and other related conjectures in $C^*$-algebras for hyperbolic groups, see [ConM] [Yu1-3] [MineY] [MinMS].

For the Poisson boundary of hyperbolic groups, see [Kai1-2]. For related results on random walks on groups, see [KaiW] [Woe].

For other applications of the Gromov boundary of hyperbolic groups and boundary actions to $C^*$-algebras, see [Eme]. For $K$-groups of $C^*$-algebras of actions on boundary of buildings, see [Rob].

The pioneering work on hyperbolic groups is [Gro3]. See also [Gro6]. For a more thorough discussion, see the book [GhH]. The book [Oh] covers hyperbolic groups and many other related topics. Some other references on the Gromov boundary include [Ger2] [KapK1-2] [CroK] [Sele1-2] [Fuj2] [Bou1].

The action of hyperbolic groups on the Gromov boundary is closely related to the notion of convergence groups. See [Bow3] for a survey and also [GehM2] [Bis1-2]. See [Gab] for a characterization of convergence subgroups of Homeo($S^1$) as Fuchsian groups and applications to the Seifert fibered space conjecture.

CHAPTER 20

# Mapping class groups and outer automorphism groups of free groups

We have discussed several generalizations of arithmetic subgroups, such as $S$-arithmetic subgroups, tree lattices, building lattices, CAT(0)-groups, hyperbolic groups. In this section, we discuss two more important classes of groups. They occur naturally in topology, complex analysis, algebraic geometry and combinational group theory. As pointed out in the introduction of this book, together with arithmetic subgroups, they form three important classes of groups having similar structures and acting on similar spaces. Results and methods from one class can lead to results for other classes.

A general reference for mapping class groups is [Iv2], and a survey on outer automorphism groups of free groups is [Vogt1]. An article emphasizing similarities between these three classes is [BridV]. See also [HainL] for the point of view of algebraic geometry.

## 20.1. Mapping class groups

The arithmetic subgroup $SL(2,\mathbb{Z}) \subset SL(n,\mathbb{R})$ is the most basic arithmetic subgroup. One natural generalization is the family of subgroups $SL(n,\mathbb{Z}) \subset SL(n,\mathbb{R})$, $n \geq 2$. Another natural generalization is the mapping class group of surfaces of genus $g \geq 2$.

Let $S = S_g$ be a closed and oriented surface of genus $g$. Let $\mathrm{Diff}(S)$ be the group of all diffeomorphisms of $S$, and $\mathrm{Diff}^+(S)$ the subgroup of orientation preserving diffeomorphisms. Denote the identity component of $\mathrm{Diff}(S)$ by $\mathrm{Diff}^0(S)$, which is also contained in $\mathrm{Diff}^+(S)$. (We can also use the groups of homeomorphisms of $S$ instead of diffeomorphisms to define the mapping class groups).

Then the component group $\mathrm{Diff}(S)/\mathrm{Diff}^0(S) = \pi_0(\mathrm{Diff}(S))$ is called the extended mapping class group of $S$, often denoted by $Mod_S$ or $\Gamma_g$. Each element represents an isotropy class of diffeomorphisms of $S$. The group $\mathrm{Diff}^+(S)/\mathrm{Diff}^0(S)$ is called the mapping class group, denoted by $Mod_S^+$ or $\Gamma_g^+$.

When $S$ is the torus $\mathbb{Z}^2\backslash\mathbb{R}^2$, i.e., $g = 1$, then $Mod_S^+ = SL(2,\mathbb{Z})$; on the other hand, the extended mapping class group $Mod_S = GL(2,\mathbb{Z})$.

We can also similarly define mapping class groups for surfaces with punctures and boundary. If $S_{g,p}$ is a surface of genus $g$ with $p$ punctures, then its mapping class group is often denoted by $\Gamma^+_{g,p}$, and the extended mapping class group by $\Gamma_{g,p}$.

For any connected orientable smooth manifold $M$, we can also define its mapping class groups by

$$Mod_M = \mathrm{Diff}(M)/\mathrm{Diff}^0(M), \quad \text{and} \quad Mod^+_M = \mathrm{Diff}^+(M)/\mathrm{Diff}^0(M).$$

When $M$ is a closed 1-dimensional manifold, i.e., $M = S^1$, the unit circle, then $Mod_M = \mathbb{Z}/2\mathbb{Z}$, which is generated by the reflection with respect to a diameter of the unit disc. For $M = \mathbb{Z}^n\backslash\mathbb{R}^n$, $Mod^+_M = SL(n,\mathbb{Z})$, and $Mod_M = GL(n,\mathbb{Z})$.

For surfaces $S$, the structure of $Mod_S$ has been studied intensively and understood relatively well. In fact, these mapping class groups enjoy many properties similar to those of arithmetic groups, in particular of $SL(2,\mathbb{Z})$, and methods used to study them are often similar. For example, an element $\gamma$ of $SL(2,\mathbb{R})$ and hence of $SL(2,\mathbb{Z})$ belongs to one of the following three types: elliptic, parabolic or hyperbolic, depending on if $\gamma$ fixes a point in $\mathbb{H}$, only one boundary point $\partial\mathbb{H} = \mathbb{R} \cup \{i\infty\}$, or excatly two points in $\partial\mathbb{H}$. A vast generalization of this to elements of general $Mod_S$ has been obtained by Thurston using its extended action on the Thurston compactification of the Teichmüller space of $S$ and is important to three dimensional hyperbolic geometry. See [Th1-2] [FatLP].

See the next subsection for similarities between cohomology groups of arithmetic subgroups and mapping class groups.

On the other hand, for general higher dimensional manifolds $M$, it is difficult to understand $Mod_M$, and little has been done.

## 20.2. Teichmüller spaces of Riemann surfaces

Actions of arithmetic groups on symmetric spaces are crucial for understanding arithmetic subgroups. The analogue of symmetric spaces for $Mod_S$ is the Teichmüller space $T_S$ of the surface $S$.

Basically the idea is that the Teichmüller space $T_S$ parametrizes marked special structures on $S$, and $Mod_S$ acts on $T_S$ by changing the markings.

For example, when the genus $g$ of $S$ is at least 2, $S$ admits hyperbolic metrics (the sectional curvature is normalized to be $-1$), and $T_S$ parametrizes marked hyperbolic metrics on $S$.

When $g = 1$, the only constant curvature metrics on $S$ are the flat ones. Then $T_S$ parametrizes the marked flat metrics on $S$ of total area 1. (Recall that in this case scaling does not change the zero curvature.)

For a higher dimension manifold $M$, if it admits some special metrics or structures, we can also define its Teichmüller space $T_M$. For example, for $M = \mathbb{Z}^n\backslash\mathbb{R}^n$, we can consider the marked flat metrics on $M$ of total area 1. It turns out that $T_M$ is equal to the symmetric space associated with the Lie group $SL(n,\mathbb{R})$, i.e., the symmetric space $SL(n,\mathbb{R})/SO(n)$ of positive definite matrices of determinant 1.

In order to use the action of the mapping class group on the Teichmüller space to study the group, it is important that the Teichmüller space is contractible, for example, for the purpose to construct various classifying (or universal) spaces. In most of the above cases, the Teichmüller spaces are diffeomorphic to some $\mathbb{R}^n$.

A comprehensive survey on the mapping class groups $Mod_S$ and Teichmüller spaces $T_S$ is given in [Iv2] together with extensive references. See also [Schu] for a survey on Teichmüller spaces and a generalization to higher dimensional Kähler manifolds. For some higher Teichmüller spaces, see [FoG]. For actions of mapping class groups on representation (or character) varieties and related results, see [Wie] [BurILW] and references there.

For the Poisson boundary of the mapping class groups and Teichmüller spaces, see [KaiM1-2].

## 20.3. Topology of moduli spaces of Riemann surfaces and the mapping class groups

We concentrate on the case of mapping class groups of surfaces $S_g$, $g \geq 2$. The quotient $Mod_S^+ \backslash T_S$ is the moduli space of hyperbolic metrics on $S$, denoted by $\mathcal{M}_g$. (In fact, the quotient by $Mod_S^+$ removes the marking.) It is also the moduli space of complex structures on $S$. In fact, in each complex structure on $S$, there is a unique compatible hyperbolic on $S$, where the compatibility means that in the local holomorphic coordinate $z$ of the complex structure, the hyperbolic metric is conformal to $|dz|^2$. The space $\mathcal{M}_g$ is also the moduli space of smooth projective curves over $\mathbb{C}$ of genus $g$. Hence it is an important space in algebraic geometry and complex analysis.

A basic question about the moduli space is to compute the homology groups of $\mathcal{M}_g$. Since $Mod_S^+$ acts properly on $T_S$ and $T_S$ is contractible, it follows that

$$H^*(\mathcal{M}_g, \mathbb{Q}) = H^*(Mod_S^+, \mathbb{Q}).$$

(Note that $Mod_S^+$ contains nontrivial torsion elements, and hence its action on $T_S$ has fixed points and $\mathcal{M}_g$ is an orbifold. Consequently, $Mod_S^+ \backslash T_S$ is not a $K(Mod_S^+, 1)$-space. The equality $H^*(\mathcal{M}_g, \mathbb{Z}) = H^*(Mod_S^+, \mathbb{Z})$ does not hold due to these fixed points.)

Methods from cohomology of arithmetic subgroups can be used to understand $H^*(Mod_S^+, \mathbb{Q})$. Many other properties can also be shown. For example, Euler number and lower degree cohomology groups of the mapping class groups (or the moduli spaces) are similarly determined as in the case of arithmetic subgroups; and the cohomological dimension of mapping class groups can be computed explicitly, and the mapping class groups are also virtual duality groups of dimension $4g - 5$ but not virtual Poincaré duality groups. See [Hare] for an exposition (summary) and references, and also [IvJ] for the non-Poincaré duality property.

For homotopy groups of the moduli spaces (or mapping class groups) and the Mumford conjecture on the homology groups of stable mapping class groups, see [MadW] and its references.

## 20.4. Compactifications of Teichmüller spaces

Besides many other applications, compactifications of symmetric and locally symmetric spaces are crucial for understanding cohomology groups of arithmetic groups and showing the arithmetic groups are virtual duality groups.

Similar constructions work for Teichmüller spaces. The analogue of the Borel-Serre partial compacctification of symmetric spaces can be defined for Teichmüller spaces $T_S$, whose boundary components are parametrized by the *curve complex* of $S$, an analogue of the spherical Tits building $\Delta_{\mathbb{Q}}(\mathbf{G})$ of semisimple linear algebraic groups $\mathbf{G}$ defined over $\mathbb{Q}$. See [Harv] and [Iv2-3]. ([Iv2] contains detailed properties and applications of the curve complex. For some recent applications in three dimensional hyperbolic geometry and topology, see [Mins] and the references there.)

Briefly, the Tits building $\Delta_{\mathbb{Q}}(\mathbf{G})$ of a linear semisimple algebraic Lie group $\mathbf{G}$ defined over $\mathbb{Q}$ is an infinite simplicial complex with simplexes $\sigma_{\mathbf{P}}$ corresponding to proper $\mathbb{Q}$-parabolic subgroups $\mathbf{P}$ of $\mathbf{G}$ such that

(1) $\sigma_{\mathbf{P}}$ is a vertex, i.e., a simplex of dimension 0 if and only if $\mathbf{P}$ is a proper maximal $\mathbb{Q}$-parabolic subgroup of $\mathbf{G}$.
(2) For every pair of $\mathbb{Q}$-parabolic subgroups $\mathbf{P}_1$ and $\mathbf{P}_2$, $\sigma_{\mathbf{P}_1}$ is a face of $\sigma_{\mathbf{P}_2}$ if and only if $\mathbf{P}_1$ contains $\mathbf{P}_2$. In particular, vertices $\sigma_{\mathbf{P}_1}, \cdots,$ $\sigma_{\mathbf{P}_{k+1}}$ are the vertices of a $k$-simplex $\sigma_{\mathbf{P}}$ if and only if $\mathbf{P} = \mathbf{P}_1 \cap \cdots \cap \mathbf{P}_{k+1}$.

(This Tits building is related to partial compactifications of the symmetric space $X = G/K$ such as the Borel-Serre partial compactification $\overline{X}^{BS}$, where $G = \mathbf{G}(\mathbb{R})$. If we consider $G = \mathbf{G}(\mathbb{R})$ as a semisimple Lie group and all its real parabolic subgroups, then we also get another spherical Tits building $\Delta(G)$, which is useful for compactifications of the symmetric space $X$. In fact, as mentioned before, its underlying space can be identified with the sphere at infinity $X(\infty)$. See [GuiJT]. If the real rank of $\mathbf{G}$ is not equal to the $\mathbb{Q}$-rank of $\mathbf{G}$, the rational Tits building $\Delta_{\mathbb{Q}}(\mathbf{G})$ is not contained in $\Delta(G)$ as a subcomplex.)

Assume that $S$ is a closed surface. For the curve complex $\mathcal{C}(S)$ of the surface $S$, every vertex corresponds to a simple closed curve. Several such curves form the vertices of a simplex if they are disjoint and not homotopic to each other.

There is also a truncation $T_S(\varepsilon)$ of $T_S$ by removing hyperbolic metrics on $S$ (or marked hyperbolic surfaces) containing geodesics with length less than $\varepsilon$, when $\varepsilon$ is sufficiently small (see [Iv4].) This subspace $T_S(\varepsilon)$ is clearly invariant under $Mod_S$ and can be shown to be a cofinite universal space for proper actions of $Mod_S$. This is similar to the truncation of symmetric spaces in §10.8. There is another similarity. The compactness of $Mod_S \backslash T_S(\varepsilon)$ follows from the Mumford compactness criterion for subsets of the moduli space $\mathcal{M}_g$ (see [Iv4] and [Buse] for example), which is similar to the Mahler compactness criterion for subsets of the space of lattices (and hence also for subsets of locally symmetric spaces associated with arithmetic subgroups. See [Bore4].)

It can be shown that $T_S(\varepsilon)$ is a cofinite universal space for proper actions of $Mod_S$. From this, it follows that the rational Novikov conjecture in algebraic $K$-theory holds for $Mod_S$. Other finiteness properties of $Mod_S$ such as $FP_\infty$ and finite generation of $H^i(Mod_S, \mathbb{Z})$ in every degree also follow.

A compactification of the Teichmüller space gives an ideal boundary. The Poisson boundary of a group and a space is motivated by the random walks and harmonic functions. For the Poisson boundary of the mapping class groups and Teichmüller spaaces, see [KaiM1-2].

In constructing compactifications of locally symmetric spaces from compactifications of symmetric spaces, the notion of rational boundary components is needed. The reduction theory for arithmetic subgroups is also used crucially. There is a similar reduction theory for $Mod_S$ acting on $T_S$. See [Buse] and [Ji10].

The Teichmüller space $T_S$ admits many different compactifications such as the Thurston compactification, the Teichmüller compactification [Ke], the Bers compactification [KeT] [Abi], and the CAT(0)-space compactification of the Weil-Petersson completion of $T_S$ [Wolp3]. But only the Thurston compactification admits the continuous extended action of $Mod_S$.

It turns out that the curve complex sits in the boundary of Thurston compactification as rational boundary components. Then a compacctification of $\mathcal{M}_S$ (also denoted by $\mathcal{M}_g$ if $S$ is a closed surface of genus $g$) can be constructed by following the usual procedure of compactifications of locally symmetric spaces:

(1) Pick out rational boundary components of the Thurston compactification of $T_S$ using the reduction theory.
(2) Add these rational boundary components (the curve complex) at infinity using the Fenchel-Nielson coordinates (or rather partial Fenchel-Nielson coordinates) to obtain a partial compactification of $T_S$.
(3) Use the reduction theory of $Mod_S$ to show that the quotient by $Mod_S$ of the partial compactification is a compact and Hausdorff space.

The boundary of this compactification of $\mathcal{M}_g$ is a finite simplicial complex, which is the quotient of the curve complex by $Mod_S$. This compactification is different from the well-known Deligne-Mumford compactificattion of $\mathcal{M}_g$ by stable curves. See [Ji10].

## 20.5. Symmetry of Teichmüller spaces

In this book, we have tried to emphasize similarities between Teichmüller spaces and symmetric spaces. On the other hand, it is also important to point out some differences between them.

As it is well-known and follows from the definition, a symmetric space $X$ is a homogeneous space $X = G/K$ and admits a $G$-invariant Riemannian metric. In particular, $X$ with an invariant (or canonical) metric admits positive dimensional families of symmetry (or isometry.) If $X$ is a Hermitian symmetric space, then its group of holomorphic automorphisms also acts transitively on $X$. (Note that by definition, a symmetric space is a Hermitian symmetric space if it admits a $G$-invariant complex structure.)

A natural question is to understand the symmetry of the Teichmüller space $T_S$. It is known that $T_S$ is a complex manifold, and $Mod_S^+$ acts on it by biholomorphic maps. A famous result of Royden [Roy] says that $Mod_S^+$ is the whole group of biholomorphic automorphisms of $T_S$.

The Teichmüller space $T_S$ admits several $Mod_S$-invariant metrics. The first one is the Teichmüller metric, which is also equal to the Kobayashi metric when $T_S$ is considered as a bounded domain in $\mathbb{C}^n$ [Roy]. It is shown in [Roy] [EarK] that every isometry of the Teichmüller metric is induced by an element of the extended mapping class group $Mod_S$.

The Teichmüller space $T_S$ also admits a Kähler metric, the Weil-Petersson metric, which is also invariant under $Mod_S$. Its isometry group is shown to be equal to $Mod_S$ in [MazW] (a simplification in [Wolp3]), by using a fact about the automorphism group of the curve complex of $S$. That the automorphism of the curve complex is equal to $Mod_S$ when the genus of S is at least 2 can be found in [Iv7] and [Luo]. The genus 1 and 0 cases except the 2-holed torus was established independently in [Korkl] and [Luo]. The most difficult case of the 2-holed torus in which the automorphism is not equal to the mapping class group was established in [Luo]. Full details of the proofs in all cases are provided in [Luo].

There are also Kähler-Einstein metric, the Ricci metric and the perturbed Ricci metric and McMullen metric etc on $T_S$. For recent work on comparison of these metrics and other related results such as the volume of the moduli spaces, see [LiuSY1] [LiuSY2] [Mir1] [Mir2] [Wolp3] [Wolp6] [Ye5] [Mcm].

CHAPTER 21

# Outer automorphism group of free groups and the outer spaces

We have discussed several ways to realize the group $SL(2,\mathbb{Z})$ or $GL(2,\mathbb{Z})$. There are two more ways from the point of view of combinatorial group theory.

For any group $\Gamma$, let $Out(\Gamma)$ be the group of outer automorphisms of $\Gamma$. Specifically, let $Aut(\Gamma)$ be the group of all automorphisms of $\Gamma$ and $Inn(\Gamma)$ the normal subgroup given by conjugation by elements of $\Gamma$. Then

$$Out(\Gamma) = Aut(\Gamma)/Inn(\Gamma).$$

The first realization in the combinatorial group theory is given by the identification

$$GL(2,\mathbb{Z}) = Out(\mathbb{Z}^2).$$

Let $F_2$ be the free group on two generators. Then the second realization in the combinatorial group theory is that

$$GL(2,\mathbb{Z}) = Out(F_2).$$

In fact, the abelization of $F_2$ is equal to $\mathbb{Z}^2$. This gives a map from $Out(F_2)$ to $Out(\mathbb{Z}^2) = GL(2,\mathbb{Z})$, which turns out to be an isomorphism. It is natural to try to generalize such relations to free groups $F_n$ on more generators.

## 21.1. Outer automorphism group of free groups

Let $F_n$ be the free group on $n$ generators. Its automorphism group $Aut(F_n)$ and the outer automorphism group $Out(F_n)$ are fundamental objects in the theory of combinatorial groups. One reason is that they keep track of generators of $F_n$, very much as $GL(n,\mathbb{Z})$ keeps track of bases of the lattice $\mathbb{Z}^n$. In fact, the vectors $e_1 = (1,0,\dots,0),\dots,e_n = (0,\dots,0,1)$ give the standard basis of $\mathbb{Z}^n$, and any other basis is the image of $e_1,\dots,e_n$ under some element $\gamma \in GL(n,\mathbb{Z})$. (If we fix an orientation, we can use elements of the subgroup $SL(n,\mathbb{Z})$.) Some basic books are [MagKS] [ChaW].

## 21.2. Outer spaces

The analogue of the Teichmüller space is the outer space $\hat{X}_n$, which is defined to be the moduli space of marked normalized metric graphs with the fundamental group equal to $F_n$. There is also a reduced outer space, denoted by $X_n$ in this book.

To motivate these spaces, we note that a closed surface $S_g$ with $g \geq 1$ is a classifying space for its fundamental group, i.e., an aspherical manifold. Similarly, $\mathbb{Z}^n \backslash \mathbb{R}^n$ is a classifying space for $\mathbb{Z}^n$. The Teichmüller spaces of $S$ and $\mathbb{Z}^n \backslash \mathbb{R}^n$ represent marked special metrics on these spaces (hyperbolic metrics on the former, and flat metrics on the latter).

For the free group $F_n$, a classifying space is given by a connected graph with $n$ loops. It turns out that it is best to consider connected graphs with no vertices of valence 1 or 2. A metric on such a graph is completely determined by edge lengths, which are normalized to have the total length of all edges of the graph equal to 1. Then the outer space $\hat{X}_n$ is the space of all marked and normalized metric graphs with fundamental group equal to $F_n$ and the length of every nontrivial loop in the graphs is positive.

If we require that the graphs do not contain any bridges, i.e., separating edges, then we get the reduced outer space $X_n$ mentioned above. Note that collapsing of separating edges of a graph will not change the topology of the graph.

For each marked graph, all possible metrics on it fill out a simplex, with some faces of the simplex possibly missing. This arises due to the condition that the total length of every loop is positive. Hence the outer space $\hat{X}_n$ and the reduced outer space $X_n$ are infinite simplicial complexes.

For $n = 2$, $X_n$ can be identified with the upper half plane $\mathbb{H}$ with the simplexes corresponding to ideal triangles with rational boundary points, i.e., Farey tessellation of $\mathbb{H}$ mentioned earlier. In this case, $X_n$ is clearly contractible. In fact, $X_n$ is contractible for all $n$ [CuV].

$Out(F_n)$ acts properly on both the outer space $\hat{X}_n$ and the reduced outer space $X_n$, and these are spaces are equivariantly homotopic to each other. For various reasons, it is simpler to consider the reduced space $X_n$.

## 21.3. Compactifications of outer spaces

The quotient of $Out(F_n) \backslash X_n$ is noncompact, since metric graphs can degenerate when the total length of some loops goes to 0, but it can be retracted to compact subsets. For example, the spine $K_n$ of the simplicial complex $X_n$ is an equivariant retraction with respect to $Out(F_n)$ such that the quotient $Out(F_n) \backslash K_n$ is compact. It can also be shown that $K_n$ is a cofinite universal space for proper actions of $Out(F_n)$. This implies that $Out(F_n)$ enjoys many finiteness properties such as $FP_\infty$ as in the cases of arithmetic subgroups and mapping class groups.

For some other purposes, for example, to prove that $Out(F_n)$ is a virtual duality group of dimension $2n - 3$, we need a Borel-Serre type compactification of $X_n$, which was constructed in [BestF1]. It can also be shown that $Out(F_n)$ is not a virtual Poincaré duality group. See [Ji10].

We pointed out earlier in §10.8 that the Borel-Serre compactification of symmetric spaces can also be realized by a truncated subspace. A similar truncation $T_S(\varepsilon)$ for the Teichmüller space $T_S$ also gives a Borel-Serre type compactification of $T_S$.

For $T_S$, the construction is simple. The truncated space $T_S(\varepsilon)$ is defined to consist of hyperbolic metrics on the surface $S$ containing no geodesic with length less than $\varepsilon$. For the reduced outer space $X_n$, the same construction by picking out marked metric graphs such that every loop on them has length at least $\varepsilon$ will not work. In fact, we need to impose lower bounds on the lengths of unions of loops, the so-called core subgraphs. One reason is that in Riemannian manifolds, if two geodesic are tangent at one point, in particular share one segment, then they must be the same. This is not true with loops in graphs. Depending on the numbers of loops contained in the core subgraphs, we give a suitable lower bound on the length of core subgraphs. (More specifically, if a core subgraph contains $n$ loops, the lower bound is $3^{n-1}\varepsilon$. Therefore, for one loop, the lower bound is $\varepsilon$, and for a core subgraph of two loops, the lower bound is $3\varepsilon$, etc.) This gives a truncated subpace $X_n(\varepsilon)$.

The boundary of this space $X_n(\varepsilon)$ is parametrized by a simplicial complex similar to the Tits building for arithmetic groups, and the curve complex for mapping class groups.

See the surveys [Vogt1] [Vogt2] [Best3] [BridV] and many references there for many properties of outer automorphism groups and outer spaces.

## 21.4. Outer automorphism group of non-free groups

One point of the above discussions is that the outer automorphism group of the free group $F_2$ is similar to arithmetic groups. A natural problem is to study automorphism and outer automorphism groups of other classes of groups.

As mentioned earlier, the outer automorphism group of the group $\mathbb{Z}^n$ is equal to $GL(n, \mathbb{Z})$, an arithmetic group. The automorphism groups of nilpotent groups have been extensively studied. It turns out that the outer automorphism groups of polycyclic-by-finite groups are arithmetic groups, though the automorphism groups are not so. See [BauG] [PicR] [Seg1-2] and the references there.

If $\Gamma$ is an discrete subgroup of a semisimple Lie group $G$ enjoying the Mostow strong rigidity, then every automorphism of $\Gamma$ is induced by an automorphism group of $G$. Hence there is a surjective map from $Out(G)$ to $Out(\Gamma)$. Since $Out(G)$ is finite, $Out(\Gamma)$ is also finite.

The only exception to the Mostow rigidity is the case of hyperbolic surfaces $\Gamma\backslash\mathbb{H}$, i.e., $\Gamma \subset SL(2, \mathbb{R})$. Assume that $\Gamma\backslash\mathbb{H}$ is a surface $S_{g,p}$ of genus $g$ with $p$ punctures, then the outer automorphism group of $\Gamma$ is related to the mapping class group $\mathrm{Mod}_{g,p}$ of $S_{g,p}$. See [Vogt1] and the references there.

For results on automorphism groups of the mapping class groups and its subgroups of finite index, see [IvM] [Kork2] and references there.

If $\Gamma$ is a hyperbolic group, some information about its automorphism groups is known. One result in [Pau1] says that if $\Gamma$ is a hyperbolic group and satisfies the property (T), then $Out(\Gamma)$ is finite. See also [Pau2] [Fuj2] [Sel1-2] for other results.

# References


[Abe1] H. Abels, *Finiteness properties of certain arithmetic groups in the function field case*, Israel J. Math. 76 (1991) 113–128.

[Abe2] H. Abels, *Finite presentability of S-arithmetic groups. Compact presentability of solvable groups*, Lecture Notes in Mathematics, vol. 1261, Springer-Verlag, 1987. vi+178 pp.

[Abe3] H. Abels, *Reductive groups as metric spaces*, in *Groups: topological, combinatorial and arithmetic aspects*, pp. 1–20, in London Math. Soc. Lecture Note Ser., 311, Cambridge Univ. Press, 2004.

[Abe4] H. Abels, *Properly discontinuous groups of affine transformations: a survey*, Geom. Dedicata 87 (2001) 309–333.

[Abe5] H. Abels, *Proximal linear maps*, to appear in Pure and Applied Math Quarterly (4) 1 (2008).

[AbeM] H. Abels, G. Margulis, *Coarsely geodesic metrics on reductive groups*, in *Modern dynamical systems and applications*, pp. 163–183, Cambridge Univ. Press, 2004.

[AbeMS1] H. Abels, G. Margulis, G. Soifer, *On the Zariski closure of the linear part of a properly discontinuous group of affine transformations*, J. Differential Geom. 60 (2002) 315–344.

[AbeMS2] H. Abels, G. Margulis, G. Soifer, *Semigroups containing proximal linear maps*, Israel J. Math. 91 (1995) 1–30.

[AbeMS3] H. Abels, G. Margulis, G. Soifer, *The Auslander conjecture for groups leaving a form of signature $(n-2,2)$ invariant*, Israel J. Math. 148 (2005) 11–21.

[AbeMS4] H. Abels, G. Margulis, G. Soifer, *Properly discontinuous groups of affine transformations with orthogonal linear part*, C. R. Acad. Sci. Paris Ser. I Math., 324 (1997) 253–258.

[AbLP] M. Abért, A. Lubotzky, L. Pyber, *Bounded generation and linear groups*, Internat. J. Algebra Comput. 13 (2003) 401–413.

[AbNS] M. Abért, N. Nikolov, B. Szegedy, *Congruence subgroup growth of arithmetic groups in positive characteristic*, Duke Math. J. 117 (2003) 367–383.

[Abi] W. Abikoff, *Degenerating families of Riemann surfaces*, Ann. of Math. 105 (1977) 29–44.

[Abr1] P. Abramenko, *Twin buildings and applications to S-arithmetic groups*, Lecture Notes in Mathematics, vol. 1641, Springer-Verlag, 1996. x+123 pp.

[Abr2] P. Abramenko, *Finiteness properties of Chevalley groups over $F_q[t]$*, Israel J. Math. 87 (1994) 203–223.

[Ada] C. Adams, *The noncompact hyperbolic 3-manifold of minimal volume*, Proc. Amer. Math. Soc. 100 (1987) 601–606.

[Adam] J. Adams, *Lectures on Lie groups*, W. A. Benjamin, Inc., 1969 xii+182 pp.

[AdaS] S. Adams, R. Spatzier, *Kazhdan groups, cocycles and trees*, Amer. J. Math. 112 (1990) 271–287.

[AdeD] A. Adem, J. Davis, *Topics in transformation groups*, in *Handbook of geometric topology*, pp. 1–54, North-Holland, 2002.

[AdeR] A. Adem, Y. Ruan, *Twisted orbifold $K$-theory*, Comm. Math. Phys. 237 (2003) 533–556.

[AdM] S. Adian, J. Mennicke, *On bounded generation of* $\mathrm{SL}_n(Z)$, Internat. J. Algebra Comput. 2 (1992) 357–365.

[Ag1] I. Agol, *Finiteness of arithmetic Kleinian reflection groups*, math.GT/0512560.

[Ag2] I. Agol, *Tameness of hyperbolic 3-manifolds*, math.GT/0405568.

[AgBSW] I. Agol, M. Belolipetsky, P. Storm, K. Whyte, *Finiteness of arithmetic hyperbolic reflection groups*, math.GT/0612132.

[Ak1] D. Akhiezer, *Lie group actions in complex analysis*, Aspects of Mathematics, E27. Friedr. Vieweg & Sohn, 1995. viii+201 pp.

[Ak2] D. Akhiezer, *Homogeneous complex manifolds*, in *Several complex variables. IV. Algebraic aspects of complex analysis*, edited by S. G. Gindikin and G. M. Khenkin, Encyclopaedia of Mathematical Sciences, 10. Springer-Verlag, Berlin, 1990, pp. 195–244.

[Alb] P. Albuquerque, *Patterson-Sullivan theory in higher rank symmetric spaces*, Geom. Funct. Anal. 9 (1999) 1–28.

[AleVS] D. Alekseevskij, E. Vinberg, A. Solodovnikov, *Geometry of spaces of constant curvature*, in *Geometry, II*, pp. 1–138, Encyclopaedia Math. Sci., 29, Springer, 1993.

[Alex] V. Alexeev, *Complete moduli in the presence of semiabelian group action*, Ann. of Math. (2) 155 (2002) 611–708.

[All] N.D. Allan, *The problem of the maximality of arithmetic groups*, in *Algebraic groups and discontinuous subgroups*, Proc. Symp. Pure Math IX, (A. Borel and G.D. Mostow eds), AMS 1966.

[AllCT1] D. Allcock, J. Carlson, D. Toledo, *The complex hyperbolic geometry of the moduli space of cubic surfaces*, J. Algebraic Geom. 11 (2002) 659–724.

[AllCT2] D. Allcock, J. Carlson, D. Toledo, *Non-Arithmetic uniformization of Some Real Moduli Spaces*, Geometriae Dedicata 122 (2006) 159–169.

[Alp] R. Alperin, *Locally compact groups acting on trees and property $T$*, Monatsh. Math. 93 (1982) 261–265.

[AmG] G. D'Ambra, M. Gromov, *Lectures on transformation groups: geometry and dynamics*, in *Surveys in differential geometry*, pp. 19–111, 1991.

[Ana] N. Anantharaman, *Entropy and the localization of eigenfunctions*, to appear in Ann. of Math.

[AnaN] N. Anantharaman, S. Nonnenmacher, *Entropy of semiclassical measures of the Walsh-quantized baker's map*, Ann. Henri Poincaré 8 (2007) 37–74.

[AnHS] M. Ando, M. Hopkins, N. Strickland, *Elliptic spectra, the Witten genus and the theorem of the cube*, Invent. Math. 146 (2001) 595–687.

[Ap1] B. Apanasov, *Conformal geometry of discrete groups and manifolds*, Walter de Gruyter, 2000.

[Ap2] B. Apanasov, *Discrete groups in space and uniformization problems*, Kluwer Academic Publishers, 1991.

[ApX] B. Apanasov, X. Xie, *Geometrically finite complex hyperbolic manifolds*, Internat. J. Math. 8 (1997) 703–757.

[ApX2] B. Apanasov, X. Xie, *Discrete actions on nilpotent Lie groups and negatively curved spaces*, Differential Geom. Appl. 20 (2004) 11–29.

[Arm] M. Armstrong, *Groups and symmetry*, Undergraduate Texts in Mathematics, Springer-Verlag, 1988, xii+186 pp.

[Arn] V. Arnold, *Sur la géométrie différentielle des groupes de Lie de dimension infinie et ses applications á l'hydrodynamique des fluides parfaits*, Ann. Inst. Fourier (Grenoble) 16 (1966) 319–361.

[ArnK] V. Arnold, B. Khesin, *Topological methods in hydrodynamics*, Applied Mathematical Sciences, 125. Springer-Verlag, 1998.

[Art1] J. Arthur, *An introduction to the trace formula*, in *Harmonic analysis, the trace formula, and Shimura varieties*, pp. 1–263, Clay Math. Proc., 4, Amer. Math. Soc., Providence, RI, 2005.

[Art2] J. Arthur, *The principle of functoriality*, Bull. Amer. Math. Soc. (N.S.) 40 (2003) 39–53.

[Art3] J. Arthur, *Harmonic analysis and group representations*, Notices Amer. Math. Soc. 47 (2000) 26–34.

[Art4] J. Arthur, *Automorphic representations and number theory*, in *1980 Seminar on Harmonic Analysis*, pp. 3–51, CMS Conf. Proc., 1, Amer. Math. Soc., 1981.

[ArtG] J. Arthur, S. Gelbart, *Lectures on automorphic L-functions*, in *L-functions and arithmetic*, pp. 1–59, London Math. Soc. Lecture Note Ser., 153, Cambridge Univ. Press, 1991.

[As1] A. Ash, *Small-dimensional classifying spaces for arithmetic subgroups of general linear groups*, Duke Math. J. 51 (1984) 459–468.

[As2] A. Ash, *Cohomology of congruence subgroups* $\mathrm{SL}(n, Z)$, Math. Ann. 249 (1980) 55–73.

[As3] A. Ash, *On eutactic forms*, Canad. J. Math. 29 (1977) 1040–1054.

[As4] A. Ash, *Deformation retracts with lowest possible dimension of arithmetic quotients of self-adjoint homogeneous cones*, Math. Ann. 225 (1977) 69–76.

[As5] A. Ash, *A note on minimal modular symbols*, Proc. Amer. Math. Soc. 96 (1986) 394–396.

[AsB] A. Ash, B. Borel, *Generalized modular symbols*, in *Cohomology of arithmetic groups and automorphic forms*, pp. 57–75, Lecture Notes in Math., vol. 1447, Springer, 1990.

[AsG1] A. Ash, R. Gross, *Fearless symmetry. Exposing the hidden patterns of numbers*, Princeton University Press, 2006. xx+272 pp.

[AsG2] A. Ash, R. Gross, *Generalized non-abelian reciprocity laws: a context for Wiles' proof*, Bull. London Math. Soc. 32 (2000) 385–397.

[AsGM] A. Ash, P. Gunnells, M. McConnell, *Cohomology of congruence subgroups of* $\mathrm{SL}_4(\mathbb{Z})$, J. Number Theory 94 (2002) 181–212.

[AsM] A. Ash, M. McConnell, *Cohomology at infinity and the well-rounded retract for general linear groups*, Duke Math. J. 90 (1997) 549–576.

[AsMRT] A. Ash, D. Mumford, M. Rapoport, Y. Tai, *Smooth compactifications of locally symmetric varieties*, Math. Sci. Press, Brookline, 1975.

[AsR] A. Ash, L. Rudolph, *The modular symbol and continued fractions in higher dimensions*, Invent. Math. 55 (1979) 241–250.

[At] M. Atiyah, *Elliptic operators, discrete groups and von Neumann algebras*, pp. 43–72. Asterisque, No. 32–33, Soc. Math. France, 1976.

[AtS] M. Atiyah, W. Schmid, *A geometric construction of the discrete series for semisimple Lie groups*, Invent. Math. 42 (1977) 1–62.

[AuBG] K. Aubert, E. Bombieri, D. Goldfeld, *Number theory, trace formulas and discrete groups, Symposium in honor of Atle Selberg held at the University of Oslo*, Academic Press, 1989. xxii+510 pp.

[Aus] L. Auslander, *The structure of complete locally affine manifolds*, Topology 3 (1964) 131–139.

[BabP1] M. Babillot, M. Peigné, *Asymptotic laws for geodesic homology on hyperbolic manifolds with cusps*, Bull. Soc. Math. France 134 (2006) 119–163.

[BabP2] M. Babillot, M. Peigné, *Homologie des géodésiques fermées sur des variétés hyperboliques avec bouts cuspidaux*, Ann. Sci. École Norm. Sup. (4) 33 (2000) 81–120.

[BadFGM] U. Bader, A. Furman, T. Gelander, N. Monod, *Property (T) and rigidity for actions on Banach spaces*, Acta Math. 198 (2007) 57–105. (arXiv:math/0506361.)

[BadFS] U. Bader, A. Furman, A. Shaker, *Superrigidity, Weyl groups, and actions on the Circle*, arXiv:math/0605276.

[BadS] U. Bader, Y. Shalom, *Factor and normal subgroup theorems for lattices in products of groups*, Invent. Math. 163 (2006) 415–454.

[BaeN] T. Bailey and A. Knapp, *Representation theory and automorphic forms*, Proceedings of Symposia in Pure Mathematics, 61. American Mathematical Society, 1997. viii+479 pp.

[BaiB] W. Baily, A. Borel, *Compactification of arithmetic quotients of bounded symmetric domains*, Ann. of Math. 84 (1966) 442–528.

[BakK] A. Baklouti, F. Khlif, *Proper actions on some exponential solvable homogeneous spaces*, Internat. J. Math., 16 (2005) 941–955.

[Bal1] W. Ballmann, *Lectures on spaces of nonpositive curvature*, Birkhäuser Verlag, 1995.

[Bal2] W. Ballmann, *Manifolds of nonpositive sectional curvature and manifolds without conjugate points*, in *Proceedings of the International Congress of Mathematicians*, (Berkeley, Calif., 1986) pp. 484–490, Amer. Math. Soc., 1987.

[Bal3] W. Ballmann, *Nonpositively curved manifolds of higher rank*, Ann. of Math. 122 (1985) 597–609.

[BalBE] W. Ballmann, M. Brin, P. Eberlein, *Structure of manifolds of nonpositive curvature. I*, Ann. of Math. 122 (1985) 171–203.

[BalBS] W. Ballmann, M. Brin, R. Spatzier, *Structure of manifolds of nonpositive curvature. II*, Ann. of Math. 122 (1985) 205–235.

[BalGS] W. Ballmann, M. Gromov, V. Schroeder, *Manifolds of nonpositive curvature*, in *Progress in Mathematics*, vol. 61, Birkhäuser Boston, 1985.

[BaoT] D. Bao, T. Ratiu, *On a maximal torus in the volume-preserving diffeomorphism group of the finite cylinder*, Differential Geom. Appl. 7 (1997) 193–210.

[BarLR1] A. Bartels, W. Lück, H. Reich, *On the Farrell-Jones Conjecture and its applications*, preprint, arXiv:math/0703548.

[BarLR2] A. Bartels, W. Lück, H. Reich, *The K-theoretic Farrell-Jones Conjecture for hyperbolic groups*, preprint, arXiv:math/0701434.

[BarLR3] A. Bartels, W. Lück, H. Reich, *Equivariant covers for hyperbolic groups*, preprint, arXiv:math/0609685.

[BarHPV] W. Barth, K. Hulek, C. Peters, A. Van de Ven, *Compact complex surfaces*, Second edition, Springer-Verlag, 2004. xii+436 pp.

[BasK] H. Bass, R. Kulkarni, *Uniform tree lattices*, J. Amer. Math. Soc. 3 (1990) 843–902.

[BasLS] H. Bass, M. Lazard, J.P. Serre, *Sous-groupes d'indice fini dans $SL(n, Z)$*, Bull. Amer. Math. Soc. 70 (1964) 385–392.

[BasL] H. Bass, A. Lubotzky, *Tree lattices*, in *Progress in Mathematics*, vol. 176, Birkhäuser, 2001. xiv+233 pp.

[BasMS] H. Bass, J. Milnor, J.P. Serre, *Solution of the congruence subgroup problem for $\mathrm{SL}_n\,(n \geq 3)$ and $\mathrm{Sp}_{2n}\,(n \geq 2)$*, Inst. Hautes Études Sci. Publ. Math. No. 33 (1967) 59–137.

[BauG] O. Baues, F. Grunewald, *Automorphism groups of polycyclic-by-finite groups and arithmetic groups*, Publ. Math. Inst. Hautes Études Sci. No. 104 (2006) 213–268.

[BauCH] P. Baum, A. Connes, N. Higson, *Classifying space for proper actions and $K$-theory of group $C^*$-algebras*, in *$C^*$-algebras*, pp. 240–291, Contemp. Math., 167, Amer. Math. Soc., 1994.

[Bav1] C. Bavard, *Théorie de Voronoii géom'etrique. Propriétés de finitude pour les familles de réseaux et analogues*, Bull. Soc. Math. France 133 (2005) 205–257.

[Bav2] C. Bavard, *Systole et invariant d'Hermite*, J. Reine Angew. Math. 482 (1997) 93–120.

[Bav3] C. Bavard, *Classes minimales de réseaux et rétractions géométriques équivariantes dans les espaces symétriques*, J. London Math. Soc. (2) 64 (2001) 275–286.

[Bea] A. Beardon, *The geometry of discrete groups*, Corrected reprint of the 1983 original, Graduate Texts in Mathematics, vol. 91, Springer-Verlag, 1995. xii+337 pp.

[Beh1] H. Behr, *Higher finiteness properties of $S$-arithmetic groups in the function field case I*, in *Groups: topological, combinatorial and arithmetic aspects*, pp. 27–42, London Math. Soc. Lecture Note Ser., 311, Cambridge Univ. Press, 2004.

[Beh2] H. Behr, *Arithmetic groups over function fields. I. A complete characterization of finitely generated and finitely presented arithmetic subgroups of reductive algebraic groups*, J. Reine Angew. Math. 495 (1998) 79–118.

[BehM] J. Behrstock, Y. Minsky, *Dimension and rank for mapping class groups*, math.GT/0512352, to appear in Annals of Mathematics.

[BehN] J. Behrstock, W. Neumann, *Quasi-isometric classification of graph manifold groups*, math.GT/0604042.

[BeiB] J. Beineke, D. Bump, *A summation formula for divisor functions associated to lattices*, Forum Math. 18 (2006) 869–906.

[Bek] M. Bekka, *On the full $C^*$-algebras of arithmetic groups and the congruence subgroup problem*, Forum Math. 11 (1999) 705–715.

[Bele1] I. Belegradek, *Nonnegative curvature, symmetry and fundamental group*, Geom. Dedicata 106 (2004) 169–184.

[Bele2] I. Belegradek, *Pinching surface groups in complex hyperbolic plane*, Geom. Dedicata 97 (2003) 45–54.

[Bele3] I. Belegradek, *On Mostow rigidity for variable negative curvature*, Topology 41 (2002) 341–361.

[Bele4] I. Belegradek, *Aspherical manifolds, relative hyperbolicity, simplicial volume and assembly maps*, Algebr. Geom. Topol. 6 (2006) 1341–1354.

[BeleK] I. Belegradek, V. Kapovitch, *Classification of negatively pinched manifolds with amenable fundamental groups*, Acta Math. 196 (2006) 229–260.

[Belo1] M. Belolipetsky, *Counting maximal arithmetic subgroups*, arXiv:math.GR/0501198.

[Belo2] M. Belolipetsky, *On volumes of arithmetic quotients of* $\mathrm{SO}(1,n)$, Ann. Sc. Norm. Super. Pisa Cl. Sci. (5) 3 (2004) 749–770.

[BeloL] M. Belolipetsky, A. Lubotzky, *Finite groups and hyperbolic manifolds*, Invent. Math. 162 (2005) 459–472.

[BeneP] R. Benedetti, C. Petronio, *Lectures on hyperbolic geometry*, Universitext, Springer-Verlag, 1992. xiv+330 pp.

[Beno1] Y. Benoist, *Convexes divisibles. II*, Duke Math. J. 120 (2003) 97–120.

[Beno2] Y. Benoist, *Propriétés asymptotiques des groupes liniaires*, Geom. Funct. Anal. 7 (1997) 1–47.

[Beno3] Y. Benoist, *Actions propres sur les espaces homogènes réductifs*, Ann. of Math. 144 (1996) 315–347.

[Beno4] Y. Benoist, *Une nilvariété non affine*, J. Differential Geom. 41 (1995) 21–52.

[Ber] P. Bérard, *On the wave equation on a compact Riemannian manifold without conjugate points*, Math. Z. 155 (1977) 249–276.

[Bera] L. Berard-Bergery, *Les variétés riemanniennes homogènes simplement connexes de dimension impaire courbure strictement positive*, J. Math. Pures Appl. (9) 55 (1976) 47–67.

[Berg1] M. Berger, *A panoramic view of Riemannian geometry*, Springer-Verlag, 2003. xxiv+824 pp.

[Berg2] M. Berger, *Les espaces symétriques non compacts*, Ann. Sci. École Norm. Sup., 74 (1957) 85–177.

[Berge1] N. Bergeron, *Lefschetz properties for arithmetic real and complex hyperbolic manifolds*, Int. Math. Res. Not. 2003, no. 20, 1089–1122.

[Berge2] N. Bergeron, *Sur la cohomologie et le spectre des variétés localement symétriques*, Ensaios Mateméticos [Mathematical Surveys], 11. Sociedade Brasileira de Matemática, Rio de Janeiro, 2006. 105 pp.

[BergeC] N. Bergeron, L. Clozel, *Spectre automorphe des variétés hyperboliques et applications topologiques*, Astérisque No. 303 (2005), xx+218 pp.

[BergeG] N. Bergeron, T. Gelander, *A note on local rigidity*, Geom. Dedicata 107 (2004) 111–131.

[BernG] J. Bernstein and S. Gelbart, *An introduction to the Langlands program*, Birkhäuser, 2003. x+281 pp.

[BernR] J. Bernstein, A. Reznikov, *Analytic continuation of representations and estimates of automorphic forms*, Ann. of Math. 150 (1999) 329–352.

[Bes] A. Besse, *Einstein manifolds*, Springer-Verlag, 1987. xii+510 pp.

[BesCG1] G. Besson, G. Courtois, S. Gallot, *Entropies et rigiditis des espaces localement symitriques de courbure strictement nigative*, Geom. Funct. Anal. 5 (1995) 731–799.

[BesCG2] G. Besson, G. Courtois, S. Gallot, *Minimal entropy and Mostow's rigidity theorems*, Ergodic Theory Dynam. Systems 16 (1996) 623–649.

[Best1] M. Bestvina, *Local homology properties of boundaries of groups*, Michigan Math. J. 43 (1996) 123–139.

[Best2] M. Bestvina, *$\mathbb{R}$-trees in topology, geometry, and group theory*, in *Handbook of geometric topology*, pp. 55–91, North-Holland, 2002.

[Best3] M. Bestvina, *The topology of* $\mathrm{Out}(F_n)$, Proc. of ICM, Vol. II (Beijing, 2002), 373–384, Higher Ed. Press, 2002.

[BestF1] M. Bestvina, M. Feighn, *The topology at infinity of* $\mathrm{Out}(F_n)$, Invent. Math. 140 (2000) 651–692.

[BestF2] M. Bestvina, M. Feighn, *Proper actions of lattices on contractible manifolds*, Invent. Math. 150 (2002) 237–256.

[BesFH1] M. Bestvina, M. Feighn, M. Handel, *Laminations, trees, and irreducible automorphisms of free groups*, Geom. Funct. Anal. 7 (1997) 215–244.

[BesFH2] M. Bestvina, M. Feighn, M. Handel, *The Tits alternative for* $\mathrm{Out}(F_n)$*. II. A Kolchin type theorem*, Ann. of Math. (2) 161 (2005) 1–59.

[BesFH3] M. Bestvina, M. Feighn, M. Handel, *The Tits alternative for* $\mathrm{Out}(F_n)$*. I. Dynamics of exponentially-growing automorphisms*, Ann. of Math. (2) 151 (2000) 517–623.

[BestM] M. Bestvina, G. Mess, *The boundary of negatively curved groups*, J. Amer. Math. Soc. 4 (1991) 469–481.

[Bieb1] L. Bieberbach, *Über die Bewegungsgruppen der Euklidischen Räume*, Math. Ann. 70 (1911) 297–336.

[Bieb2] L. Bieberbach, *Über die Bewegungsgruppen der Euklidischen Räume (Zweite Abhandlung.) Die Gruppen mit einem endlichen Fundamentalbereich*, Math. Ann. 72 (1912) 400–412.

[Bier1] R. Bieri, *Homological dimension of discrete groups*, Second edition, *Queen Mary College Mathematical Notes*, Queen Mary College, Department of Pure Mathematics, London, 1981. iv+198 pp.

[Bier2] R. Bieri, *The geometric invariants of a group. A survey with emphasis on the homotopical approach*, in *Geometric group theory*, Vol. 1, pp. 24–36, London Math. Soc. Lecture Note Ser., 181, Cambridge Univ. Press, 1993.

[BierE] R. Bieri, B. Eckmann, *Groups with homological duality generalizing Poincaré duality*, Invent. Math. 20 (1973) 103–124.

[BiG1] R. Bieri, R. Geoghegan, *Controlled topology and group actions*, in *Groups: topological, combinatorial and arithmetic aspects*, pp. 43–63, London Math. Soc. Lecture Note Ser., 311, Cambridge Univ. Press, 2004.

[BiG2] R. Bieri, R. Geoghegan, *Topological properties of* $\mathrm{SL}_2$ *actions on the hyperbolic plane*, Geom. Dedicata 99 (2003) 137–166.

[BiG3] R. Bieri, R. Geoghegan, *Connectivity properties of group actions on non-positively curved spaces*, Mem. Amer. Math. Soc. 161 (2003), no. 765, xiv+83 pp.

[BiORS] J. Birget, A. Olshanskii, E. Rips, M. Sapir, *Isoperimetric functions of groups and computational complexity of the word problem*, Ann. of Math. 156 (2002) 467–518.

[BirRS] J. Birget, E. Rips, M. Sapir, *Isoperimetric and isodiametric functions of groups*, Ann. of Math. 156 (2002) 345–466.

[BirS] J. Birman, C. Series, *Geodesics with bounded intersection number on surfaces are sparsely distributed*, Topology 24 (1985) 217–225.

[Bis1] C. Bishop, *Divergence groups have the Bowen property*, Ann. of Math. 154 (2001) 205–217.

[Bis2] C. Bishop, *Geometric exponents and Kleinian groups*, Invent. Math. 127 (1997) 33–50.

[BisP] C. Bishop, P. Jones, *Hausdorff dimension and Kleinian groups*, Acta Math. 179 (1997) 1–39.

[BisS] C. Bishop, T. Steger, *Three rigidity criteria for* $\mathrm{PSL}(2, R)$, Bull. Amer. Math. Soc. 24 (1991) 117–123.

[BlFG] D. Blasius, J. Franke, F. Grunewald, *Cohomology of $S$-arithmetic subgroups in the number field case*, Invent. Math. 116 (1994) 75–93.

[BloFR1] A. Bloch, H. Flaschka, T. Ratiu, *Schur-Horn-Kostant convexity theorem for the diffeomorphism group of the annulus*, Invent. Math. 113 (1993) 511–529.

[BloFR2] A. Bloch, H. Flaschka, T. Ratiu, *The Toda PDE and the geometry of the diffeomorphism group of the annulus*, in *Mechanics day (Waterloo, ON, 1992)*, pp. 57–92, Fields Inst. Commun., 7, Amer. Math. Soc., 1996.

[BloW] J. Block, S. Weinberger, *Arithmetic manifolds of positive scalar curvature*, J. Differential Geom. 52 (1999) 375–406.

[BoHM] M. Bökstedt, W. Hsiang, I. Madsen, *The cyclotomic trace and algebraic $K$-theory of spaces*, Invent. Math. 111 (1993) 465–539.

[BolCS] J. Boland, C. Connell, J. Souto, *Volume rigidity for finite volume manifolds*, Amer. J. Math. 127 (2005) 535–550.

[BonK1] M. Bonk, B. Kleiner, *Conformal dimension and Gromov hyperbolic groups with 2-sphere boundary*, Geom. Topol. 9 (2005) 219–246.

[BonK2] M. Bonk, B. Kleiner, *Rigidity for quasi-Fuchsian actions on negatively curved spaces*, Int. Math. Res. Not. 61 (2004) 3309–3316.

[BonK3] M. Bonk, B. Kleiner, *Rigidity for quasi-Mbius group actions*, J. Differential Geom. 61 (2002) 81–106.

[Borc] R. Borcherds, *What is Moonshine?* in *Proceedings of the International Congress of Mathematicians*, Vol. I Doc. Math. 1998, Extra Vol. I, pp. 607–615.

[Bore1] A. Borel, *Density properties for certain subgroups of semisimple groups without compact factors*, Ann. of Math. 72 (1960), 179–188.

[Bore2] A. Borel, *Density and maximality of arithmetic subgroups*, J. Reine Angew. Math. 224 (1966) 78–89.

[Bore3] A. Borel, *Reduction theory for arithmetic groups*, Proc. of Symp. in Pure Math., Vol. 9, 1969, pp. 20–25.

[Bore4] A. Borel, *Introduction aux groupes arithmétiques*, Hermann, Paris, 1969.

[Bore5] A. Borel, *Sous-groupes discrets de groupes semi-simples*, Séminaire Bourbaki, Exp. 358 (1968/69), Lect. Notes Math. 179 (1971) 199–216.

[Bore6] A. Borel, *Some metric properties of arithmetic quotients of symmetric spaces and an extension Theorem*, J. of Diff. Geom. 6 (1972) 543–560.

[Bore7] A. Borel, *Stable real cohomology of arithmetic groups*, Ann. Sci. Ec. Norm. Super. 7 (1974) 235–272.

[Bore8] A. Borel, *Linear algebraic groups*, 2nd enlarged edition, GTM 126, Springer 1991.

[Bore9] A. Borel, *Automorphic forms on $SL_2(R)$*, Cambridge Tracts in Mathematics 130, Cambridge University Press, 1997.

[Bore10] A. Borel, *Semisimple groups and Riemannian symmetric spaces*, Texts & Readings in Math. vol. 16, Hindustan Book Agency, New Delhi, 1998.

[Bore11] A. Borel, *Essays in the history of Lie groups and algebraic groups*, History of Mathematics 21, A.M.S. 2001.

[Bore12] A. Borel, *Values of zeta-functions at integers, cohomology and polylogarithms*, in *Current trends in mathematics and physics*, pp. 1–44, Narosa, New Delhi, 1995.

[Bore13] A. Borel, *Linear algebraic groups*, in *Algebraic Groups and Discontinuous Subgroups*, Proc. Sympos. Pure Math., vol. 9 (1965), pp. 3–19.

[Bore14] A. Borel, *Some finiteness properties of adele groups over number fields*, Inst. Hautes Études Sci. Publ. Math. 16 (1963) 5–30.

[Bore15] A. Borel, *Arithmetic properties of linear algebraic groups*, 1963 Proc. Internat. Congr. Mathematicians (Stockholm, 1962) pp. 10–22.

[Bore16] A. Borel, *Compact Clifford-Klein forms of symmetric spaces*, Topology 2 (1963) 111–122.

[Bore17] A. Borel, *Commensurability classes and volumes of hyperbolic 3-manifolds*, Ann. Scuola Norm. Sup. Pisa Cl. Sci. 8 (1981) 1–33.

[Bore18] A. Borel, *Introduction to the cohomology of arithmetic groups*, in *Lie groups and automorphic forms*, pp. 51–86, AMS/IP Stud. Adv. Math., 37, Amer. Math. Soc., 2006.

[Bore19] A. Borel, *Cohomologie de* $\mathrm{SL}_n$ *et valeurs de fonctions zeta aux points entiers*, Ann. Scuola Norm. Sup. Pisa Cl. Sci. (4) 4 (1977) 613–636.

[Bore20] A. Borel, *Cohomologie réelle stable de groupes $S$-arithmétiques classiques*, C. R. Acad. Sci. Paris Sér. A-B, 274 (1972), A1700–A1702.

[Bore21] A. Borel, *Œuvres: collected papers. Vol. I, II, III, IV*, Springer-Verlag, 1983 & 2001.

[Bore22] A. Borel, *On the work of M. S. Raghunathan*, in *Algebraic groups and arithmetic*, pp. 1–24, Tata Inst. Fund. Res., Mumbai, 2004.

[BoreC1] A. Borel, W. Casselman, *Automorphic forms, representations, and L-functions. Part 1 and 2*, Proc. of Symp. in Pure Math, XXXIII. American Mathematical Society, 1979, x+322 pp, vii+382 pp.

[BoreC2] A. Borel, W. Casselman, *$L^2$-cohomology of locally symmetric manifolds of finite volume*, Duke Math. J. 50 (1983) 625–647.

[BoreHC] A. Borel, Harish-Chandra, *Arithmetic subgroups of algebraic groups*, Ann. of Math. 75 (1962) 485–535.

[BoreJ] A. Borel, L. Ji, *Compactifications of symmetric and locally symmetric spaces*, Birkhäuser, 2006, 479+ix.

[BoreLS] A. Borel, J.P. Labesse, J. Schwermer, *On the cuspidal cohomology of $S$-arithmetic subgroups of reductive groups over number fields*, Compositio Math. 102 (1996) 1–40.

[BoreP] A. Borel, G. Prasad, *Finiteness theorems for discrete subgroups of bounded covolume in semi-simple groups*, Inst. Hautes Études Sci. Publ. Math. No. 69 (1989) 119–171.

[BoreS1] A. Borel, J.P. Serre, *Corners and arithmetic Groups*, Comment. Math. Helv. 48 (1973) 436–491.

[BoreS2] A. Borel, J.P. Serre, *Cohomologie d'immeubles et groups S-arithmétiques*, Topology 15 (1976) 211–232.

[BoreW] A. Borel, N. Wallach, *Continuous cohomology, discrete subgroups, and representations of reductive groups*, Second edition, *Mathematical Surveys and Monographs*, vol. 67, Amer. Math. Soc., 2000, xviii+260 pp.

[BoSh] A. Borevich, I.R. Shafarevich, *Number theory*, Academic Press, 1966 x+435 pp.

[Bou1] M. Bourdon, *Cohomologie $l_p$ et produits amalgamés*, Geom. Dedicata 107 (2004) 85–98.

[Bou2] M. Bourdon, *Immeubles hyperboliques, dimension conforme et rigidité de Mostow*, Geom. Funct. Anal. 7 (1997) 245–268.

[Bou3] M. Bourdon, *Structure conforme au bord et flot godsique d'un* CAT(−1)*-espace*, Enseign. Math. (2) 41 (1995) 63–102.

[BouMV] M. Bourdon, F. Martin, A. Valette, *Vanishing and non-vanishing for the first $L^p$-cohomology of groups*, Comment. Math. Helv. 80 (2005) 377–389.

[BouP1] M. Bourdon, H. Pajot, *Quasi-conformal geometry and hyperbolic geometry*, in *Rigidity in dynamics and geometry*, pp. 1–17, Springer, 2002.

[BouP2] M. Bourdon, H. Pajot, *Rigidity of quasi-isometries for some hyperbolic buildings*, Comment. Math. Helv. 75 (2000) 701–736.

[Bow1] B. Bowditch, *Geometrical finiteness for hyperbolic groups*, J. Funct. Anal. 113 (1993) 245–317.

[Bow2] B. Bowditch, *Geometrical finiteness with variable negative curvature*, Duke Math. J. 77 (1995) 229–274.

[Bow3] B. Bowditch, *Convergence groups and configuration spaces*, in *Geometric group theory down under*, pp. 23–54, de Gruyter, 1999.

[Bowe] R. Bowen, *The equidistribution of closed geodesics*, Amer. J. Math. 94 (1972) 413–423.

[BrB] N. Brady, M. Bridson, *There is only one gap in the isoperimetric spectrum*, Geom. Funct. Anal. 10 (2000) 1053–1070.

[BrC] N. Brady, J. Crisp, *Two-dimensional Artin groups with* CAT(0) *dimension three*, Geom. Dedicata 94 (2002) 185–214.

[BrG1] E. Breuillard, T. Gelander, *Cheeger constant and algebraic entropy of linear groups*, Int. Math. Res. Not. 56 (2005) 3511–3523.

[BrG2] E. Breuillard, T. Gelander, *On dense free subgroups of Lie groups*, J. Algebra 261 (2003) 448–467.

[Bri] S. Brick, *Quasi-isometries and ends of groups*, J. Pure Appl. Algebra 86 (1993) 23–33.

[Brid] M. Bridson, *Fractional isoperimetric inequalities and subgroup distortion*, J. Amer. Math. Soc. 12 (1999) 1103–1118.

[BridH] M. Bridson, A. Haefliger, *Metric spaces of non-positive curvature*, in Grundlehren der Mathematischen Wissenschaften, 319, Springer-Verlag, 1999. xxii+643 pp.

[BridV] M. Bridson, K. Vogtmann, *Automorphism groups of free groups, surface groups and free abelian groups*, in *Problems on mapping class groups and related topics*, pp. 301–316, Proc. Sympos. Pure Math., 74, Amer. Math. Soc., 2006.

[Bro] K. Bromberg, *Rigidity of geometrically finite hyperbolic cone-manifolds*, Geom. Dedicata 105 (2004) 143–170.

[BroM] R. Brooks, E. Makover, *Riemann surfaces with large first eigenvalue*, J. Anal. Math. 83 (2001) 243–258.

[Bro1] K. Brown, *Buildings*, Springer, 1988.

[Bro2] K. Brown, *Cohomology of groups*, Springer-Verlag, 1982.

[Bro3] K. Brown, *Groups of virtually finite dimension*, in *Homological group theory*, pp. 27–70, London Math. Soc. Lecture Note Ser., 36, Cambridge Univ. Press, 1979.

[Bro4] K. Brown, *Finiteness properties of groups*, J. of Pure and Applied Algebra 44 (1987) 45–75.

[Brug1] R. Bruggeman, *Fourier coefficients of automorphic forms*, Lecture Notes in Mathematics, 865. Springer-Verlag, 1981. i+201 pp.

[Brug2] R. Bruggeman, *Families of automorphic forms*, Monographs in Mathematics, 88. Birkhäuser Boston, 1994. x+317 pp.

[BruM] R. Bruggeman, R. Miatello, *Sum formula for $SL_2$ over a totally real number field*, to appear in Memoir of AMS.

[BruMP] R. Bruggeman, R. Miatello, I. Pacharoni, *Density results for automorphic forms on Hilbert modular groups*, Geom. Funct. Anal. 13 (2003) 681–719.

[BruMW] R. Bruggeman, R. Miatello, N. Wallach, *Resolvent and lattice points on symmetric spaces of strictly negative curvature*, Math. Ann. 315 (1999) 617–639.

[BryZ] J. Brylinski, S. Zucker, *An overview of recent advances in Hodge theory*, in *Several complex variables, VI*, pp. 39–142, Encyclopaedia Math. Sci., 69, Springer, 1990.

[Bum1] D. Bump, *Automorphic forms and representations*, Cambridge University Press, Cambridge, 1997. xiv+574 pp.

[Bum2] D. Bump, *Automorphic forms on* $\mathrm{GL}(3, R)$, Lecture Notes in Mathematics, Vol. 1083, Springer-Verlag, 1984. xi+184 pp.

[BuO1] U. Bunke, M. Olbrich, *The spectrum of Kleinian manifolds*, J. Funct. Anal. 172 (2000) 76–164.

[BuO2] U. Bunke, M. Olbrich, *Group cohomology and the singularities of the Selberg zeta function associated to a Kleinian group*, Ann. of Math. 149 (1999) 627–689.

[BuO3] U. Bunke, M. Olbrich, *Gamma-cohomology and the Selberg zeta function*, J. Reine Angew. Math. 467 (1995) 199–219.

[BuO4] U. Bunke, M. Olbrich, *Selberg zeta and theta functions. A differential operator approach*, Akademie-Verlag, Berlin, 1995. 168 pp.

[BuO5] U. Bunke, M. Olbrich, *The wave kernel for the Laplacian on the classical locally symmetric spaces of rank one, theta functions, trace formulas and the Selberg zeta function*, Ann. Global Anal. Geom. 12 (1994) 357–405.

[BuO6] U. Bunke, M. Olbrich, *On quantum ergodicity for vector bundles*, Acta Appl. Math. 90 (2006) 19–41.

[BuO7] U. Bunke, M. Olbrich, *Fuchsian groups of the second kind and representations carried by the limit set*, Invent. Math. 127 (1997) 127–154.

[BurDD] D. Burde, K. Dekimpe, S. Deschamps, *The Auslander conjecture for NIL-affine crystallographic groups*, Math. Ann. 332 (2005) 161–176.

[Burg] E. Burger, *Homogeneous Diophantine approximation in $S$-integers*, Pacific J. Math. 152 (1992) 211–253.

[Burge] M. Burger, *Rigidity properties of group actions on* $\mathrm{CAT}(0)$*-spaces*, in *Proceedings of the International Congress of Mathematicians*, pp. 761–769, Birkhäuser, Basel, 1995.

[BurGLM] M. Burger, T. Gelander, A. Lubotzky, S. Mozes, *Counting hyperbolic manifolds*, Geom. Funct. Anal. 12 (2002) 1161–1173.

[BurI1] M. Burger, A. Iozzi, *Bounded Kähler class rigidity of actions on Hermitian symmetric spaces*, Ann. Sci. École Norm. Sup. 37 (2004) 77–103.

[BurI2] M. Burger, A. Iozzi, *Rigidity in dynamics and geometry*, Springer-Verlag, 2002. xiv+492 pp.

[BurILW] M. Burger, A. Iozzi, F. Labourie, A. Wienhard, *Maximal representations of surface groups: symplectic Anosov structures*, Pure Appl. Math. Q. 1 (2005) 543–590.

[BurM1] M. Burger, N. Monod, *Bounded cohomology of lattices in higher rank Lie groups*, J. Eur. Math. Soc. 1 (1999) 199–235.

[BurM2] M. Burger, N. Monod, *Continuous bounded cohomology and applications to rigidity theory*, Geom. Funct. Anal. 12 (2002) 219–280.

[BurMo1] M. Burger, S. Mozes, *Lattices in product of trees*, Inst. Hautes Études Sci. Publ. Math. No. 92 (2000) 151–194.

[BurMo2] M. Burger, S. Mozes, *Groups acting on trees: from local to global structure*, Inst. Hautes Études Sci. Publ. Math. No. 92 (2000) 113–150.

[BurMo3] M. Burger, S. Mozes, CAT(-1)*-spaces, divergence groups and their commensurators*, J. Amer. Math. Soc. 9 (1996) 57–93.

[BurS1] M. Burger, V. Schroeder, *Amenable groups and stabilizers of measures on the boundary of a Hadamard manifold*, Math. Ann. 276 (1987) 505–514.

[BurS2] M. Burger, V. Schroeder, *Volume, diameter and the first eigenvalue of locally symmetric spaces of rank one*, J. Differential Geom. 26 (1987) 273–284.

[BuS1] K. Burns, R. Spatzier, *Manifolds of nonpositive curvature and their buildings*, Inst. Hautes Études Sci. Publ. Math. 65 (1987) 35–59.

[BuS2] K. Burns, R. Spatzier, *On topological Tits buildings and their classification*, Inst. Hautes Études Sci. Publ. Math. 65 (1987) 5–34.

[Buse] P. Buser, *Geometry and spectra of compact Riemann surfaces*, Progress in Mathematics, 106. Birkhäuser, 1992. xiv+454 pp.

[Bux] K. Bux, *Finiteness properties of soluble $S$-arithmetic groups: a survey*, in *Groups: topological, combinatorial and arithmetic aspects*, pp. 64–92, London Math. Soc. Lecture Note Ser., 311, Cambridge Univ. Press, 2004.

[BuW1] K. Bux, K. Wortman, *Finiteness properties of arithmetic groups over function fields*, Invent. Math. 167 (2007) 355–378.

[BuW2] K. Bux, K. Wortman, *A geometric proof that* $\mathrm{SL}_2(\mathbb{Z}[t,t^{-1}])$ *is not finitely presented*, Algebr. Geom. Topol. 6 (2006) 839–852.

[CalaM] E. Calabi and L. Markus, *Relativistic space forms*, Ann. of Math. 75 (1962) 63–76.

[CaleG] D. Calegari, D. Gabai, *Shrinkwrapping and the taming of hyperbolic 3-manifolds*, J. Amer. Math. Soc. 19 (2006) 385–446.

[Can] R. Canary, *Ends of hyperbolic* 3*-manifolds*, J. Amer. Math. Soc. 6 (1993) 1–35.

[CanMT] R. Canary, Y. Minsky, E. Taylor, *Spectral theory, Hausdorff dimension and the topology of hyperbolic 3-manifolds*, J. Geom. Anal. 9 (1999) 17–40.

[CapS] Sylvain E. Cappell, Julius L. Shaneson, *Some Problems in Number Theory I: The Circle Problem*, arXiv:math/0702613.

[Car1] L. Carbone, *Non-uniform lattices on uniform trees*, Mem. Amer. Math. Soc. 152 (2001), no. 724, xii+127 pp.

[Car2] L. Carbone, *The Tree lattice existence theorems*, Comptes Rendus de l'Academie des Sciences. Serie I, Mathematique, 335 (2002) 223–228.

[Car3] L. Carbone, *Congruence subgroups of lattices in rank 2 Kac-Moody groups over finite fields*, Preprint, 2007.

[CarER] L. Carbone, M. Ershov, G. Ritter, *Abstract simplicity of complete Kac-Moody groups over finite fields*, Submitted 2006.

[CarG1] L. Carbone, H. Garland, *Lattices in Kac-Moody groups*, Math. Res. Lett. 6 (1999) 439–447.

[CarG2] L. Carbone, H. Garland, *Existence of lattices in Kac-Moody groups over finite fields*, Commun. Contemp. Math. 5 (2003) 813–867.

[CarbR] L. Carbone, G. Rosenberg, *Lattices on nonuniform trees*, Geom. Dedicata 98 (2003) 161–188.

[CarMP] J. Carlson, S. Müller-Stach, C. Peters, *Period mappings and period domains*, Cambridge Studies in Advanced Mathematics, 85. Cambridge University Press, 2003. xvi+430 pp.

[CarP] G. Carlsson, E. Pedersen, *Controlled algebra and the Novikov conjectures for K- and L-theory*, Topology 34 (1995) 731–758.

[CartK] D. Carter, G. Keller, *Bounded elementary generation of* $\mathrm{SL}_n(\mathcal{O})$, Amer. J. Math. 105 (1983) 673–687.

[CartMS1] D. Cartwright, A. Mantero, T. Steger, A. Zappa, *Groups acting simply transitively on the vertices of a building of type* 2. *I*, Geom. Dedicata 47 (1993) 143–166.

[CartMS2] D. Cartwright, A. Mantero, T. Steger, A. Zappa, *Groups acting simply transitively on the vertices of a building of type* 2. *II. The cases* $q = 2$ *and* $q = 3$, Geom. Dedicata 47 (1993) 167–223.

[CartMSt] D. Cartwright, W. Motkowski, T. Steger, *Property (T) and* $\tilde{A}_2$ *groups*, Ann. Inst. Fourier (Grenoble) 44 (1994) 213–248.

[CartS] D. Cartwright, T. Steger, *A family of* $\tilde{A}_n$*-groups*, Israel J. Math. 103 (1998) 125–140.

[CartW] D. Cartwright, W. Woess, *Isotropic random walks in a building of type* $\tilde{A}_d$, Math. Z. 247 (2004) 101–135.

[Cass] W. Casselman, *Geometric rationality of Satake compactifications*, in *Algebraic groups and Lie groups*, Cambridge Univ. Press, 1997.

[Cass1] J. Cassels, *An introduction to the geometry of numbers*, Corrected reprint of the 1971 edition, in *Classics in Mathematics*, Springer-Verlag, 1997. viii+344 pp.

[Cass2] J. Cassels, *Rational quadratic forms*, in *London Mathematical Society Monographs*, 13. Academic Press, Inc., 1978. xvi+413 pp.

[CassF] J. Cassels, A. Fröhlich, *Algebraic number theory*, Academic Press, 1986. xviii+366 pp.

[Cat1] E. Cattani, *On the partial compactification of the arithmetic quotient of a period matrix domain*, Bull. Amer. Math. Soc. 80 (1974) 330–333.

[Cat2] E. Cattani, *Mixed Hodge structures, compactifications and monodromy weight filtration*, in *Topics in transcendental algebraic geometry*, pp. 75–100, Ann. of Math. Stud., vol. 106, Princeton Univ. Press, 1984.

[Cat3] E. Cattani, *Homogeneous spaces and Hodge theory, Conference on differential geometry on homogeneous spaces*, Rend. Sem. Mat. Univ. Politec. Torino 1983, Special Issue, 1–16 (1984).

[CatK] E. Cattani, A. Kaplan, *Extension of period mappings for Hodge structures of weight two*, Duke Math. J. 44 (1977) 1–43.

[CatKS1] E. Cattani, A. Kaplan, W. Schmid, *Some remarks on* $L^2$ *and intersection cohomologies*, in *Hodge theory*, 32–41, Lecture Notes in Math., vol. 1246, Springer, 1987.

[CatKS2] E. Cattani, A. Kaplan, W. Schmid, *Variations of polarized Hodge structure: asymptotics and monodromy*, in *Hodge theory*, pp. 16–31, Lecture Notes in Math., vol. 1246, Springer, 1987.

[CatKS3] E. Cattani, A. Kaplan, W. Schmid, $L^2$ *and intersection cohomologies for a polarizable variation of Hodge structure*, Invent. Math. 87 (1987) 217–252.

[Ch1] C.L. Chai, *Arithmetic minimal compactification of the Hilbert-Blumenthal moduli spaces*, Ann. of Math. 131 (1990) 541–554.

[Ch2] C.L. Chai, *Arithmetic compactification of the Siegel moduli space*, in *Theta functions—Bowdoin 1987*, Part 2, pp. 19–44, Proc. Sympos. Pure Math., vol. 49, Part 2, Amer. Math. Soc., 1989.

[Ch3] C.L. Chai, *Siegel moduli schemes and their compactifications over* $\mathbb{C}$, in *Arithmetic geometry*, pp. 231–251, Springer, 1986.

[Ch4] C.L. Chai, *Compactification of Siegel moduli schemes*, London Mathematical Society Lecture Note Series, vol. 107, Cambridge University Press, 1985. xvi+326 pp.

[ChaW] B. Chandler, W. Magnus, *The history of combinatorial group theory. A case study in the history of ideas*, Studies in the History of Mathematics and Physical Sciences, 9. Springer-Verlag, 1982. viii+234 pp.

[ChaWe] S. Chang, S. Weinberger, *Topological nonrigidity of nonuniform lattices*, Comm. Pure Appl. Math. 60 (2007) 282–290.

[ChaD1] R. Charney, M. Davis, *Finite* $K(\pi,1)$*s for Artin groups*, in *Prospects in topology*, pp. 110–124, Ann. of Math. Stud., 138, Princeton Univ. Press, 1995.

[ChaD2] R. Charney, M. Davis, *The* $K(\pi,1)$*-problem for hyperplane complements associated to infinite reflection groups*, J. Amer. Math. Soc. 8 (1995) 597–627.

[ChaL] R. Charney, R. Lee, *Cohomology of the Satake compactification*, Topology 22 (1983) 389–423.

[ChaM] C. Chatterjee, D. Morris, *Divergent torus orbits in homogeneous spaces of* $\mathbb{Q}$*-rank two*, Israel J. Math. 152 (2006) 229–243.

[ChaR] I. Chatterji, K. Ruane, *Some geometric groups with rapid decay*, Geom. Funct. Anal. 15 (2005) 311–339.

[CheCM] J. Cheeger, T. Colding, W. Minicozzi, *Linear growth harmonic functions on complete manifolds with nonnegative Ricci curvature*, Geom. Funct. Anal. 5 (1995) 948–954.

[CheG1] J. Cheeger, M. Gromov, *Bounds on the von Neumann dimension of* $L_2$*-cohomology and the Gauss-Bonnet theorem for open manifolds*, J. Diff. Geom. (1985) 1–39.

[CheG2] J. Cheeger, M. Gromov, $L_2$*-cohomology and group cohomology*, Topology 25 (1986) 189–215.

[CheG3] J. Cheeger, M. Gromov, *On the characteristic numbers of complete manifolds of bounded curvature and finite volume*, in *Differential geometry and complex analysis*, pp. 115–154, Springer, 1985.

[CheJV] P. Cherix, M. Cowling, P. Jolissaint, P. Julg, A. Valette, *Groups with the Haagerup property. Gromovs a-T-menability*, Progress in Mathematics, 197. Birkhuser Verlag, 2001. viii+126 pp.

[Chi] T. Chinburg, *Volumes of hyperbolic manifolds*, J. Differential Geom. 18 (1983) 783–789.

[ClaW] B. Clair, K. Whyte, *Growth of Betti numbers*, Topology 42 (2003) 1125–1142.

[ClO] L. Clozel, J.P. Otal, *Unique ergodicité des correspondances modulaires*, in *Essays on geometry and related topics*, Vol. 1, 2, pp. 205–216, Monogr. Enseign. Math., 38, Enseignement Math., Geneva, 2001.

[ClOU] L. Clozel, H. Oh, E. Ullmo, *Hecke operators and equidistribution of Hecke points*, Invent. Math. 144 (2001) 327–351.

[ClU1] L. Clozel, E. Ullmo, *Équidistribution des points de Hecke*, in *Contributions to automorphic forms, geometry, and number theory*, pp. 193–254, Johns Hopkins Univ. Press, 2004.

[ClU2] L. Clozel, E. Ullmo, *Équidistribution de sous-variétés spéciales*, Ann. of Math. 161 (2005) 1571–1588.

[CoJPS] J. Coates, L. Ji, G. Prasad, Y.T. Siu, S.T. Yau, *Special issue dedicated to the memory of Professor Armand Borel, 1923–2003*, Asian J. Math. 8 (2004), no. 4.

[Cog1] J. Cogdell, *Lectures on L-functions, converse theorems, and functoriality for* $\mathrm{GL}_n$, in *Lectures on automorphic L-functions*, pp. 1–96, Fields Inst. Monogr., 20, Amer. Math. Soc., 2004.

[Cog2] J. Cogdell, *Converse theorems, functoriality, and applications*, Pure Appl. Math. Q. 1 (2005) 341–367.

[CogKM] J. Cogdell, H. Kim, M. Murty, *Lectures on automorphic L-functions*, Fields Institute Monographs, 20. American Mathematical Society, 2004. xii+283 pp.

[CogSP] J. Cogdell, I. Piatetski-Shapiro, *Converse theorems, functoriality, and applications to number theory*, in *Proceedings of the International Congress of Mathematicians*, Vol. II, pp. 119–128, Higher Ed. Press, Beijing, 2002.

[Coh] D. Cohen, *Groups of cohomological dimension one*, Lecture Notes in Mathematics, Vol. 245. Springer-Verlag, 1972. v+99 pp.

[Cohn] H. Cohn, *A classical invitation to algebraic numbers and class fields. With two appendices by Olga Taussky: "Artin's 1932 Gttingen lectures on class field theory" and "Connections between algebraic number theory and integral matrices"*, Universitext. Springer-Verlag, 1978. xiii+328 pp.

[CoKu] H. Cohn, A. Kumar, *The densest lattice in twenty-four dimensions*, Electron. Res. Announc. Amer. Math. Soc. 10 (2004) 58–67.

[ColM1] T. Colding, W. Minicozzi, *Harmonic functions with polynomial growth*, J. Differential Geom. 46 (1997) 1–77.

[ColM2] T. Colding, W. Minicozzi, *Weyl type bounds for harmonic functions*, Invent. Math. 131 (1998) 257–298.

[ColM3] T. Colding, W. Minicozzi, *Liouville theorems for harmonic sections and applications*, Comm. Pure Appl. Math. 51 (1998) 113–138.

[ColM4] T. Colding, W. Minicozzi, *Harmonic functions on manifolds*, Ann. of Math. 146 (1997) 725–747.

[Con1] C. Connell, *A characterization of homogeneous spaces with positive hyperbolic rank*, Geom. Dedicata 93 (2002) 205–233.

[Con2] C. Connell, *Asymptotic harmonicity of negatively curved homogeneous spaces and their measures at infinity*, Comm. Anal. Geom. 8 (2000) 575–633.

[Con3] C. Connell, *Minimal Lyapunov exponents, quasiconformal structures, and rigidity of non-positively curved manifolds*, Ergodic Theory Dynam. Systems 23 (2003) 429–446.

[ConF1] C. Connell, B. Farb, *Minimal entropy rigidity for lattices in products of rank one symmetric spaces*, Comm. Anal. Geom. 11 (2003) 1001–1026.

[ConF2] C. Connell, B. Farb, *Some recent applications of the barycenter method in geometry*, in *Topology and geometry of manifolds*, pp. 19–50, Proc. Sympos. Pure Math., 71, Amer. Math. Soc., 2003.

[ConF3] C. Connell, B. Farb, *The degree theorem in higher rank*, J. Differential Geom. 65 (2003) 19–59.

[ConMT] G. Conner, M. Mihalik, S. Tschantz, *Homotopy of ends and boundaries of CAT(0) groups*, Geom. Dedicata 120 (2006) 1–17.

[ConM] A. Connes, H. Moscovici, *Cyclic cohomology, the Novikov conjecture and hyperbolic groups*, Topology 29 (1990) 345–388.

[ConS] J.H. Conway, N. Sloane, *Sphere packings, lattices and groups*, Third edition, Springer-Verlag, 1999. lxxiv+703 pp.

[ConGS] J.H. Conway, C. Goodman-Strauss, N. Sloane, *Recent progress in sphere packing*, in *Current developments in mathematics, 1999*, pp. 37–76, Int. Press, 1999.

[ConzG1] J. Conze, Y. Guivarch, *Densité d'orbites d'actions de groupes linéaires et propriétés d'équidistribution de marches aléatoires*, in *Rigidity in dynamics and geometry*, pp. 39–76, Springer, 2002.

[ConzG2] J. Conze, Y. Guivarch, *Limit sets of groups of linear transformations*, Ergodic theory and harmonic analysis, Sankhya Ser. A 62 (2000) 367–385.

[Coo] M. Coornaert, *Mesures de Patterson-Sullivan sur le bord d'un espace hyperbolique au sens de Gromov*, Pacific J. Math. 159 (1993) 241–270.

[Cor1] K. Corlette, *Archimedean superrigidity and hyperbolic geometry*, Ann. of Math. 135 (1992) 165–182.

[Cor2] K. Corlette, *Hausdorff dimensions of limit sets. I*, Invent. Math. 102 (1990) 521–541.

[CorK] G. Cornelissen, F. Kato, *The p-adic icosahedron*, Notices Amer. Math. Soc. 52 (2005) 720–727.

[Cox] D. Cox, *Primes of the form $x^2 + ny^2$. Fermat, class field theory and complex multiplication*, A Wiley-Interscience Publication, John Wiley & Sons, 1989. xiv+351 pp.

[Cr] J. Crisp, *On the* CAT(0) *dimension of 2-dimensional Bestvina-Brady groups*, Algebr. Geom. Topol. 2 (2002) 921–936.

[Cro1] C. Croke, *Rigidity for surfaces of nonpositive curvature*, Comment. Math. Helv. 65 (1990) 150–169.

[Cro2] C. Croke, *Rigidity theorems in Riemannian geometry*, in *Geometric methods in inverse problems and PDE control*, pp. 47–72, IMA Vol. Math. Appl., 137, Springer, 2004.

[CroK] C. Croke, B. Kleiner, *Spaces with nonpositive curvature and their ideal boundaries*, Topology 39 (2000) 549–556.

[CuM] M. Culler, J. Morgan, *Group actions on R-trees*, Proc. London Math. Soc. 55 (1987) 571–604.

[CuV] M. Culler, K. Vogtmann, *Moduli of graphs and automorphisms of free groups*, Invent. Math. 84 (1986) 91–119.

[DalK] F. Dal'Bo, I. Kim, *Marked length rigidity for symmetric spaces*, Comment. Math. Helv. 77 (2002) 399–407.

[Dan1] S. Dani, *Flows on homogeneous spaces: a review*, in *Ergodic theory of $\mathbb{Z}_d$ actions*, pp. 63–112, London Math. Soc. Lecture Note Ser., 228, Cambridge Univ. Press, 1996.

[Dan2] S. Dani, *On orbits of unipotent flows on homogeneous spaces*, Ergodic Theory Dynam. Systems 4 (1984) 25–34.

[Dan3] S. Dani, *On invariant measures, minimal sets and a lemma of Margulis*, Invent. Math. 51 (1979) 239–260.

[Dan4] S. Dani, *Bounded orbits of flows on homogeneous spaces*, Comment. Math. Helv. 61 (1986) 636–660.

[Dan5] S. Dani, *Flows on homogeneous spaces and Diophantine approximation*, Proceedings of the International Congress of Mathematicians, Vol. 1, 2, pp. 780–789, Birkhäuser, 1995.

[Dan6] S. Dani, *Divergent trajectories of flows on homogeneous spaces and Diophantine approximation*, J. Reine Angew. Math. 359 (1985) 55–89.

[Dan7] S. Dani, *Invariant measures and minimal sets of horospherical flows*, Invent. Math. 64 (1981) 357–385.

[Dan8] S. Dani, *Flows on homogeneous spaces: a review*, in *Ergodic theory of $\mathbb{Z}_d$ actions*, pp. 63–112, London Math. Soc. Lecture Note Ser., 228, Cambridge Univ. Press, 1996.

[Dan9] S. Dani, *Orbits of horospherical flows*, Duke Math. J. 53 (1986) 177–188.

[DasDW] G. Daskalopoulos, S. Dostoglou, R. Wentworth, *On the Morgan-Shalen compactification of the* SL(2, $\mathbb{C}$) *character varieties of surface groups*, Duke Math. J. 101 (2000) 189–207.

[DaSV] G. Davidoff, P. Sarnak, A. Valette, *Elementary number theory, group theory, and Ramanujan graphs*, London Mathematical Society Student Texts, 55. Cambridge University Press, 2003. x+144 pp.

[Dav] J. Davis, *Manifold aspects of the Novikov conjecture*, in *Surveys on surgery theory*, Vol. 1, pp. 195–224, Ann. of Math. Stud., 145, Princeton Univ. Press, 2000.

[Dav1] M. Davis, *Nonpositive curvature and reflection groups*, in *Handbook of geometric topology*, pp.373–422, North-Holland, 2002.

[Dav2] M. Davis, *Some examples of discrete group actions on aspherical manifolds*, in *High-dimensional manifold topology*, pp. 139–150, World Sci. Publ., 2003.

[Dav3] M. Davis, *Exotic aspherical manifolds*, in *Topology of high-dimensional manifolds*, pp. 371–404, ICTP Lect. Notes, 9, 2002.

[DavJ] M. Davis, T. Januszkiewicz, *Hyperbolization of polyhedra*, J. Differential Geom. 34 (1991) 347–388.

[DavM] M. Davis, G. Moussong, *Notes on nonpositively curved polyhedra*, in *Low dimensional topology*, pp. 11–94, Bolyai Soc. Math. Stud., 8, 1999.

[Deg] D. de George, *Length spectrum for compact locally symmetric spaces of strictly negative curvature*, Ann. Sci. Icole Norm. Sup. 10 (1977) 133–152.

[DegW] D. de George, N. Wallach, *Limit formulas for multiplicities in* $L^2(\Gamma\backslash G)$, Ann. Math. 107 (1978) 133–150.

[Dei1] A. Deitmar, *Geometric zeta functions of locally symmetric spaces*, Amer. J. Math. 122 (2000) 887–926.

[Dei2] A. Deitmar, *Equivariant torsion of locally symmetric spaces*, Pacific J. Math. 182 (1998) 205–227.

[Dei3] A. Deitmar, *A prime geodesic theorem for higher rank spaces*, Geom. Funct. Anal. 14 (2004) 1238–1266.

[Dei4] A. Deitmar, *A prime geodesic theorem for higher rank spaces II: singular geodesics*, http://arxiv.org/abs/math.DG/0410553, to appear in Rocky Mountain J. Math.

[Dei5] A. Deitmar, *A Lefschetz formula for higher rank*, arXiv:math/0505403, to appear in The Journal of Fixed Point Theory and Applications.

[Dei6] A. Deitmar, *Higher torsion zeta functions*, Adv. Math. 110 (1995) 109–128.

[DeiP] A. Deitmar, M. Pavey, *A prime geodesic theorem for SL(4)*, arXiv:math/0605048, Annals of Global Analysis and Geometry.

[DeitH] A. Deitmar, J. Hilgert, *Cohomology of arithmetic groups with infinite dimensional coefficient spaces*, Doc. Math. 10 (2005) 199–216.

[DeitH1] A. Deitmar, W. Hoffmann, *Asymptotics of class numbers*, Invent. Math. 160 (2005) 647–675.

[DeitH2] A. Deitmar, W. Hoffmann, *On limit multiplicities for spaces of automorphic forms*, Canad. J. Math. 51 (1999) 952–976.

[DeitH3] A. Deitmar, W. Hoffmann, *Spectral estimates for towers of noncompact quotients*, Canad. J. Math. 51 (1999) 266–293.

[Dek] K. Dekimpe, *Any virtually polycyclic group admits a NIL-affine crystallographic action*, Topology 42 (2003) 821–832.

[DekI] K. Dekimpe, P. Igodt, *Polycyclic-by-finite groups admit a bounded-degree polynomial structure*, Invent. Math. 129 (1997) 121–140.

[DeL] J. De Loera, *The many aspects of counting lattice points in polytopes*, Math. Semesterber. 52 (2005) 175–195.

[DeM] P. Deligne, G. Mostow, *Commensurabilities among lattices in* $\mathrm{PU}(1,n)$, Annals of Mathematics Studies, 132, Princeton University Press, 1993. viii+183 pp.

[DiaT] A. Diaconu, Y. Tian, *Twisted Fermat curves over totally real fields*, Ann. of Math. 162 (2005) 1353–1376.

[Din] J. Ding, *A proof of a conjecture of C. L. Siegel*, J. Number Theory 46 (1994) 1–11.

[Dod] J. Dodziuk, *de Rham-Hodge theory for $L^2$-cohomology of infinite coverings*, Topology 16 (1977) 157–165.

[Dol1] I. Dolgachev, *Reflection groups in algebraic geometry*, math.AG/0610938.

[Dol2] I. Dolgachev, *Abstract configurations in algebraic geometry*, in *The Fano Conference*, pp. 423–462, Univ. Torino, 2004.

[DolK] I. Dolgachev, S. Kondo, *Moduli spaces of K3 surfaces and complex ball quotients*, math.AG/0511051.

[DolGK] I. Dolgachev, B. van Geemen, S. Kondo, *A complex ball uniformization of the moduli space of cubic surfaces via periods of K3 surfaces*, J. Reine Angew. Math. 588 (2005) 99–148.

[Don] H. Donnelly, *Quantum unique ergodicity*, Proc. Amer. Math. Soc. 131 (2003) 2945–2951.

[Dr] T. Drumm, *Linear holonomy of Margulis space-times*, J. Differential Geom. 38 (1993) 679–690.

[Dru] C. Drutu, *Filling in solvable groups and in lattices in semisimple groups.* Topology **43** (2004) 983–1033.

[DruS] C. Drutu, M. Sapir, *Tree-graded spaces and asymptotic cones of groups*, Topology 44 (2005) 959–1058.

[DuG] J. Duistermaat, V. Guillemin, *The spectrum of positive elliptic operators and periodic bicharacteristics*, Invent. Math. 29 (1975) 39–79.

[DuKV] J. Duistermaat, J. Kolk, V. Varadarajan, *Spectra of compact locally symmetric manifolds of negative curvature*, Invent. Math. 52 (1979) 27–93.

[DuT] N. Dunfield, W. Thurston, *Finite covers of random 3-manifolds*, Invent. Math. 166 (2006) 457–521.

[Dun] M. Dunwoody, *The accessibility of finitely presented groups*, Invent. Math. 81 (1985) 449–457.

[DunS] M. Dunwoody, M. Sageev, *JSJ-splittings for finitely presented groups over slender groups*, Invent. Math. 135 (1999) 25–44.

[DyJ] J. Dymara, T. Januszkiewicz, *Cohomology of buildings and their automorphism groups*, Invent. Math. 150 (2002) 579–627.

[EarK] C. Earle, I. Kra, *On isometries between Teichmller spaces*, Duke Math. J. 41 (1974) 583–591.

[Eb] P. Eberlein, *Geometry of nonpositively curved manifolds*, Chicago Lectures in Mathematics, 1996.

[EbHS] P. Eberlein, U. Hamenstadt, V. Schroeder, *Manifolds of nonpositive curvature*, in *Differential geometry: Riemannian geometry*, pp. 179–227, Proc. Sympos. Pure Math., 54, Part 3, Amer. Math. Soc., 1993.

[EbM] D. Ebin, J. Marsden, *Groups of diffeomorphisms and the notion of an incompressible fluid*, Ann. of Math. 92 (1970) 102–163.

[Ef] I. Efrat, *On the existence of cusp forms over function fields*, J. Reine Angew. Math. 399 (1989) 173–187.

[El] M. Elder, *Patterns theory and geodesic automatic structure for a class of groups*, arXiv:math/0611876.

[ElC] N. Elkies, H. Cohn, *New upper bounds on sphere packings I*, Ann. of Math. 157 (2003) 689–714.

[EllV1] J. Ellenberg, A. Venkatesh, *The number of extensions of a number field with fixed degree and bounded discriminant*, Ann. of Math. 163 (2006) 723–741.

[EllV2] J. Ellenberg, A. Venkatesh, *Counting extensions of function fields with bounded discriminant and specified Galois group*, in *Geometric methods in algebra and number theory*, pp. 151–168, Progr. Math., 235, Birkhäuser, 2005.

[ElsGM] J. Elstrodt, F. Grunewald, J. Mennicke, *Groups acting on hyperbolic spaces*, Springer-Verlag, 1998.

[Eme] H. Emerson, *Noncommutative Poincar duality for boundary actions of hyperbolic groups*, J. Reine Angew. Math. 564 (2003) 1–33.

[EmM] H. Emerson, R. Meyer, *Euler characteristics and Gysin sequences for group actions on boundaries*, Math. Ann. 334 (2006) 853–904.

[EepCHL] D. Epstein, J. Cannon, D. Holt, S. Levy, M. Paterson, W. Thurston, *Word processing in groups*, Jones and Bartlett Publishers, Boston, 1992. xii+330 pp.

[ErR] I. Erovenko, A. Rapinchuk, *Bounded generation of $S$-arithmetic subgroups of isotropic orthogonal groups over number fields*, J. Number Theory 119 (2006) 28–48.

[Esk1] A. Eskin, *Quasi-isometric rigidity of nonuniform lattices in higher rank symmetric spaces*, J. Amer. Math. Soc. 11 (1998) 321–361.

[Esk2] A. Eskin, *Counting problems and semisimple groups*, Proc. of ICM, Vol. II (Berlin, 1998.) Doc. Math. 1998, Extra Vol. II, pp.539–552.

[EskF1] A. Eskin, B. Farb, *Quasi-flats in $H^2 \times H^2$*, in *Lie groups and ergodic theory*, pp. 75–103, Tata Inst. Fund. Res. Stud. Math., 14, Tata Inst. Fund. Res., 1998.

[EskF2] A. Eskin, B. Farb, *Quasi-flats and rigidity in higher rank symmetric spaces*, J. Amer. Math. Soc. 10 (1997) 653–692.

[EsMaM] A. Eskin, G. Margulis, S. Mozes, *Upper bounds and asymptotics in a quantitative version of the Oppenheim conjecture*, Ann. of Math. 147 (1998) 93–141.

[EsMc] A. Eskin, C. McMullen, *Mixing, counting, and equidistribution in Lie groups*, Duke Math. J. 71 (1993) 181–209.

[EsMS] A. Eskin, S. Mozes, N. Shah, *Unipotent flows and counting lattice points on homogeneous varieties*, Ann. of Math. 143 (1996) 253–299.

[EskO] A. Eskin, H. Oh, *Representations of integers by an invariant polynomial and unipotent flows*, Duke Math. J. 135 (2006) 481–506.

[FaK] E. Falbel, P. Koseleff, *Rigidity and flexibility of triangle groups in complex hyperbolic geometry*, Topology 41 (2002) 767–786.

[FaP1] E. Falbel, J. Parker, *The moduli space of the modular group in complex hyperbolic geometry*, Invent. Math. 152 (2003) 57–88.

[FaP2] E. Falbel, J. Parker, *The geometry of the Eisenstein-Picard modular group*, Duke Math. J. 131 (2006) 249–289.

[Far] B. Farb, *The quasi-isometry classification of lattices in semisimple Lie groups*, Math. Res. Lett. 4 (1997) 705–717.

[FarbM1] B. Farb, L. Mosher, *A rigidity theorem for the solvable Baumslag-Solitar groups*, Invent. Math. 131 (1998) 419–451.

[FarbM2] B. Farb, L. Mosher, *Quasi-isometric rigidity for the solvable Baumslag-Solitar groups. II*, Invent. Math. 137 (1999) 613–649.

[FarbS] B. Farb, R. Schwartz, *The large-scale geometry of Hilbert modular groups*, J. Differential Geom. 44 (1996) 435–478.

[FarbW1] B. Farb, S. Weinberger, *Hidden symmetries and arithmetic manifolds*, in *Geometry, spectral theory, groups, and dynamics*, pp. 111–119, Contemp. Math., 387, Amer. Math. Soc., 2005.

[FarbW2] B. Farb, S. Weinberger, *Isometries, rigidity and universal covers*, to appear in Annals of Math.

[Farr] F. Farrell, *Lectures on surgical methods in rigidity*, Published for the Tata Institute of Fundamental Research, Springer-Verlag, 1996. iv+98 pp.

[FarrJ1] F. Farrell, L. Jones, *Classical aspherical manifolds*, CBMS Regional Conference Series in Mathematics, 75, American Mathematical Society, 1990. viii+54 pp.

[FarrJ2] F. Farrell, L. Jones, *Rigidity in geometry and topology*, Proceedings of the International Congress of Mathematicians, pp. 653–663, Math. Soc. Japan, 1991.

[FarrJ3] F. Farrell, L. Jones, *Isomorphism conjectures in algebraic $K$-theory*, J. Amer. Math. Soc. 6 (1993) 249–297.

[FarrJO] F. Farrell, L. Jones, P. Ontaneda, *Hyperbolic manifolds with negatively curved exotic triangulations in dimensions greater than five*, J. Differential Geom. 48 (1998) 319–322.

[FarrL] F. Farrell, J. Lafont, *EZ-structures and topological applications*, Comment. Math. Helv. 80 (2005) 103–121.

[FatLP] A. Fathi, F. Laudenbach, V. Poenaru, *Travaux de Thurston sur les surfaces*, Séminaire Orsay, Astérisque, 66–67. Société Mathématique de France, 1979. 284 pp.

[FeN] W. Fenchel, J. Nielsen, *Discontinuous groups of isometries in the hyperbolic plane*, Edited and with a preface by Asmus L. Schmidt, Biography of the authors by Bent Fuglede, de Gruyter Studies in Mathematics, 29. Walter de Gruyter & Co., Berlin, 2003. xxii+364 pp.

[Fer1] R. Feres, *An introduction to cocycle super-rigidity*, in *Rigidity in dynamics and geometry*, pp. 99–134, Springer, Berlin, 2002.

[Fer2] R. Feres, *Connection preserving actions of lattices in* $\mathrm{SL}_n(\mathbb{R})$, Israel J. Math. 79 (1992) 1–21.

[Fer3] R. Feres, *Rigid geometric structures and actions of semisimple Lie groups*, in *Rigidité groupe fondamental et dynamique*, pp. 121–167, Panor. Synthéses, 13, Soc. Math. France, Paris, 2002.

[FerK] R. Feres, A. Katok, *Ergodic theory and dynamics of $G$-spaces (with special emphasis on rigidity phenomena)*, in *Handbook of dynamical systems, Vol. 1A*, pp. 665–763, North-Holland, Amsterdam, 2002.

[FerL] R. Feres, F. Labourie, *Topological superrigidity and Anosov actions of lattices*, Ann. Sci. École Norm. Sup. 31 (1998) 5, 599–629.

[Fern] T. Fernos, *Relative property (T) and linear groups*, Les Annales de l'Institut Fourier 56 (2006) 1767–1804.

[FeRR] S. Ferry, A. Ranicki, J. Rosenberg, *Novikov conjectures, index theorems and rigidity*, vol. 1 & 2, Cambridge University Press, 1995.

[FiR] B. Fine, G. Rosenberger, *Algebraic generalizations of discrete groups. A path to combinatorial group theory through one-relator products*, Monographs and Textbooks in Pure and Applied Mathematics, 223. Marcel Dekker, 1999. x+317 pp.

[Fis1] D. Fisher, *Local Rigidity: Past, Present, Future*, to appear in MSRI volume resulting from Katok birthday conference.

[Fis2] D. Fisher, $\mathrm{Out}(F_n)$ *and the spectral gap conjecture*, Int. Math. Res. Not. 2006, Art. ID 26028, 9 pp.

[FisM1] D. Fisher, G. Margulis, *Almost isometric actions, property $(T)$, and local rigidity*, Invent. Math. 162 (2005) 19–80.

[FisM2] D. Fisher, G. Margulis, *Local rigidity for cocycles*, in *Surveys in differential geometry*, Vol. VIII, pp. 191–234, Surv. Differ. Geom., VIII, Int. Press, 2003.

[FisW] D. Fisher, K. Whyte, *When is a group action determined by its orbit structure?* Geom. Funct. Anal. 13 (2003) 1189–1200.

[FisZ] D. Fisher, R. Zimmer, *Geometric lattice actions, entropy and fundamental groups*, Comment. Math. Helv. 77 (2002) 326–338.

[Fle] M. Flensted-Jensen, *Discrete series for semisimple symmetric spaces*, Ann. of Math. 111 (1980) 253–311.

[FoG] V. Fock, A. Goncharov, *Moduli spaces of local systems and higher Teichmller theory*, Publ. Math. Inst. Hautes Études Sci. No. 103 (2006) 1–211.

[Fra] J. Franke, *Harmonic analysis in weighted $L_2$-spaces*, Ann. Sci. École Norm. Sup. 31 (1998) 181–279.

[Fre] E. Freitag, *Hilbert modular forms*, Springer-Verlag, 1990.

[Fri1] D. Fried, *Torsion and closed geodesics on complex hyperbolic manifolds*, Invent. Math. 91 (1988) 31–51.

[Fri2] D. Fried, *The zeta functions of Ruelle and Selberg*, I. Ann. Sci. École Norm. Sup. (4) 19 (1986) 491–517.

[Fri3] D. Fried, *Reduction theory over quadratic imaginary fields*, J. Number Theory, 110 (2005) 44–74.

[FrW] E. Friedlander, C. Weibel, *An overview of algebraic K-theory*, in *Algebraic K-theory and its applications*, World Sci. Publishing, 1999.

[Fuj1] K. Fujiwara, *On non-bounded generation of discrete subgroups in rank-1 Lie group*, in *Geometry, spectral theory, groups, and dynamics*, pp.153–156, Contemp. Math., 387, Amer. Math. Soc., 2005.

[Fuj2] K. Fujiwara, *On the outer automorphism group of a hyperbolic group*, Israel J. Math. 131 (2002) 277–284.

[Fuk] K. Fukaya, *Margulis' lemma in Riemannian geometry*, Sugaku Expositions 6 (1993) 201–219.

[FukY] K. Fukaya, T. Yamaguchi, *The fundamental groups of almost non-negatively curved manifolds*, Ann. of Math. 136 (1992) 253–333.

[FuM] J. Funke, J. Millson, *Cycles with local coefficients for orthogonal groups and vector-valued Siegel modular forms*, Amer. J. Math. 128 (2006) 899–948.

[Fur1] A. Furman, *Mostow-Margulis rigidity with locally compact targets*, Geom. Funct. Anal. 11 (2001) 30–59.

[Fur2] A. Furman, *Gromov's measure equivalence and rigidity of higher rank lattices*, Ann. of Math. 150 (1999) 1059–1081.

[Fur3] A. Furman, *Orbit equivalence rigidity*, Ann. of Math. 150 (1999) 1083–1108.

[Fur4] A. Furman, *On Popa's Cocycle Superrigidity Theorem*, Inter. Math. Res. Notices, 2007 (2007) 1–46.

[Furr1] H. Furstenberg, *A Poisson formula for semi-simple Lie groups*, Ann. of Math. 77 (1963) 335–386.

[Furr2] H. Furstenberg, *Boundaries of Lie groups and discrete subgroups*, in *Actes du Congrhs International des Mathimaticiens (Nice, 1970)*, Tome 2, pp. 301–306, Gauthier-Villars, 1971.

[Furr3] H. Furstenberg, *Poisson boundaries and envelopes of discrete groups*, Bull. Amer. Math. Soc. 73 (1967) 350–356.

[Furr4] H. Furstenberg, *Boundary theory and stochastic processes on homogeneous spaces. Harmonic analysis on homogeneous spaces*, Proc. Sympos. Pure Math., Vol. XXVI, pp.193–229. Amer. Math. Soc., 1973.

[Furr5] H. Furstenberg, *The unique ergodicity of the horocycle flow*, in *Recent advances in topological dynamics*, pp. 95–115. Lecture Notes in Math., Vol. 318, Springer, 1973.

[Furu] M. Furushima, *The complete classification of compactifications of $C^3$ which are projective manifolds with the second Betti number one*, Math. Ann. 297 (1993) 627–662.

[Gab] D. Gabai, *Convergence groups are Fuchsian groups*, Ann. of Math. 136 (1992) 447–510.

[GabMM] D. Gabai, G. Meyerhoff, P. Milley, *Volumes of tubes in hyperbolic 3-manifolds*, J. Differential Geom. 57 (2001) 23–46.

[Gabo] D. Gaboriau, *Invariants $l^2$ de relations d'équivalence et de groupes*, Publ. Math. Inst. Hautes Études Sci. No. 95 (2002) 93–150.

[GabP] D. Gaboriau, S. Popa, *An uncountable family of nonorbit equivalent actions of $\mathbb{F}_n$*, J. Amer. Math. Soc. 18 (2005) 547–559.

[Gan1] R. Gangolli, *zeta functions of Selberg's type for compact space forms of symmetric spaces of rank one*, Illinois J. Math. 21 (1977) 1–41.

[Gan2] R. Gangolli, *Spectra of discrete uniform subgroups of semisimple Lie groups*, in *Symmetric spaces (Short Courses, Washington Univ., St. Louis, Mo., 1969–1970)*, pp. 93–117, Dekker, 1972.

[Gan3] R. Gangolli, *The length spectra of some compact manifolds of negative curvature*, J. Differential Geom. 12 (1977) 403–424.

[GanW] R. Gangolli, G. Warner, *zeta functions of Selberg's type for some noncompact quotients of symmetric spaces of rank one*, Nagoya Math. J. 78 (1980) 1–44.

[GaR] H. Garland, M.S. Raghunathan, *Fundamental domains for lattices in $(R)$-Rank 1 semisimple Lie groups*, Ann. of Math. 92 (1970) 279–326.

[Gar1] H. Garland, *Absolute convergence of Eisenstein series on loop groups*, Duke Math. J. 135 (2006) 203–260.

[Gar2] H. Garland, *Certain Eisenstein series on loop groups: convergence and the constant term*, in *Algebraic groups and arithmetic*, pp. 275–319, Tata Inst. Fund. Res., Mumbai, 2004.

[Gar3] H. Garland, *The arithmetic theory of loop groups*, Inst. Hautes Études Sci. Publ. Math. 52 (1980) 5–136.

[Gat1] P. Garrett, *Holomorphic Hilbert modular forms*, Wadsworth & Brooks/Cole, 1990.

[Gat2] P. Garrett, *Buildings and classical groups*, Chapman & Hall,1997. xii+373 pp.

[GehM1] F. Gehring, G. Martin, *The volume of hyperbolic 3-folds with $p$-torsion, $p \geq 6$*, Quart. J. Math. Oxford Ser. (2) 50 (1999) 1–12.

[GehM2] F. Gehring, G. Martin, *Discrete convergence groups*, in *Complex analysis, I*, pp. 158–167, Lecture Notes in Math., 1275, Springer, 1987.

[Gel] T. Gelander, *Homotopy type and volume of locally symmetric manifolds*, Duke Math. J. 124 (2004) 459–515.

[GelaZ] T. Gelander, A. Zuk, *Dependence of Kazhdan constants on generating subsets*, Israel J. Math. 129 (2002) 93–98.

[Gelb1] S. Gelbart, *Automorphic forms on adèle groups*, Annals of Mathematics Studies, No. 83, Princeton University Press, 1975. x+267 pp.

[Gelb2] S. Gelbart, *Lectures on the Arthur-Selberg trace formula*, University Lecture Series, 9, American Mathematical Society, 1996. x+99 pp.

[Gelb3] S. Gelbart, *Elliptic curves and automorphic representations*, Advances in Math. 21 (1976) 235–292.

[Gelb4] S. Gelbart, *An elementary introduction to the Langlands program*, Bull. Amer. Math. Soc. 10 (1984) 177–219.

[GelbS] S. Gelbart, F. Shahidi, *Analytic properties of automorphic $L$-functions*, Perspectives in Mathematics, 6. Academic Press, 1988. viii+131 pp.

[GelGP] I.M. Gelfand, M.I. Graev, I. Pyatetskii-Shapiro, *Representation theory and automorphic functions*, Translated from the Russian by K. A. Hirsch W. B. Saunders Co., 1969 xvi+426 pp.

[GenL] S. Geninska, E. Leuzinger, *A geometric characterizaton of arthmetic Fuchsian groups*, preprint, Karlsruhe, 2006.

[Ger1] S. Gersten, *Quasi-isometry invariance of cohomological dimension*, C. R. Acad. Sci. Paris Sér. I Math. 316 (1993) 411–416.

[Ger2] S. Gersten, *Introduction to hyperbolic and automatic groups*, in *Summer School in Group Theory in Banff (1996)*, pp. 45–70, CRM Proc. Lecture Notes, 17, Amer. Math. Soc., 1999.

[Gh] É. Ghys, *Déformations des structures complexes sur les espaces homogènes de* $\mathrm{SL}(2,\mathbb{C})$, J. Riene Angew. Math 486 (1995) 113–138.

[GhH] É. Ghys and P. de la Harpe, *Sur les groupes hyperboliques d'après Mikhael Gromov*, in *Progress in Mathematics*, 83. Birkhäuser, 1990. xii+285 pp.

[GilG] H. Gillet, D. Grayson, *Volumes of symmetric spaces via lattice points*, Doc. Math. 11 (2006) 425–447.

[Gin] S. Gindikin, *Lie groups and symmetric spaces. In memory of F. I. Karpelevich*, Amer. Math. Soc. Translations, Series 2, 210. Advances in the Mathematical Sciences, 54. Amer. Math. Soc., 2003. xii+355 pp.

[Goe] E. Goetze, *Connection preserving actions of connected and discrete Lie groups*, J. Differential Geom. 40 (1994) 595–620.

[GoeS1] E. Goetze, R. Spatzier, *On Livšic's theorem, superrigidity, and Anosov actions of semisimple Lie groups*, Duke Math. J. 88 (1997) 1–27.

[GoeS2] E. Goetze, R. Spatzier, *Smooth classification of Cartan actions of higher rank semisimple Lie groups and their lattices*, Ann. of Math. 150 (1999) 743–773.

[Gold] B. Goldfarb, *Novikov conjectures for arithmetic groups with large actions at infinity*, K-Theory 11 (1997) 319–372.

[Gold] D. Goldfeld, *Automorphic forms and L-functions for the group* $\mathrm{GL}(n,\mathbb{R})$, Cambridge Studies in Advanced Mathematics, 99. Cambridge University Press, 2006. xiv+493 pp.

[GoldL] D. Goldfeld, X. Li, *Voronoi formulas on* $\mathrm{GL}(n)$, Int. Math. Res. Not. 2006, Art. ID 86295, 25 pp.

[GoldLP] D. Goldfeld, A. Lubotzky, L. Pyber, *Counting congruence subgroups*, Acta Math. 193 (2004) 73–104.

[Golm] J. Goldman, *The queen of mathematics. A historically motivated guide to number theory*, A K Peters, 1998. xxiv+525 pp.

[Goldw1] W. Goldman, *Complex hyperbolic Kleinian groups*, in *Complex geometry*, pp. 31–52, Lecture Notes in Pure and Appl. Math., 143, Dekker, 1993.

[Goldw2] W. Goldman, *Complex hyperbolic geometry*, The Clarendon Press, Oxford University Press, 1999. xx+316 pp.

[Goldw3] W. Goldman, *Nonstandard Lorentz space forms*, J. Differential Geom. 21 (1985) 301–308.

[GoldK] W. Goldman, Y. Kamishima, *The fundamental group of a compact flat Lorentz space form is virtually polycyclic*, J. Differential Geom. 19 (1984) 233–240.

[GoldM] W. Goldman, J. Millson, *Local rigidity of discrete groups acting on complex hyperbolic space*, Invent. Math. 88 (1987) 495–520.

[Goo] A. Good, *Local analysis of Selberg's trace formula*, Lecture Notes in Mathematics, 1040. Springer-Verlag, 1983. i+128 pp.

[Gord] C. Gordon, *Survey of isospectral manifolds*, in *Handbook of differential geometry*, Vol. I, pp. 747–778, North-Holland, 2000.

[GordR] C. Gordon, J. Rossetti, *Boundary volume and length spectra of Riemannian manifolds: what the middle degree Hodge spectrum doesn't reveal*, Ann. Inst. Fourier (Grenoble) 53 (2003) 2297–2314.

[Gore1] M. Goresky, *Compactifications and cohomology of modular varieties*, in *Harmonic analysis, the trace formula, and Shimura varieties*, pp. 551–582, Clay Math. Proc., 4, Amer. Math. Soc., 2005.

[Gore2] M. Goresky, *$L^2$-cohomology is intersection cohomology*, in *The zeta functions of Picard modular surfaces*, pp. 47–63, Univ. Montréal, 1992.

[GorHM] M. Goresky, G. Harder, R. MacPherson, *Weighted cohomology*, Invent. Math. 116 (1994) 139–213.

[GorHMN] M. Goresky, G. Harder, R. MacPherson, A. Nair, *Local intersection cohomology of Baily-Borel compactifications*, Compositio Math. 134 (2002) 243–268.

[GorM1] M. Goresky, R. MacPherson, *Lefschetz numbers of Hecke correspondences*, in *The zeta functions of Picard modular surfaces*, ed. by R. Langlands and D. Ramakrishnan, Univ. Montrial, 1992, pp. 465–478.

[GorM2] M. Goresky, R. MacPherson, *Intersection homology. II*, Invent. Math. 72 (1983) 77–129.

[GorM3] M. Goresky, R. MacPherson, *The topological trace formula*, J. Reine Angew. Math. 560 (2003) 77–150.

[GorM4] M. Goresky, R. MacPherson, *Local contribution to the Lefschetz fixed point formula*, Invent. Math. 111 (1993) 1–33.

[GorT] M. Goresky, Y. Tai, *Toroidal and reductive Borel-Serre compactifications of locally symmetric spaces*, Amer. J. Math. 121 (1999) 1095–1151.

[Gra1] D. Grayson, *Reduction theory using semistability. II*, Comment. Math. Helv. 61 (1986) 661–676.

[Gra2] D. Grayson, *Reduction theory using semistability*, Comment. Math. Helv. 59 (1984) 600–634.

[Gra3] D. Grayson, *Finite generation of K-groups of a curve over a finite field (after Daniel Quillen)*, in *Algebraic K-theory, Part I*, pp. 69–90, Lecture Notes in Math., 966, Springer, 1982.

[GreG1] M. Green, P. Griffiths, *On the tangent space to the space of algebraic cycles on a smooth algebraic variety*, Annals of Mathematics Studies, 157. Princeton University Press, 2005. vi+200 pp.

[GreG2] M. Green, P. Griffiths, *Algebraic cycles and singularities of normal functions II*, in *Inspired by S. S. Chern*, World Scientific, 2006, pp. 179–268.

[Gree] L. Greenberg, *Maximal groups and signatures*, in *Discontinuous groups and Riemann surfaces* pp. 207–226. Ann. of Math. Studies, No. 79, Princeton Univ. Press, 1974.

[Gren1] D. Grenier, *On the shape of fundamental domains in* $\mathrm{GL}(n, \mathbf{R})/\mathrm{O}(n)$, Pacific J. Math. 160 (1993) 53–66.

[Gren2] D. Grenier, *Fundamental domains for the general linear group*, Pacific J. Math. 132 (1988) 293–317.

[Gri] P. Griffiths, *Periods of integrals on algebraic manifolds: Summary of main results and discussion of open problems*, Bull. Amer. Math. Soc. 76 (1970) 228–296.

[GriH] P. Griffiths, J. Harris, *Principles of algebraic geometry*, Reprint of the 1978 original. Wiley Classics Library. John Wiley & Sons, 1994. xiv+813 pp.

[GriS] P. Griffiths, W. Schmid, *Recent developments in Hodge theory: a discussion of techniques and results*, in *Discrete subgroups of Lie groups and applications to moduli*, Oxford Univ. Press, 1975, pp. 31–127.

[Gro1] M. Gromov, *Asymptotic invariants of infinite groups*, in *Geometric group theory*, ed. by A. Niblo and M. Roller, Cambridge University Press, 1993, pp. 1–295.

[Gro2] M. Gromov, *Random walk in random groups*, Geom. Funct. Anal. 13 (2003) 73–146.

[Gro3] M. Gromov, *Hyperbolic groups, Essays in group theory*, pp. 75–263, Math. Sci. Res. Inst. Publ., 8, Springer, 1987.

[Gro4] M. Gromov, *Infinite groups as geometric objects*, in *Proceedings of the International Congress of Mathematicians*, pp. 385–392, Warsaw, 1984.

[Gro5] M. Gromov, *Groups of polynomial growth and expanding maps*, Inst. Hautes Études Sci. Publ. Math. No. 53 (1981) 53–73.

[Gro6] M. Gromov, *Hyperbolic manifolds, groups and actions*, in *Riemann surfaces and related topics*, pp. 183–213, Ann. of Math. Stud., 97, Princeton Univ. Press, 1981.

[Gro7] M. Gromov, *Hyperbolic manifolds (according to Thurston and Jorgensen)*, Bourbaki Seminar, Vol. 1979/80, pp. 40–53, Lecture Notes in Math., 842, Springer, 1981.

[Gro8] M. Gromov, *Volume and bounded cohomology*, Inst. Hautes Études Sci. Publ. Math. (1982) 5–99.

[Gro9] M. Gromov, *Rigid transformations groups*, in *Géométrie différentielle (Paris, 1986)*, pp. 65–139, Travaux en Cours, 33, Hermann, Paris, 1988.

[GrP] M. Gromov, P. Pansu, *Rigidity of lattices: an introduction*, in *Geometric topology: recent developments*, pp. 39–137, Lecture Notes in Math., 1504, Springer, 1991.

[GrPS] M. Gromov, Piatetski-Shapiro, *Nonarithmetic groups in Lobachevsky spaces*, Inst. Hautes Études Sci. Publ. Math. No. 66 (1988) 93–103.

[GrS] M. Gromov, R. Schoen, *Harmonic maps into singular spaces and p-adic superrigidity for lattices in groups of rank one*, Inst. Hautes Itudes Sci. Publ. Math. 76 (1992) 165–246.

[GrSh] M. Gromov, M. Shubin, *von Neumann spectra near zero*, Geom. Funct. Anal. 1 (1991) 375–404.

[GrosN] B. Gross, G. Nebe, *Globally maximal arithmetic groups*, J. Algebra 272 (2004) 625–642.

[Gru] P. Gruber, *Geometry of numbers*, in *Handbook of convex geometry*, Vol. A, B, pp. 739–763, North-Holland, 1993.

[GruL] P. Gruber, C. Lekkerkerker, *Geometry of numbers*, Second edition, North-Holland Publishing Co., 1987. xvi+732 pp.

[GrunH] F. Grunewald, S. Kühnlein, *On the proof of Humbert's volume formula*, Manuscripta Math. 95 (1998) 431–436.

[GruP1] F. Grunewald, V. Platonov, *Rigidity and automorphism groups of solvable arithmetic groups*, C. R. Acad. Sci. Paris Sér. I Math. 327 (1998) 427–432.

[GruP2] F. Grunewald, V. Platonov, *Rigidity results for groups with radical cohomology of finite groups and arithmeticity problems*, Duke Math. J. 100 (1999) 321–358.

[GruP3] F. Grunewald, V. Platonov, *Solvable arithmetic groups and arithmeticity problems*, Internat. J. Math. 10 (1999) 327–366.

[GruPiS] F. Grunewald, P. Pickel, D. Segal, *Polycyclic groups with isomorphic finite quotients*, Ann. of Math. (2) 111 (1980) 155–195.

[GruS1] F. Grunewald, D. Segal, *Some general algorithms. I. Arithmetic groups*, Ann. of Math. 112 (1980) 531–583.

[GruS2] F. Grunewald, D. Segal, *Some general algorithms. II. Nilpotent groups*, Ann. of Math. 112 (1980) 585–617.

[GruS3] F. Grunewald, D. Segal, *Decision problems concerning S-arithmetic groups*, J. Symbolic Logic 50 (1985) 743–772.

[GruS4] F. Grunewald, D. Segal, *A note on arithmetic groups*, Bull. London Math. Soc. 10 (1978) 297–302.

[GueHW] E. Guentner, N. Higson, S. Weinberger, *The Novokov conjecture for linear groups*, Publ. Math. Inst. Hautes Études Sci. 101 (2005) 243–268.

[GuiZ1] L. Guillopé, M. Zworski, *The wave trace for Riemann surfaces*, Geom. Funct. Anal. 9 (1999) 1156–1168.

[GuiZ2] L. Guillopé, M. Zworski, *Scattering asymptotics for Riemann surfaces*, Ann. of Math. (2) 145 (1997) 597–660.

[Guiv] Y. Guivarc'h, *Produits de matrices aléatoires et applications aux propriétés géométriques des sous-groupes du groupe linéaire*, Ergodic Theory Dynam. Systems 10 (1990) 483–512.

[GuiJT] Y. Guivarc'h, L. Ji, J.C. Taylor, *Compactifications of symmetric spaces*, in *Progress in Math.*, vol. 156, Birkhäuser, Boston, 1998.

[GuiS] Y. Guivarc'h, A. Starkov, *Orbits of linear group actions, random walks on homogeneous spaces and toral automorphisms*, Ergodic Theory Dynam. Systems 24 (2004) 767–802.

[Gun1] P. Gunnells, *Modular symbols and Hecke operators*, in *Algorithmic number theory*, pp. 347–357, Lecture Notes in Comput. Sci., 1838, Springer, 2000.

[Gun2] P. Gunnells, *Finiteness of minimal modular symbols for* $\mathrm{SL}_n$, J. Number Theory 82 (2000) 134–139.

[Gun3] P. Gunnells, *Symplectic modular symbols*, Duke Math. J. 102 (2000) 329–350.

[GunMM] R. Gupta, M. Murty, K. Murty, *The Euclidean algorithm for S-integers*, in *Number theory*, CMS Conf. Proc., 7, Amer. Math. Soc., 1987.

[GuP1] N. Gusevskii, J. Parker, *Representations of free Fuchsian groups in complex hyperbolic space*, Topology 39 (2000) 33–60.

[GuP2] N. Gusevskii, J. Parker, *Complex hyperbolic quasi-Fuchsian groups and Toledo's invariant*, Geom. Dedicata 97 (2003) 151–185.

[HaaM] U. Haagerup, H. Munkholm, *Simplices of maximal volume in hyperbolic n-space*, Acta Math. 147 (1981) 1–11.

[Had] J. Hadamard, *Non-Euclidean geometry in the theory of automorphic functions*, with historical introduction by J. Gray, History of Mathematics, 17. American Mathematical Society, and London Mathematical Society, 1999, xii+95 pp.

[Hai] R. Hain, *Periods of limit mixed Hodge structures*, in *Current developments in mathematics, 2002*, pp. 113–133, Int. Press, 2003.

[HaiL] R. Hain, E. Looijenga, *Mapping class groups and moduli spaces of curves*, in *Algebraic geometry—Santa Cruz 1995*, pp. 97–142, Proc. Sympos. Pure Math., 62, Part 2, Amer. Math. Soc., 1997.

[Ham1] U. Hamenstaedt, *Geometry of the mapping class groups III: Quasi-isometric rigidity*, math.GT/0512429.

[Ham2] U. Hamenstaedt, *A geometric characterization of negatively curved locally symmetric spaces*, J. Differential Geom. 34 (1991) 193–221.

[Ham3] U. Hamenstaedt, *Ergodic properties of function groups*, Geom. Dedicata 93 (2002) 163–176.

[Har1] G. Harder, *Eisenstein cohomology of arithmetic groups and its applications to number theory*, in *Proceedings of the International Congress of Mathematicians, Vol. I, II*, pp. 779–790, Math. Soc. Japan, Tokyo, 1991.

[Har2] G. Harder, *On the cohomology of discrete arithmetically defined groups*, in *Discrete subgroups of Lie groups and applications to moduli*, pp. 129–160, Oxford Univ. Press, 1975.

[Har3] G. Harder, *A Gauss-Bonnet formula for discrete arithmetically defined groups*, Ann. Sci. Icole Norm. Sup. 4 (1971) 409–455.

[Har4] G. Harder, *Chevalley groups over function fields and automorphic forms*, Ann. of Math. 100 (1974) 249–306.

[Har5] G. Harder, *Eisenstein cohomology of arithmetic groups. The case* $\mathrm{GL}_2$, Invent. Math. 89 (1987) 37–118.

[Har6] G. Harder, *Die Kohomologie S-arithmetischer Gruppen über Funktionenkörpern*, Invent. Math. 42 (1977) 135–175.

[HarLW] G. Harder, W. Li, J. Weisinger, *Dimensions of spaces of cusp forms over function fields*, J. Reine Angew. Math. 34 (1980) 73–103.

[Hare] J. Harer, *The cohomology of the moduli space of curves*, in *Theory of moduli*, pp. 138–221, Lecture Notes in Math., 1337, Springer, 1988.

[HariC] Harish-Chandra, *Automorphic forms on semisimple Lie groups*, Lect. Notes in Math. vol. 62, Springer-Verlag, 1968.

[Harp] P. de la Harpe, *Topics in geometric group theory*, Chicago University Press, 2000.

[HarpV] P. de la Harpe, A. Valette, *La propriété (T) de Kazhdan pour les groupes localement compacts*, Astérique No. 175, (1989), 158 pp.

[HarrK] M. Harris, S. Kudla, *The central critical value of a triple product L-function*, Ann. of Math. 133 (1991) 605–672.

[HarrT] M. Harris, R. Taylor, *The geometry and cohomology of some simple Shimura varieties*, Annals of Mathematics Studies, 151. Princeton University Press, 2001. viii+276 pp.

[HarrZ1] M. Harris, S. Zucker, *Boundary cohomology of Shimura varieties. II. Hodge theory at the boundary* Invent. Math. 116 (1994) 243–308.

[HarrZ2] M. Harris, S. Zucker, *Boundary cohomology of Shimura varieties. I. Coherent cohomology on toroidal compactifications*, Ann. Sci. Icole Norm. Sup. 27 (1994) 249–344.

[Harv] W. Harvey, *Boundary structure of the modular group*, in *Riemann surfaces and related topics*, pp. 245–251, Ann. of Math. Stud., 97, Princeton Univ. Press, 1981.

[Has1] K. Hashimoto, *Representations of the finite symplectic group* $\mathrm{Sp}(4, F_p)$ *in the spaces of Siegel modular forms*, in *The Selberg trace formula and related topics*, pp. 253–276, Contemp. Math., 53, Amer. Math. Soc., 1986.

[Has2] K. Hashimoto, *Explicit form of quaternion modular embeddings*, Osaka J. Math. 32 (1995) 533–546.

[Hatc] A. Hatcher, *Concordance spaces, higher simple-homotopy theory, and applications*, in *Algebraic and geometric topology*, Part 1, pp. 3–21, Proc. Sympos. Pure Math., XXXII, Amer. Math. Soc., 1978.

[Hatt1] T. Hattori, *Geometric limit sets of higher rank lattices*, Proc. London Math. Soc. 90 (2005) 689–710.

[Hatt2] T. Hattori, *Asymptotic geometry of arithmetic quotients of symmetric spaces*, Math. Z. 222 (1996) 247–277.

[Hatt3] T. Hattori, *Collapsing of quotient spaces of* $\mathrm{SO}(n)\backslash\mathrm{SL}(n, R)$ *at infinity*, J. Math. Soc. Japan 47 (1995) 193–225.

[Haw] T. Hawkins, *Emergence of the theory of Lie groups. An essay in the history of mathematics 1869–1926*, in *Sources and Studies in the History of Mathematics and Physical Sciences*, Springer-Verlag, New York, 2000. xiv+564 pp.

[Hej1] D. Hejhal, *The Selberg trace formula for* $\mathrm{PSL}(2, R)$. *Vol. I*, Lecture Notes in Mathematics, Vol. 548, Springer-Verlag, 1976. vi+516 pp.

[Hej2] D. Hejhal, *The Selberg trace formula for* $\mathrm{PSL}(2,\ R)$. *Vol. 2*, Lecture Notes in Mathematics, Vol. 1001, Springer-Verlag, 1983.

[Hej3] D. Hejhal, *Eigenvalues of the Laplacian for Hecke triangle groups*, Mem. Amer. Math. Soc. 97 (1992), no. 469, vi+165 pp.

[HejST] D. Hejhal, P. Sarnak, A. Terras, *The Selberg trace formula and related topics*, Contemporary Mathematics, vol. 53, American Mathematical Society, 1986. x+554 pp.

[Hel1] S. Helgason, *Differential geometry, Lie groups, and symmetric spaces*, Pure and Applied Math. vol. 80, Academic Press, 1978.

[Hel2] S. Helgason, *Groups and geometric analysis. Integral geometry, invariant differential operators, and spherical functions*, Pure and Applied Mathematics, 113. Academic Press, Inc., 1984. xix+654 pp.

[HePa] S. Hersonsky, F. Paulin, *Counting orbit points in coverings of negatively curved manifolds and Hausdorff dimension of cusp excursions*, Ergodic Theory Dynam. Systems 24 (2004) 803–824.

[Hig] N. Higson, *The Baum-Connes conjecture*, Proc. of ICM, Vol. II (Berlin, 1998), Doc. Math. 1998, Extra Vol. II, 637–646.

[HigR] N. Higson, J. Roe, *Amenable group actions and the Novikov conjecture*, J. Reine Angew. Math. 519 (2000) 143–153.

[HilO] J. Hilgert, G. Olafsson, *Causal symmetric spaces. Geometry and harmonic analysis*, Perspectives in Mathematics, 18. Academic Press, 1997. xvi+286 pp.

[Hirz1] F. Hirzebruch, *The Hilbert modular group, resolution of the singularities at the cusps and related problems*, Siminaire Bourbaki, 23hme annie (1970/1971), Exp. No. 396, pp. 275–288. Lecture Notes in Math., Vol. 244, Springer, 1971.

[Hirz2] F. Hirzebruch, *Hilbert modular surfaces*, Enseignement Math. 19 (1973) 183–281.

[Hirz3] F. Hirzebruch, *Chern numbers of algebraic surfaces: an example*, Math. Ann. 266 (1984) 351–356.

[Hirz4] F. Hirzebruch, *The Hilbert modular group for the field $\mathbb{Q}(\sqrt{5})$, and the cubic diagonal surface of Clebsch and Klein*, Russian Math. Surveys 31 (1976) 96–110.

[Hoc] G. Hochschild, *The structure of Lie groups*, Holden-Day, Inc., San Francisco-London-Amsterdam 1965, ix+230 pp.

[HodK] C. Hodgson, S. Kerckhoff, *Rigidity of hyperbolic cone-manifolds and hyperbolic Dehn surgery*, J. Differential Geom. 48 (1998) 1–59.

[Hof] W. Hoffmann, *Cuspidal functions and Diophantine approximation*, Math. Ann. 291 (1991) 357–372.

[Hol1] R. Holzapfel, *Geometry and arithmetic around Euler partial differential equations*, Reidel Publishing Company, 1986.

[Hol2] R. Holzapfel, *The ball and some Hilbert problems*, Birkhäuser, 1995.

[HoLW] S. Hoory, N. Linial, A. Wigderson, *Expander graphs and their applications*, Bull. Amer. Math. Soc. (N.S.) 43 (2006) 439–561.

[Hop] M. Hopkins, *Algebraic topology and modular forms*, Proc. of the International Congress of Mathematicians, Vol. I (Beijing, 2002), pp.291–317, Higher Ed. Press, 2002.

[Hor] I. Horozov, *Euler characteristics of arithmetic groups*, Math. Res. Lett. 12 (2005) 275–291.

[How] R. Howe, *Discrete groups in geometry and analysis, Papers from the conference in honor of G. D. Mostow on his sixtieth birthday held at Yale University*, 1984. Ed. by Roger Howe. Progress in Mathematics, 67. Birkhäuser Boston, 1987. xii+210 pp.

[Hub] A. Huber, *On subharmonic functions and differential geometry in the large*, Comment. Math. Helv. 32 (1957) 13–72.

[HuKW] K. Hulek, C. Kahn, S. Weintraub, *Moduli spaces of abelian surfaces: compactification, degenerations, and theta functions*, Walter de Gruyter & Co. 1993. xii+347 pp.

[Hum1] J. Humphreys, *Reflection groups and Coxeter groups*, Cambridge University Press, 1990.

[Hum2] J. Humphreys, *Arithmetic groups*, Lecture Notes in Mathematics, 789, Springer, 1980, vii+158 pp.

[Hum3] J. Humphreys, *Linear algebraic groups*, in *Graduate Texts in Mathematics*, No. 21. Springer-Verlag, 1975. xiv+247 pp.

[Hun] B. Hunt, *The geometry of some special arithmetic quotients*, Lect. Notes in Math., vol. 1637, Springer-Verlag, 1996.

[Hw1] J. Hwang, *On the number of complex hyperbolic manifolds of bounded volume*, Internat. J. Math. 16 (2005) 863–873.

[Hw2] J. Hwang, *On the volumes of complex hyperbolic manifolds with cusps*, Internat. J. Math. 15 (2004) 567–572.

[HwM1] J. Hwang, N. Mok, *Rigidity of irreducible Hermitian symmetric spaces of the compact type under Kähler deformation*, Invent. Math. 131 (1998) 393–418.

[HwM2] J. Hwang, N. Mok, *Characterization and deformation-rigidity of compact irreducible Hermitian symmetric spaces of rank $\geq 2$ among Fano manifolds*, in *Algebra and geometry*, pp.15–46, Int. Press, 1998.

[Iha] Y. Ihara, *On discrete subgroups of the two by two projective linear group over* $\mathfrak{p}$*-adic fields*, J. Math. Soc. Japan 18 (1966) 219–235.

[IoW] A. Iozzi, D. Witte, *Tessellations of homogeneous spaces of classical groups of real rank two*, Geom. Dedicata, 103 (2004) 115–191.

[IsK] M. Ishida, F. Kato, *The strong rigidity theorem for non-Archimedean uniformization*, Tohoku Math. J. 50 (1998) 537–555.

[Iv1] N. Ivanov, *Action of Möbius transformations on homeomorphisms: stability and rigidity*, Geom. Funct. Anal. 6 (1996) 79–119.

[Iv2] N. Ivanov, *Mapping class groups*, in *Handbook of geometric topology*, pp. 523–633, North-Holland, 2002.

[Iv3] N. Ivanov, *Attaching corners to Teichmller space*, Leningrad Math. J. 1 (1990) 1177–1205.

[Iv4] N. Ivanov, *Complexes of curves and Teichmller spaces*, Math. Notes 49 (1991) 479–484.

[Iv5] N. Ivanov, *Algebraic properties of the Teichmller modular group*, Dokl. Akad. Nauk SSSR 275 (1984) 786–789.

[Iv6] N. Ivanov, *Isometries of Teichmller spaces from the point of view of Mostow rigidity*, in *Topology, ergodic theory, real algebraic geometry*, pp. 131–149, Amer. Math. Soc. Transl. Ser. 2, 202, Amer. Math. Soc., 2001.

[Iv7] N. Ivanov, *Automorphism of complexes of curves and of Teichmller spaces*, Internat. Math. Res. Notices 14 (1997) 651–666.

[IvJ] N. Ivanov, L. Ji, *Infinite topology of curve complex and non-Poincaré duality of Teichmüller modular groups*, preprint.

[IvM] N. Ivanov, J. McCarthy, *On injective homomorphisms between Teichmller modular groups. I*, Invent. Math. 135 (1999) 425–486.

[Ivr] V. Ivrii, *Microlocal analysis and precise spectral asymptotics*, Springer-Verlag, 1998. xvi+731 pp.

[Iw1] H. Iwaniec, *Spectral methods of automorphic forms*, Second edition, Graduate Studies in Mathematics, 53. American Mathematical Society, 2002.

[Iw2] H. Iwaniec, *Topics in classical automorphic forms*, Graduate Studies in Mathematics, 17, American Mathematical Society, 1997. xii+259 pp.

[IwS] H. Iwaniec, P. Sarnak, $L^\infty$ *norms of eigenfunctions of arithmetic surfaces*, Ann. of Math. 141 (1995) 301–320.

[Iz] H. Izeki, *Limit sets of Kleinian groups and conformally flat Riemannian manifolds*, Invent. Math. 122 (1995) 603–625.

[IzN] H. Izeki, S. Nayatani, *Combinatorial harmonic maps and discrete-group actions on Hadamard spaces*, Geom. Dedicata 114 (2005) 147–188.

[Jac] H. Jacquet, *A guide to the relative trace formula*, in *Automorphic representations, L-functions and applications: progress and prospects*, pp. 257–272, Ohio State Univ. Math. Res. Inst. Publ., 11, de Gruyter, 2005.

[JacL] H. Jacquet, R. Langlands, *Automorphic forms on* GL(2), Lecture Notes in Mathematics, Vol. 114. Springer-Verlag, 1970. vii+548 pp.

[Jak1] D. Jakobson, *Quantum limits on flat tori*, Ann. of Math. 145 (1997) 235–266.

[Jak2] D. Jakobson, *Quantum unique ergodicity for Eisenstein series on* $\mathrm{PSL}_2(Z)\backslash\mathrm{PSL}_2(R)$, Ann. Inst. Fourier 44 (1994) 1477–1504.

[Jan] T. Januszkiewicz, *Hyperbolizations*, in *Group theory from a geometrical viewpoint*, pp. 464–490, World Sci. Publ., 1991.

[JaS1] T. Januszkiewicz, J. Świątkowski, *Hyperbolic Coxeter groups of large dimension*, Comment. Math. Helv. 78 (2003) 555–583.

[JaS2] T. Januszkiewicz, J. Świątkowski, *Simplicial nonpositive curvature*, Publ. Math. Inst. Hautes Études Sci. No. 104 (2006) 1–85.

[JaS3] T. Januszkiewicz, J. Świątkowski, *Hyperbolic Coxeter groups of large dimension*, Comment. Math. Helv. 78 (2003) 555–583.

[Ji1] L. Ji, *Asymptotic dimension and the integral K-theoretic Novikov conjecture for arithmetic groups*, J. Differential Geom. 68 (2004) 535–544.

[Ji2] L. Ji, *The trace class conjecture for arithmetic groups*, J. Diff. Geom. 48 (1998) 165–203.

[Ji3] L. Ji, *The greatest common quotient of Borel–Serre and the toroidal compactifications*, Geometric and Functional Analysis 8 (1998) 978–1015.

[Ji4] L. Ji, *Metric compactifications of locally symmetric spaces*, International J. of Math. 9 (1998) 465–491.

[Ji5] L. Ji, *Buildings and their applications in geometry and topology*, Asian Journal of Math. 10 (2006) 11–80.

[Ji6] L. Ji, *Lectures on locally symmetric spaces and arithmetic groups*, in *Lie groups and automorphic forms*, pp. 87–146, AMS/IP Stud. Adv. Math., 37, Amer. Math. Soc., 2006.

[Ji7] L. Ji, *The integral Novikov conjectures for linear groups containing torsion elements*, to appear in the Journal of Topology, 2008.

[Ji8] L. Ji, *Introduction to symmetric spaces and their compactifications. Lie theory*, pp. 1–67, Progr. Math., 229, Birkhäuser 2005.

[Ji9] L. Ji, *A summary of the work of Gregory Margulis*, Pure and Applied Mathematics Quarterly (Special Issue in honor of Gregory Margulis, Part 2 of 2) 4 (2008) 169.

[Ji10] L. Ji, *Arithmetic groups, mapping class groups, related groups, and their associated spaces*, preprint, 2007.

[Ji11] L. Ji, *From symmetric spaces to buildings, curve complexes and outer spaces*, preprint 2007.

[JiLW] L. Ji, P. Li, J. Wang, *Ends of locally symmetric spaces with maximal bottom of spectrum*, preprint, 2007.

[JiM] L. Ji, R. MacPherson, *Geometry of compactifications of locally symmetric spaces*, Ann. Inst. Fourier (Grenoble) 52 (2002) 457–559.

[JiMSS] L. Ji, K. Murty, L. Saper, J. Scherk, *The Congruence subgroup kernel and the fundamental group of Satake compactifications*, preprint 2007.

[JiZ1] L. Ji, M. Zworski, *Scattering matrices and scattering geodesics*, Ann. Scient. de Éc. Norm. Sup. 34 (2001), 441–469.

[JiZ2] L. Ji, M. Zworski, *Correction and supplements to "Scattering matrices and scattering geodesics"*, Ann. Scient. Éc. Norm. Sup. 35 (2002) 897–901.

[JiZ3] L. Ji, M. Zworski, *The remainder estimate in spectral accumulation for degenerating hyperbolic surfaces*, J. Funct. Anal. 114 (1993) 412–420.

[JiaPS] D. Jiang, I. Piatetski-Shapiro, *Arithmeticity of discrete subgroups and automorphic forms*, Geom. Funct. Anal. 8 (1998) 586–605.

[Jon] V. Jones, *Ten problems*, in *Mathematics: frontiers and perspectives*, pp. 79–91, Amer. Math. Soc., 2000.

[Jos] J. Jost, *Nonpositive curvature: geometric and analytic aspects*, Birkhäuser Verlag, 1997. viii+108 pp.

[JosL] J. Jost, J. Li, *Finite energy and totally geodesic maps from locally symmetric spaces of finite volume*, Calc. Var. Partial Differential Equations 4 (1996) 409–420.

[JoY1] J. Jost, S.T. Yau, *Harmonic maps and superrigidity*, in *Differential Geometry*, Proc. Symp in Pure Math., vol. 54, Part I, 1993, pp. 245–280.

[JoY2] J. Jost, S.T. Yau, *Applications of quasilinear PDE to algebraic geometry and arithmetic lattices*, in *Algebraic geometry and related topics*, pp. 169–193, Internat. Press, 1993.

[JoY3] J. Jost, S.T. Yau, *Harmonic mappings and Kähler manifolds*, Math. Ann. 262 (1983) 145–166.

[JoY4] J. Jost, S.T. Yau, *Harmonic maps and rigidity theorems for spaces of nonpositive curvature*, Comm. Anal. Geom. 7 (1999) 681–694.

[Ju] A. Juhl, *Cohomological theory of dynamical zeta functions*, Progress in Mathematics, 194. Birkhäuser, 2001. x+709 pp.

[Kac] M. Kac, *Can one hear the shape of a drum?* Amer. Math. Monthly 73 (1966), no. 4, part II, 1–23.

[Kacv] V. Kac, *Infinite-dimensional Lie algebras*, Third edition. Cambridge University Press, 1990. xxii+400 pp.

[Kai1] V. Kaimanovich, *Boundary amenability of hyperbolic spaces*, in *Discrete geometric analysis*, pp. 83–111, Contemp. Math., 347, Amer. Math. Soc., 2004.

[Kai2] V. Kaimanovich, *The Poisson formula for groups with hyperbolic properties*, Ann. of Math. 152 (2000) 659–692.

[Kai3] V. Kaimanovich, *Ergodic properties of the horocycle flow and classification of Fuchsian groups*, J. Dynam. Control Systems 6 (2000) 21–56.

[KaiM1] V. Kaimanovich, H. Masur, *The Poisson boundary of the mapping class group*, Invent. Math. 125 (1996) 221–264.

[KaiM2] V. Kaimanovich, H. Masur, *The Poisson boundary of Teichmller space*, J. Funct. Anal. 156 (1998) 301–332.

[KaiW] V. Kaimanovich, W. Woess, *Boundary and entropy of space homogeneous Markov chains*, Ann. Probab. 30 (2002) 323–363.

[KalS] B. Kalinin, R. Spatzier, *On the classification of Cartan actions*, Geom. Funct. Anal. 17 (2007) 468–490.

[KamY] J. Kaminker, G. Yu, *Boundary amenability of groups and positive scalar curvature*, $K$-Theory 18 (1999) 93–97.

[Kami] Y. Kamishima, *Cusp Cross-Sections of Hyperbolic Orbifolds by Heisenberg Nilmanifolds I*, Geometriae Dedicata 122 (2006) 33–49.

[Kan] S. Kan, *On rigidity of Grauert tubes over homogeneous Riemannian manifolds*, J. Reine Angew. Math. 577 (2004) 213–233.

[Kap] M. Kapovich, *Hyperbolic manifolds and discrete groups*, Progress in Math., vol.183, Birkhäuser, 2001.

[KapK1] M. Kapovich, B. Kleiner, *The weak hyperbolization conjecture for 3-dimensional CAT(0)-groups*, Groups Geom. Dyn. 1 (2007) 61–79.

[KapK2] M. Kapovich, B. Kleiner, *Hyperbolic groups with low-dimensional boundary*, Ann. Sci. École Norm. Sup. (4) 33 (2000) 647–669.

[KapKL] M. Kapovich, B. Kleiner, B. Leeb, *Quasi-isometries and the de Rham decomposition*, Topology 37 (1998) 1193–1211.

[Kar] M. Karoubi, *K-theory. An introduction*, Springer-Verlag, 1978. xviii+308 pp.

[Kat1] F. Kato, *Non-Archimedean orbifolds covered by Mumford curves*, J. Algebraic Geom. 14 (2005) 1–34.

[Kat2] F. Kato, *On Fake Projective Planes*, preprint.

[KatO] F. Kato, H. Ochiai, *Arithmetic structure of CMSZ fake projective planes*, J. Algebra 305 (2006) 1166–1185.

[KatU1] K. Kato, S. Usui, *Borel-Serre spaces and spaces of SL(2)-orbits*, in *Algebraic geometry 2000*, Azumino (Hotaka), 321–382, Adv. Stud. Pure Math., 36, Math. Soc. Japan, Tokyo, 2002.

[KatU2] K. Kato, S. Usui, *Logarithmic Hodge structures and classifying spaces*, in *The arithmetic and geometry of algebraic cycles*, pp. 115–130, CRM Proc. Lecture Notes, 24, Amer. Math. Soc., 2000.

[KatU3] K. Kato, S. Usui, *Classifying spaces of degenerating polarized Hodge structures*, preprint.

[KatU4] K. Kato, S. Usui, *Borel-Serre spaces and spaces of* SL(2)*-orbits*, in *Algebraic geometry 2000*, pp. 321–382, Adv. Stud. Pure Math., 36, Math. Soc. Japan, 2002.

[KatoL] A. Katok, J. Lewis, *Global rigidity results for lattice actions on tori and new examples of volume-preserving actions*, Israel J. Math. 93 (1996) 253–280.

[KatoLZ] A. Katok, J. Lewis, R. Zimmer, *Cocycle superrigidity and rigidity for lattice actions on tori*, Topology 35 (1996) 27–38.

[KatoS1] A. Katok, R. Spatzier, *First cohomology of Anosov actions of higher rank abelian groups and applications to rigidity*, Inst. Hautes Études Sci. Publ. Math. No. 79 (1994) 131–156.

[KatoS2] A. Katok, R. Spatzier, *Invariant measures for higher-rank hyperbolic abelian actions*, Ergodic Theory Dynam. Systems 16 (1996) 751–778.

[Katok] S. Katok, *Fuchsian groups*, University of Chicago Press, 1992.

[KatoU] S. Katok, I. Ugarcovici, *Symbolic dynamics for the modular surface and beyond*, Bull. Amer. Math. Soc. 44 (2007) 87–132.

[KatsS] A. Katsuda, T. Sunada, *Homology and closed geodesics in a compact Riemann surface*, Amer. J. Math. 110 (1988) 145–155.

[KatzS1] N. Katz, P. Sarnak, *Random matrices, Frobenius eigenvalues, and monodromy*, American Mathematical Society Colloquium Publications, 45. American Mathematical Society, 1999. xii+419 pp.

[KatzS2] N. Katz, P. Sarnak, *Zeroes of zeta functions and symmetry*, Bull. Amer. Math. Soc. (N.S.) 36 (1999) 1–26.

[Kaw] T. Kawabe, *On the discontinuity of affine motions and Auslanders conjecture*, in *Topics in almost Hermitian geometry and related fields*, pp.128–139, World Sci. Publ., 2005.

[Kaz1] D. Kazhdan, *Connection of the dual space of a group with the structure of its closed subgroups*, Funct. Anal. Appl. 1 (1967) 63–65.

[Kaz2] D. Kazhdan, *On arithmetic varieties*, in *Lie groups and their representations*, (Proc. Summer School, Bolyai János Math. Soc., Budapest, 1971), pp. 151–217, Halsted, New York, 1975.

[KazM] D. Kazhdan, G. Margulis, *A proof of Selberg's hypothesis. (Russian)* Mat. Sb. (N.S.) 117 (1968) 163–168.

[Ke] S. Kerckhoff, *The asymptotic geometry of Teichmller space*, Topology 19 (1980) 23–41.

[KeT] S. Kerckhoff, W. Thurston, *Noncontinuity of the action of the modular group at Bers' boundary of Teichmller space*, Invent. Math. 100 (1990) 25–47.

[Ki] P. Kiernan, *On the compactifications of arithmetic quotients of symmetric spaces*, Bull. Amer. Math. Soc. 80 (1974) 109–110.

[KiK1] P. Kiernan, S. Kobayashi, *Comments on Satake compactification and Picard theorem*, J. Math. Soc. Japan 28 (1976) 577–580.

[KiK2] P. Kiernan, S. Kobayashi, *Satake compactification and extension of holomorphic mappings*, Invent. Math. 16 (1972) 237–248.

[Kim1] I. Kim, *Rigidity on symmetric spaces*, Topology 43 (2004) 393–405.

[Kim2] I. Kim, *Marked length rigidity of rank one symmetric spaces and their product*, Topology 40 (2001) 1295–1323.

[Kim3] I. Kim, *Ergodic theory and rigidity on the symmetric space of non-compact type*, Ergodic Theory Dynam. Systems 21 (2001) 93–114.

[Kim4] I. Kim, *Length spectrum in rank one symmetric space is not arithmetic*, Proc. Amer. Math. Soc. 134 (2006) 3691–3696.

[Kim5] I. Kim, *Isospectral finiteness on hyperbolic 3-manifolds*, Comm. Pure Appl. Math. 59 (2006) 617–625.

[KirW] F. Kirchheimer, J. Wolfart, *Explizite Präsentation gewisser Hilbertscher Modulgruppen durch Erzeugende und Relationen*, J. Reine Angew. Math. 315 (1980) 139–173.

[Kl] S. Kleiman, *The development of intersection homology theory*, in *A century of mathematics in America*, Part II, pp. 543–585, Hist. Math., 2, Amer. Math. Soc., 1989. (An updated version of this paper has appeared in Pure and Applied Mathematics Quarterly, 3 (2007) 225–282.)

[Kle] D. Kleinbock, *Some applications of homogeneous dynamics to number theory*, in *Smooth ergodic theory and its applications*, pp.639–660, Proc. Sympos. Pure Math., 69, Amer. Math. Soc., 2001.

[KleM1] D. Kleinbock, G. Margulis, *Logarithm laws for flows on homogeneous spaces*, Invent. Math. 138 (1999) 451–494.

[KleM2] D. Kleinbock, G. Margulis, *Flows on homogeneous spaces and Diophantine approximation on manifolds*, Ann. of Math. 148 (1998) 339–360.

[KleSS] D. Kleinbock, N. Shah, A. Starkov, *Dynamics of subgroup actions on homogeneous spaces of Lie groups and applications to number theory*, in *Handbook of dynamical systems*, Vol. 1A, pp. 813–930, North-Holland, 2002.

[Klei] B. Kleiner, *The asymptotic geometry of negatively curved spaces: uniformization, geometrization and rigidity*, in *International Congress of Mathematicians*, Vol. II, 743–768, Eur. Math. Soc., Zurich, 2006.

[KlL1] B. Kleiner, B. Leeb, *Groups quasi-isometric to symmetric spaces*, Comm. Anal. Geom. 9 (2001) 239–260.

[KlL2] B. Kleiner, B. Leeb, *Rigidity of quasi-isometries for symmetric spaces and Euclidean buildings*, Inst. Hautes Études Sci. Publ. Math. 86 (1997) 115–197.

[KlL3] B. Kleiner, B. Leeb, *Quasi-isometries and the de Rham decomposition*, Topology 37 (1998) 1193–1211.

[KlL4] B. Kleiner, B. Leeb, *Rigidity of invariant convex sets in symmetric spaces*, Invent. Math. 163 (2006) 657–676.

[Klig1] B. Klingler, *Sur la rigiditi de certains groupes fondamentaux, l'arithmiticiti des riseaux hyperboliques complexes, et les "faux plans projectifs"*, Invent. Math. 153 (2003) 105–143.

[Klig2] B. Klingler, *Un théorème de rigidité non-métrique pour les variétés localement symétriques hermitiennes*, Comment. Math. Helv. 76 (2001) 200–217.

[KnL1] A. Knightly, C. Li, *Traces of Hecke operators*, in *Mathematical Surveys and Monographs*, vol. 133, American Mathematical Society, 2006. x+378 pp.

[KnL2] A. Knightly, C. Li, *A relative trace formula proof of the Petersson trace formula*, Acta Arith. 122 (2006) 297–313.

[KobN] R. Kobayashi, I. Naruki, *The structure of the icosahedral modular group*, in *Automorphic forms and geometry of arithmetic varieties*, pp. 365–372, Adv. Stud. Pure Math., 15, Academic Press, 1989.

[KobO] S. Kobayashi, T. Ochiai, *Satake compactification and the great Picard theorem*, J. Math. Soc. Japan 23 (1971) 340–350.

[Kob1] T. Kobayashi, *On discontinuous group actions on non-Riemannian homogeneous spaces*, to appear in Sugaku Expositions, Amer. Math. Soc., math.DG/0603319.

[Kob2] T. Kobayashi, *Discontinuous groups for non-Riemannian homogeneous spaces*, in *Mathematics unlimited—2001 and beyond*, pp.723–747, Springer, 2001.

[Kob3] T. Kobayashi, *Discontinuous groups and Clifford-Klein forms of pseudo-Riemannian homogeneous manifolds*, in *Algebraic and analytic methods in representation theory*, pp. 99–165, Perspect. Math., 17, Academic Press, 1997.

[Kob4] T. Kobayashi, *Proper action on a homogeneous space of reductive type*, Math. Ann. 285 (1989) 249–263.

[Kob5] T. Kobayashi, *Discontinuous groups acting on homogeneous spaces of reductive type*, In *Proceedings of ICM-90 satellite conference on Representation Theory of Lie Groups and Lie Algebras*, Fuji-Kawaguchiko, 1990 (T. Kawazoe, T. Oshima and S. Sano eds.), World Scientific, 1992, pp. 59–75.

[Kob6] T. Kobayashi, *A necessary condition for the existence of compact Clifford–Klein forms of homogeneous spaces of reductive type*, Duke Math. J. 67 (1992) 653–664.

[Kob7] T. Kobayashi, *On discontinuous groups acting on homogeneous spaces with noncompact isotropy subgroups*, J. Geom. Phys. 12 (1993) 133–144.

[Kob8] T. Kobayashi, *Harmonic analysis on homogeneous manifolds of reductive type and unitary representation theory*, in *Selected papers on harmonic analysis, groups, and invariants*, pp. 1–31, Amer. Math. Soc. Transl. Ser. 2, vol. 183, Amer. Math. Soc., 1998.

[Kob9] T. Kobayashi, *Criterion for proper actions on homogeneous spaces of reductive groups*, J. Lie Theory 6 (1996) 147–163.

[Kob10] T. Kobayashi, *Deformations of compact Clifford–Klein forms of indefinite-Riemannian homogeneous manifolds*, Math. Ann. 310 (1998) 395–409.

[Kob11] T. Kobayashi, *Discrete decomposability of the restriction of* $A_{\mathfrak{q}}(\lambda)$ *with respect to reductive subgroups II — micro-local analysis and asymptotic* $K$*-support*, Ann. of Math. 147 (1998) 709–729.

[Kob12] T. Kobayashi, *Theory of discrete decomposable branching laws of unitary representations of semisimple Lie groups and some applications*, Sugaku Expositions 18 (2005) 1–37.

[KobD] T. Kobayashi, T. Oda, *A vanishing theorem for modular symbols on locally symmetric spaces*, Comment. Math. Helv. 73 (1998) 45–70.

[KobO] T. Kobayashi, K. Ono, *Note on Hirzebruch's proportionality principle*, J. Fac. Sci. Univ. Tokyo Sect. IA Math. 37 (1990) 71–87.

[KobY] T. Kobayashi, T. Yoshino, *Compact Clifford-Klein forms of symmetric spaces—revisited*, Pure Appl. Math. Q. 1 (2005) 591–663.

[Kobl] N. Koblitz, *Introduction to elliptic curves and modular forms*, in *Graduate Texts in Mathematics*, Vol. 97, Springer-Verlag, 1984. viii+248 pp.

[Koj] S. Kojima, *Deformations of hyperbolic* 3*-cone-manifolds*, J. Differential Geom. 49 (1998) 469–516.

[Kon] S. Kondo, *Niemeier lattices, Mathieu groups, and finite groups of symplectic automorphisms of* $K3$ *surfaces, With an appendix by Shigeru Mukai*, Duke Math. J. 92 (1998) 593–603.

[Kor1] A. Koranyi, *A survey of harmonic functions on symmetric spaces*, in *Harmonic analysis in Euclidean spaces*, Part 1, pp. 323–344, Proc. Sympos. Pure Math., vol. XXXV, Amer. Math. Soc., 1979.

[Kor2] A. Koranyi, *Admissible limit sets of discrete groups on symmetric spaces of rank one*, in *Topics in Geometry: in memory of Joseph D'Atri*, ed. by S. Gindikin, Birkhäuser, 1996, pp. 231–240.

[Kork1] M. Korkmaz, *Automorphisms of complexes of curves on punctured spheres and on punctured tori*, Topology Appl. 95 (1999) 85–111.

[Kork2] M. Korkmaz, *Problems on homomorphisms of mapping class groups*, in *Problems on mapping class groups and related topics*, pp. 81–89, Proc. Sympos. Pure Math., 74, Amer. Math. Soc., 2006.

[Kos] B. Kostant, *Lie algebra cohomology and the generalized Borel-Weil theorem*, Ann. of Math. 74 (1961) 329–387.

[Kr] I. Kra, *Automorphic forms and Kleinian groups*, W. A. Benjamin, 1972. xiv+464 pp.

[KrLu] M. Kreck, Matthias, W. Lück, *The Novikov conjecture*, in *Geometry and algebra*, Oberwolfach Seminars, 33, Birkhäuser Verlag, 2005. xvi+267 pp.

[Ku] T. Kubota, *Elementary theory of Eisenstein series*, Kodansha Ltd., Tokyo; Halsted Press [John Wiley & Sons], 1973. xi+110 pp.

[KuM1] S. Kudla, J. Millson, *Geodesic cyclics and the Weil representation. I. Quotients of hyperbolic space and Siegel modular forms*, Compositio Math. 45 (1982) 207–271.

[KuM2] S. Kudla, J. Millson, *Intersection numbers of cycles on locally symmetric spaces and Fourier coefficients of holomorphic modular forms in several complex variables*, Inst. Hautes Études Sci. Publ. Math. No. 71 (1990) 121–172.

[KudR1] S. Kudla, S. Rallis, *A regularized Siegel-Weil formula: the first term identity*, Ann. of Math. 140 (1994) 1–80.

[KudR2] S. Kudla, S. Rallis, *On the Weil-Siegel formula. II. The isotropic convergent case*, J. Reine Angew. Math. 391 (1988) 65–84.

[KudR3] S. Kudla, S. Rallis, *On the Weil-Siegel formula*, J. Reine Angew. Math. 387 (1988) 1–68.

[KudRY] S. Kudla, M. Rapoport, T. Yang, *Modular forms and special cycles on Shimura curves*, Ann of Math Studies, 161. Princeton University Press, 2006. x+373 pp.

[KugI] M. Kuga, S. Ihara, *Family of families of abelian varieties*, in *Algebraic number theory*, pp. 129–142. Japan Soc. Promotion Sci., Tokyo, 1977.

[KugS] M. Kuga, I. Satake, *Abelian varieties attached to polarized $K_3$-surfaces*, Math. Ann. 169 (1967) 239–242.

[Kuh] S. Kühnlein, *On torsion in the cohomology of certain $S$-arithmetic groups*, Manuscripta Math. 83 (1994) 99–110.

[LaM] J.P. Labesse, W. Müller, *Weak Weyl's law for congruence subgroups*, Asian J. Math. 8 (2004) 733–745.

[LabS1] J.P. Labesse, J. Schwermer, *Cohomology of arithmetic groups and automorphic forms*, Lecture Notes in Mathematics, Vol. 1447, Springer-Verlag, vi+358 pp.

[LabS2] J.P. Labesse, J. Schwermer, *On liftings and cusp cohomology of arithmetic groups*, Invent. Math. 83 (1986) 383–401.

[Labo] F. Labourie, *Quelques résultats récents sur les espaces localement homogènes compacts*, in *Manifolds and geometry* (Pisa, 1993), Sympos. Math., XXXVI, Cambridge Univ. Press, 1996, pp. 267–283.

[Labo2] F. Labourie, *Fuchsian affine actions of surface groups*, J. Differential Geom. 59 (2001) 15–31.

[LabZ] F. Labourie, R. Zimmer, *On the non-existence of cocompact lattices for $SL(n)/SL(m)$*, Math. Res. Lett. 2 (1995) 75–77.

[Laf] J. Lafont, *Rigidity result for certain three-dimensional singular spaces and their fundamental groups*, Geom. Dedicata 109 (2004) 197–219.

[LafP] J. Lafont, S. Prassidis, *Roundness properties of groups*, Geom. Dedicata 117 (2006) 137–160.

[LafS] J. Lafont, B. Schmid, *Simplicial volumes of locally symmetric spaces of non-compact type*, Acta Math. 197 (2006) 129–143.

[Lal] S. Lalley, *Renewal theorems in symbolic dynamics, with applications to geodesic flows, non-Euclidean tessellations and their fractal limits*, Acta Math. 163 (1989) 1–55.

[Lan1] J. Landsberg, *On the infinitesimal rigidity of homogeneous varieties*, Compositio Math. 118 (1999) 189–201.

[Lan2] J. Landsberg, *Griffiths-Harris rigidity of compact Hermitian symmetric spaces*, J. Differential Geom. 74 (2006) 395–405.

[Land] E. Landvogt, *A compactification of the Bruhat-Tits building*, Lecture Notes in Mathematics, Vol. 1619, 1996, viii+152 pp.

[Lang1] R. Langlands, *The trace formula and its applications: an introduction to the work of James Arthur*, Canad. Math. Bull. 44 (2001) 160–209.

[Lang2] R. Langlands, *Where stands functoriality today?*, in *Representation theory and automorphic forms*, pp. 457–471, Proc. Sympos. Pure Math., 61, Amer. Math. Soc., 1997.

[Lang3] R. Langlands, *On the functional equations satisfied by Eisenstein series*, Lecture Notes in Mathematics, Vol. 544. Springer-Verlag, 1976. v+337 pp.

[Lang4] R. Langlands, *Problems in the theory of automorphic forms*, in *Lectures in modern analysis and applications, III*, pp. 18–61. Lecture Notes in Math., Vol. 170, Springer, 1970.

[LanR] R. Langlands, D. Ramakrishnan, *The zeta functions of Picard modular surfaces*, les publications CRM, Montreal, 1992.

[Lar] M. Larsen, *Arithmetic compactification of some Shimura surfaces*, in *The zeta functions of Picard modular surfaces*, pp. 31–45, Montreal, 1992.

[Lau1] G. Laumon, *Cohomology of Drinfeld modular varieties. Part I. Geometry, counting of points and local harmonic analysis*, Cambridge Studies in Advanced Mathematics, 41. Cambridge University Press, 1996. xiv+344 pp.

[Lau2] G. Laumon, *Cohomology of Drinfeld modular varieties. Part II. Automorphic forms, trace formulas and Langlands correspondence*, With an appendix by Jean-Loup Waldspurger, Cambridge Studies in Advanced Mathematics, 56. Cambridge University Press, 1997. xii+366 pp.

[Lau3] G. Laumon, *Sur la cohomologie a supports compacts des varietes de Shimura pour* $\mathrm{GSp}(4)_Q$, Compositio Math. 105 (1997) 267–359.

[Led1] F. Ledrappier, *Ergodic properties of some linear actions*, J. Math. Sci. (New York) 105 (2001) 1861–1875.

[Led2] F. Ledrappier, *Applications of dynamics to compact manifolds of negative curvature*, Proc. of the International Congress of Mathematicians, Vol. 1, 2, pp. 1195–1202, Birkhäuser, 1995.

[LedP1] F. Ledrappier, M. Pollicott, *Ergodic properties of linear actions of* $(2 \times 2)$-*matrices* Duke Math. J. 116 (2003) 353–388.

[LedP2] F. Ledrappier, M. Pollicott, *Distribution results for lattices in* $SL(2, \mathbb{Q}_p)$, Bull. Braz. Math. Soc. (N.S.) 36 (2005) 143–176.

[LeeS1] R. Lee, J. Schwermer, *Geometry and arithmetic cycles attached to* $\mathrm{SL}_3(Z)$. *I*, Topology 25 (1986) 159–174.

[LeeS2] R. Lee, J. Schwermer, *Cohomology of arithmetic subgroups of* $\mathrm{SL}_3$ *at infinity*, J. Reine Angew. Math. 330 (1982) 100–131.

[LeeW1] R. Lee, S. Weintraub, *The Siegel modular variety of degree two and level four*, Mem. Amer. Math. Soc. 133 (1998), no. 631, viii, pp. 1–58.

[LeeW2] R. Lee, S. Weintraub, *Topology of the Siegel spaces of degree two and their compactifications*, in *Proceedings of the 1986 topology conference*, Topology Proc. 11 (1986) 115–175.

[LeeW3] R. Lee, S. Weintraub, *Cohomology of* $\mathrm{Sp}_4(Z)$ *and related groups and spaces*, Topology 24 (1985) 391–410.

[LeeW4] R. Lee, S. Weintraub, *Cohomology of a Siegel modular variety of degree* 2, in *Group actions on manifolds*, pp. 433–488, Contemp. Math., vol. 36, Amer. Math. Soc., 1985.

[LeeZ1] R. Lee, R. Szczarba, *The group* $K_3(Z)$ *is cyclic of order forty-eight*, Ann. of Math. 104 (1976) 31–60.

[LeeZ2] R. Lee, R. Szczarba, *On the homology and cohomology of congruence subgroups*, Invent. Math. 33 (1976) 15–53.

[LeeZ3] R. Lee, R. Szczarba, *On the torsion in* $K_4(Z)$ *and* $K_5(Z)$, Duke Math. J. 45 (1978) 101–129.

[Leeb1] B. Leeb, *Examples of discrete groups without small actions on metric trees: arithmetic groups and infinitely generated groups*, in *Low-dimensional topology*, pp. 71–74, Int. Press, 1994.

[Leeb2] B. Leeb, *A characterization of irreducible symmetric spaces and Euclidean buildings of higher rank by their asymptotic geometry*, Bonner Mathematische Schriften, vol. 326, 2000, ii+42 pp.

[Leh] J. Lehner, *Discontinuous groups and automorphic functions*, Amer. Math. Soc., Math. Surveys, No. VIII, 1964.

[LepM] J. Lepowsky, R. Moody, *Hyperbolic Lie Algebras and quasi-regular cusps on Hilbert modular surfaces*, Math. Ann. 245 (1979) 63–88.

[Leu1] E. Leuzinger, *Tits geometry, arithmetic groups, and the proof of a conjecture of Siegel*, J. Lie Theory 14 (2004) 317–338.

[Leu2] E. Leuzinger, *Geodesic rays in locally symmetric spaces*, Differential Geom. Appl. 6 (1996) 55–65.

[Leu3] E. Leuzinger, *Kazhdan's property (T), $L^2$-spectrum and isoperimetric inequalities for locally symmetric spaces*, Comment. Math. Helv. 78 (2003) 116–133.

[Leu4] E. Leuzinger, *An exhaustion of locally symmetric spaces by compact submanifolds with corners*, Invent. Math. 121 (1995) 389–410.

[Leu5] E. Leuzinger, *On the Gauss-Bonnet formula for locally symmetric spaces of noncompact type*, Enseign. Math. 42 (1996) 201–214.

[Leu6] E. Leuzinger, *Critical exponents of discrete groups and $L^2$-spectrum*, Proc. Amer. Math. Soc. 132 (2004) 919–927.

[LeuP] E. Leuzinger and Ch. Pittet. *Isoperimetric inequalities for lattices in semisimple Lie groups of rank 2*, Geom. Funct. Anal. 6 (1996) 489–511.

[Lev1] J. Levy, *A truncated integral of the Poisson summation formula*, Canad. J. Math. 53 (2001) 122–160.

[Lev2] J. Levy, *A truncated Poisson formula for groups of rank at most two*, Amer. J. Math. 117 (1995) 1371–1408.

[Lew] J. Lewis, *Eigenfunctions on symmetric spaces with distribution-valued boundary forms*, J. Funct. Anal. 29 (1978) 287–307.

[LeZa] J. Lewis, D. Zagier, *Period functions for Maass wave forms. I.*, Ann. of Math. 153 (2001) 191–258.

[Li] J. Li, *Nonvanishing theorems for the cohomology of certain arithmetic quotients*, J. Reine Angew. Math. 428 (1992) 177–217.

[LiM] J. Li, J. Millson, *On the first Betti number of a hyperbolic manifold with an arithmetic fundamental group*, Duke Math. J. 71 (1993) 365–401.

[LiS1] J. Li, J. Schwermer, *On the Eisenstein cohomology of arithmetic groups*, Duke Math. J. 123 (2004) 141–169.

[LiS2] J. Li, J. Schwermer, *Automorphic representations and cohomology of arithmetic groups*, in *Challenges for the 21st century* (Singapore, 2000), World Sci. Publishing, 2001, pp. 102–137.

[LiS3] J. Li, J. Schwermer, *Constructions of automorphic forms and related cohomology classes for arithmetic subgroups of $G_2$*, Compositio Math. 87 (1993) 45–78.

[LiP] P. Li, *Lectures on harmonic functions*, preprint of a book.

[LiT1] P. Li, L. Tam, *Harmonic functions and the structure of complete manifolds*, J. Differential Geom. 35 (1992) 359–383.

[LiT2] P. Li, L. Tam, *Linear growth harmonic functions on a complete manifold*, J. Differential Geom. 29 (1989) 421–425.

[LiT3] P. Li, L. Tam, *Positive harmonic functions on complete manifolds with nonnegative curvature outside a compact set*, Ann. of Math. (2) 125 (1987) 171–207.

[LiW1] P. Li, J. Wang, *Counting dimensions of L-harmonic functions*, Ann. of Math. 152 (2000) 645–658.

[LiW2] P. Li, J. Wang, *Counting massive sets and dimensions of harmonic functions*, J. Differential Geom. 53 (1999) 237–278.

[LiW3] P. Li, J. Wang, *Complete manifolds with positive spectrum*, J. Differential Geom. 58 (2001) 501–534.

[LiW4] P. Li, J. Wang, *Complete manifolds with positive spectrum. II*, J. Differential Geom. 62 (2002) 143–162.

[LiW5] P. Li, J. Wang, *Comparison theorem for Khler manifolds and positivity of spectrum*, J. Differential Geom. 69 (2005) 43–74.

[LiW6] P. Li, J. Wang, *Connectedness at infinity of certain Kähler manifolds and locally symmetric spaces*, preprint, 2007.

[LiWi1] W. Li, *Number theory with applications*, World Scientific Publishing Co., 1996. xii+229 pp.

[LiWi2] W. Li, *Eisenstein series and decomposition theory over function fields*, Math. Ann. 240 (1979) 115–139.

[Lim1] S. Lim, *Minimal volume entropy on graphs*, to appear in Trans. AMS. (http://www.arxiv.org/abs/math.GR/0506216).

[Lim2] S. Lim, *Counting overlattices in automorphism groups of trees*, Geom. Dedicata 118 (2006) 1–21.

[Lin1] E. Lindenstrauss, *Invariant measures and arithmetic quantum unique ergodicity*, Ann. of Math. 163 (2006) 165–219.

[Lin2] E. Lindenstrauss, *On quantum unique ergodicity for* $\Gamma\backslash\mathbb{H}\times\mathbb{H}$, Internat. Math. Res. Notices 17 (2001) 913–933.

[LinV] E. Lindenstrauss, A. Venkatesh, *Existence and Weyl's law for spherical cusp forms*, arXiv:math.NT/0503724.

[Link1] G. Link, *Ergodicity of generalised Patterson-Sullivan measures in higher rank symmetric spaces*, Math. Z. 254 (2006) 611–625.

[Link2] G. Link, *Hausdorff dimension of limit sets of discrete subgroups of higher rank Lie groups*, Geom. Funct. Anal. 14 (2004) 400–432.

[Link3] G. Link, *Geometry and Dynamics of Discrete Isometry Groups of Higher Rank Symmetric Spaces*, Geometriae Dedicata vol 122 (2006) 51–75.

[Lip] R. Lipsman, *Proper actions and a compactness condition*, J. Lie Theory 5 (1995) 25–39.

[Liu1] K. Liu, *On elliptic genera and theta-functions*, Topology 35 (1996) 617–640.

[Liu2] K. Liu, *Modular forms and topology*, in *Moonshine, the Monster, and related topics*, pp. 237–262, Contemp. Math., 193, Amer. Math. Soc., 1996.

[LiuSY1] K. Liu, X. Sun, S.T. Yau, *Canonical metrics on the moduli space of Riemann surfaces*, I. J. Differential Geom. 68 (2004) 571–637.

[LiuSY2] K. Liu, X. Sun, S.T. Yau, *Canonical metrics on the moduli space of Riemann surfaces. II*, J. Differential Geom. 69 (2005) 163–216.

[Liuy] Y. Liu, *Density of the commensurability group of uniform tree lattices*, Journal of Algebra, 165 (1994) 346–359.

[LiuY] J. Liu, Y. Ye, *Petersson and Kuznetsov trace formulas*, in *Lie groups and automorphic forms*, pp.147–168, AMS/IP Stud. Adv. Math., 37, Amer. Math. Soc., 2006.

[LLGS] E. Lluis-Puebla, J. Loday, H. Gillet, C. Soule, V. Snaith, *Higher algebraic* $K$*-theory: an overview*, Lecture Notes in Mathematics, 1491, Springer-Verlag, 1992. x+164 pp.

[Lod] J. Loday, *Cyclic homology*, Second edition, Springer-Verlag, Berlin, 1998. xx+513 pp.

[Loo1] E. Looijenga, *Compactificatioms defined by arrangements I: The ball quotient case*, Duke Math. J. 118 (2003) 151–187.

[Loo2] E. Looijenga, *Compactificatioms defined by arrangements II: locally symmetric varieties of type IV*, Duke Math. J. 119 (2003) 527–588.

[Loo3] E. Looijenga, $L^2$*-cohomology of locally symmetric varieties*, Compositio Math. 67 (1988) 3–20.

[Loo4] E. Looijenga, *Uniformization by Lauricella functions–an overview of the theory of Deligne-Mostow*, math.CV/0507534.

[Loo5] E. Looijenga, *New compactifications of locally symmetric varieties*, in *Proceedings of the 1984 Vancouver conference in algebraic geometry*, pp. 341–364, CMS Conf. Proc., 6, Amer. Math. Soc., 1986.

[LooR] E. Looijenga, M. Rapoport, *Weights in the local cohomology of a Baily-Borel compactification*, in *Complex geometry and Lie theory*, pp. 223–260, Proc. Sympos. Pure Math., 53, Amer. Math. Soc., 1991.

[LooS] E. Looijenga, R. Swierstra, *The period map for cubic threefolds*, math.AG/0608279.

[Lot] J. Lott, *Heat kernels on covering spaces and topological invariants*, J. Differential Geom. 35 (1992) 471–510.

[LotL] J. Lott, W. Lück, *$L^2$-topological invariants of 3-manifolds*, Invent. Math. 120 (1995) 15–60.

[Lub1] A. Lubotzky, *Discrete groups, expanding graphs and invariant measures*, Progress in Mathematics, 125. Birkhäuser Verlag, 1994. xii+195 pp.

[Lub2] A. Lubotzky, *Lattices in rank one Lie groups over local fields*, Geom. Funct. Anal. 1 (1991) 406–431.

[Lub3] A. Lubotzky, *Eigenvalues of the Laplacian, the first Betti number and the congruence subgroup problem*, Ann. of Math. 144 (1996) 441–452.

[Lub4] A. Lubotzky, *Tree lattices and lattices in Lie groups*, in *Combinatorial and Geometric Group Theory*, ed A. Duncan, N. Gilbert and J. Howie, LMS, Lecture Note Series 204, Cambridge University Press, 1995, pp. 217–232.

[LubMR] A. Lubotzky, S. Mozes, M. Raghunathan, *The word and Riemannian metrics on lattices of semisimple groups*, Inst. Hautes Études Sci. Publ. Math. No. 91 (2000) 5–53.

[LubN] A. Lubotzky, N. Nikolov, *Subgroup growth of lattices in semisimple Lie groups*, Acta Math. 193 (2004) 105–139.

[LubPS1] A. Lubotzky, R. Phillips, P. Sarnak, *Ramanujan graphs*, Combinatorica 8 (1988) 261–277.

[LubPS2] A. Lubotzky, R. Phillips, P. Sarnak, *Hecke operators and distributing points on the sphere. I*, in *Frontiers of the mathematical sciences: 1985 (New York, 1985)*, Comm. Pure Appl. Math. 39 (1986), no. S, suppl., S149–S186.

[LubPS3] A. Lubotzky, R. Phillips, P. Sarnak, *Hecke operators and distributing points on $S^2$. II*, Comm. Pure Appl. Math. 40 (1987) 401–420.

[LubS] A. Lubotzky, D. Segal, *Subgroup growth*, Progress in Mathematics, 212, Birkhäuser Verlag, 2003, xxii+453 pp.

[LubV] A. Lubotzky, T. Venkataramana, *A group theoretical characterisation of $S$-arithmetic groups in higher rank semi-simple groups*, Geom. Dedicata 90 (2002) 1–28.

[LubZ1] A. Lubotzky, R. Zimmer, *Variants of Kazhdan's property for subgroups of semisimple groups*, Israel J. Math. 66 (1989) 289–299.

[LubZ2] A. Lubotzky, R. Zimmer, *Arithmetic structure of fundamental groups and actions of semisimple Lie groups*, Topology 40 (2001) 851–869.

[LubZ3] A. Lubotzky, R. Zimmer, *A canonical arithmetic quotient for simple Lie group actions*, in *Lie groups and ergodic theory*, pp. 131–142, Tata Inst. Fund. Res. Stud. Math., 14, Tata Inst. Fund. Res., Bombay, 1998.

[Luc1] W. Lück, *$L^2$-invariants: theory and applications to geometry and $K$-theory*, Springer-Verlag, 2002, xvi+595 pp.

[Luc2] W. Lück, *Survey on classifying spaces of familes of subgroups*, in *infinite groups: geometric, combinatorial and dynamical aspects*, pp. 269–322, Progr. Math., 248, Birkhäuser, 2005.

[Luc3] W. Lück, *$L^2$-invariants and their applications to geometry, group theory and spectral theory*, in *Mathematics unlimited—2001 and beyond*, pp/ 859–871, Springer, 2001.

[Luc4] W. Lück, *Approximating $L^2$-invariants by their finite-dimensional analogues*, Geom. Funct. Anal. 4 (1994) 455–481.

[LucR] W. Lück, H. Reich, *The Baum-Connes and the Farrell-Jones conjectures in $K$- and $L$-theory*, in *Handbook of $K$-theory*, pp. 703–842, Springer, 2005.

[LucS] W. Lück, T. Schick,*$L^2$-torsion of hyperbolic manifolds of finite volume*, Geom. Funct. Anal. 9 (1999) 518–567.

[Luo] F. Luo, *Automorphisms of the complex of curves*, Topology 39 (2000) 283–298.

[LuRS] W. Luo, Z. Rudnick, P. Sarnak, *On Selberg's eigenvalue conjecture*, Geom. Funct. Anal. 5 (1995) 387–401.

[LuSa1] W. Luo, P. Sarnak, *Quantum variance for Hecke eigenforms*, Ann. Sci. école Norm. Sup. 37 (2004) 769–799.

[LuSa2] W. Luo, P. Sarnak, *Number variance for arithmetic hyperbolic surfaces*, Comm. Math. Phys. 161(1994) 419–432.

[LuSa3] W. Luo, P. Sarnak, *Quantum ergodicity of eigenfunctions on* $\mathrm{PSL}_2 Z)\backslash\mathbb{H}^2$, Inst. Hautes Études Sci. Publ. Math. No. 81 (1995) 207–237.

[Luri] J. Lurie, *A survey of elliptic cohomology*, preprint, 2006.

[LyS] R. Lyndon, P. Schupp, *Combinatorial group theory*, Springer-Verlag, 1977, xiv+339 pp.

[Maa] H. Maass, *Lectures on modular functions of one complex variable*, Tata Institute of Fundamental Research Lectures on Mathematics and Physics, 29, 1983. iii+262 pp.

[MaclR] C. Maclachlan, A. Reid, *The arithmetic of hyperbolic 3-manifolds*, Graduate Texts in Mathematics, Vol. 219, Springer-Verlag, 2003.

[MacPM1] R. MacPherson, M. McConnell, *Explicit reduction theory for Siegel modular threefolds*, Invent. Math. 111 (1993) 575–625.

[MacPM2] R. MacPherson, M. McConnell, *Classical projective geometry and modular varieties*, in *Algebraic analysis, geometry, and number theory*, pp. 237–290, Johns Hopkins Univ. Press, 1989.

[MadW] I. Madsen, M. Weiss, *The stable mapping class group and stable homotopy theory*, European Congress of Mathematics, pp. 283–307, Eur. Math. Soc., 2005.

[Mag] W. Magnus, *Noneuclidean tesselations and their groups*, Academic Press, 1974.

[MagKS] W. Magnus, A. Karrass, D. Solitar, *Combinatorial group theory. Presentations of groups in terms of generators and relations*, Dover Publications, 2004. xii+444 pp.

[Mah1] J. Mahnkopf, *Cohomology of arithmetic groups, parabolic subgroups and the special values of $L$-functions on* $\mathrm{GL}_n$, J. Inst. Math. Jussieu 4 (2005) 553–637.

[Mah2] J. Mahnkopf, *Eisenstein cohomology and the construction of $p$-adic analytic $L$-functions*, Compositio Math. 124 (2000) 253–304.

[Mah3] J. Mahnkopf, *Modular symbols and values of $L$-functions on* $\mathrm{GL}_3$, J. Reine Angew. Math. 497 (1998) 91–112.

[Mai1] S. Maillot, *Quasi-isometries of groups, graphs and surfaces*, Comment. Math. Helv. 76 (2001) 29–60.

[Mai2] S. Maillot, *Large-scale conformal rigidity in dimension three*, Math. Ann. 337 (2007) 613–630.

[Mal] A. Malcev, *On one class of homogeneous spaces*, Amer. Math. Soc. Transl. 39 (1951) 276–307.

[Man] J. Manin, *Parabolic points and zeta functions of modular curves*, Izv. Akad. Nauk SSSR Ser. Mat. 36 (1972) 19–66.

[ManM] Y. Manin, M. Marcolli, *Continued fractions, modular symbols, and noncommutative geometry*, Selecta Math. (N.S.) 8 (2002) 475–521.

[Marc1] M. Marcolli, *Limiting modular symbols and the Lyapunov spectrum*, J. Number Theory 98 (2003) 348–376.

[Marc2] M. Marcolli, *Arithmetic noncommutative geometry*, University Lecture Series, 36. American Mathematical Society, 2005. xii+136 pp.

[Mard] A. Marden, *The geometry of finitely generated kleinian groups*, Ann. of Math. 99 (1974) 383–462.

[Marg1] G. Margulis, *Discrete subgroups of semisimple Lie groups*, Springer-Verlag, 1991. x+388 pp.

[Marg2] G. Margulis, *Diophantine approximation, lattices and flows on homogeneous spaces*, in *A panorama of number theory or the view from Baker's garden*, pp. 280–310, Cambridge Univ. Press, 2002.

[Marg3] G. Margulis, *Orbits of group actions and values of quadratic forms at integral points*, in *Festschrift in honor of I. I. Piatetski-Shapiro on the occasion of his sixtieth birthday*, Part II, pp. 127–150, Israel Math. Conf. Proc., 3, 1990.

[Marg4] G. Margulis, *Certain applications of ergodic theory to the investigation of manifolds of negative curvature*, Funkcional. Anal. i Prilozhen. 3 (1969) 89–90.

[Marg5] G. Margulis, *On some aspects of the theory of Anosov systems*, With a survey by Richard Sharp: *Periodic orbits of hyperbolic flows*, Springer Monographs in Mathematics, Springer, 2004. vi+139 pp.

[Marg6] G. Margulis, *Problems and conjectures in rigidity thoery*, in *Mathematics: Frontiers and Perspectives*, Amer. Math. Soc., 2000, pp. 161–174.

[Marg7] G. Margulis, *On the decomposition of discrete subgroups into amalgams*, Selecta Math. Soviet. 1 (1981) 197–213.

[MargQ] G. Margulis, N. Qian, *Rigidity of weakly hyperbolic actions of higher real rank semisimple Lie groups and their lattices*, Ergodic Theory Dynam. Systems 21 (2001) 121–164.

[MargR] G. Margulis, J. Rohlfs, *On the proportionality of covolumes of discrete subgroups*, Math. Ann. 275 (1986) 197–205.

[Mas1] B. Maskit, *Kleinian groups*, Springer-Verlag, 1998.

[Mas2] B. Maskit, *On Poincaré's theorem for fundamental polygons*, Advances in Math. 7 (1971) 219–230.

[Mat] D. Matsnev, *The Baum-Connes conjecture and group actions on affine buildings*, Thesis at Penn State University, 2005.

[May] D. Mayer, *The thermodynamic formalism approach to Selberg's zeta function for* $\mathrm{PSL}(2, Z)$, Bull. Amer. Math. Soc. 25 (1991) 55–60.

[MazW] H. Mazur, M. Wolf, *The Weil-Petersson isometry group*, Geom. Dedicata 93 (2002) 177–190.

[Mc] J. McCarthy, *A "Tits-alternative" for subgroups of surface mapping class groups*, Trans. Amer. Math. Soc. 291 (1985) 583–612.

[McC1] M. McConnell, *Cell decompositions of Satake compactifications*, in *Geometric combinatorics (Kotor, 1998)*, Publ. Inst. Math. (Beograd) (N.S.) 66 (80) (1999) 46–90.

[McC2] M. McConnell, *Classical projective geometry and arithmetic groups*. Math. Ann. 290 (1991) 441–462.

[Mck] H. McKean, *Selberg's trace formula as applied to a compact Riemann surface*, Comm. Pure Appl. Math. 25 (1972) 225–246.

[Mcm] C. McMullen, *The moduli space of Riemann surfaces is Kahler hyperbolic*, Ann. of Math. 151 (2000) 327–357.

[McR] D. McReynolds, *Peripheral separability and cusps of arithmetic hyperbolic orbifolds*, Algebr. Geom. Topol. 4 (2004) 721–755.

[McS] G. McShane, *Simple geodesics and a series constant over Teichmüller space*, Invent. Math. 132 (1998) 607–632.

[McSR] G. McShane, I. Rivin, *A norm on homology of surfaces and counting simple geodesics*, Internat. Math. Res. Notices 2 (1995) 61–69.

[MeiS] D. Meintrup, T. Schick, *A model for the universal space for proper actions of a hyperbolic group*, New York J. Math. 8 (2002) 1–7.

[Men] J. Mennicke, *Finite factor groups of the unimodular group*, Ann. of Math. 81 (1965) 31–37.

[Mer] L. Merel, *Intersections sur des courbes modulaires*, Manuscripta Math. 80 (1993) 283–289.

[Mey] C. Meyer, *Modular Calabi-Yau threefolds*, Fields Institute Monographs, 22. American Mathematical Society, 2005. x+195 pp.

[Meye] R. Meyerhoff, *Sphere-packing and volume in hyperbolic* 3*-space*, Comment. Math. Helv. 61 (1986) 271–278.

[MiaR] R. Miatello, J. Rossetti, *Length spectra and p-spectra of compact flat manifolds*, J. Geom. Anal. 13 (2003) 631–657.

[MiaW1] R. Miatello, N. Wallach, *The resolvent of the Laplacian on locally symmetric spaces*, J. Differential Geom. 36 (1992) 663–698.

[MiaW2] R. Miatello, N. Wallach, *Kuznetsov formulas for products of groups of R-rank one*, in *Festschrift in honor of I. I. Piatetski-Shapiro on the occasion of his sixtieth birthday, Part II*, pp. 305–320, Israel Math. Conf. Proc., 3, 1990.

[MiaW3] R. Miatello, N. Wallach, *Kuznetsov formulas for real rank one groups*, J. Funct. Anal. 93 (1990) 171–206.

[MihR] M. Mihalik, K. Ruane, CAT(0) *groups with non-locally connected boundary*, J. London Math. Soc. 60 (1999) 757–770.

[MilW1] S. Miller, W. Schmid, *Distributions and analytic continuation of Dirichlet series*, J. Funct. Anal. 214 (2004) 155–220.

[MilW2] S. Miller, W. Schmid, *The highly oscillatory behavior of automorphic distributions for* SL(2), Lett. Math. Phys. 69 (2004) 265–286.

[MilW3] S. Miller, W. Schmid, *Summation formulas, from Poisson and Voronoi to the present*, in *Noncommutative harmonic analysis*, pp. 419–440, Progr. Math., 220, Birkhäuser, 2004.

[MilW4] S. Miller, W. Schmid, *Automorphic Distributions, L-functions, and Voronoi Summation for GL(3)*, Ann of Math 164 (2006) 423–488.

[MilW5] S. Miller, W. Schmid, *The Rankin–Selberg method for automorphic distributions*, to appear in *Representation Theory and Automorphic Forms*, Progr. Math., Birkhäuser.

[Mill1] J. Millson, *On the first Betti number of a constant negatively curved manifold*, Ann. of Math. 104 (1976) 235–247.

[Mill2] J. Millson, *Intersection numbers of cycles on locally symmetric spaces and Fourier coefficients of holomorphic modular forms in several complex variables*, in *Theta functions*, Part 2, pp. 129–142, Proc. Sympos. Pure Math., 49, Part 2, Amer. Math. Soc., 1989.

[MilR] J. Millson, M. Raghunathan, *Geometric construction of cohomology for arithmetic groups. I*, Proc. Indian Acad. Sci. Math. Sci. 90 (1981) 103–123.

[Miln1] J. Milne, *Canonical models of (mixed) Shimura varieties and automorphic vector bundles*, in *Automorphic forms, Shimura varieties, and L-functions*, Vol. I, ed. by L. Clozel and J. Milne, Academic Press, 1990, pp. 283–414.

[Miln2] J. Milne, *Shimura varieties and motives*, in *Motives*, Proc. Sympos. Pure Math., 55, Part 2, Amer. Math. Soc., 1994, pp. 447–523.

[Miln3] J. Milne, *Introduction to Shimura varieties*, in *Harmonic analysis, the trace formula, and Shimura varieties*, pp. 265–378, Clay Math. Proc., 4, Amer. Math. Soc., 2005.

[Miln4] J. Milne, *Algebraic Groups and Arithmetic Groups*, preprint of a book (available at http://www.jmilne.org/math/CourseNotes/aag.pdf).

[Milno1] J. Milnor, *Hilbert's problem 18: on crystallographic groups, fundamental domains, and on sphere packing*, in *Mathematical developments arising from Hilbert problems*, pp. 491–506. Proc. Sympos. Pure Math., Vol. XXVIII, Amer. Math. Soc., 1976.

[Milno2] J. Milnor, *Eigenvalues of the Laplace operator on certain manifolds*, Proc. Nat. Acad. Sci. U.S.A. 51 (1964) 542.

[Milno3] J. Milnor, *Hyperbolic geometry: the first 150 years*, Bull. Amer. Math. Soc. 6 (1982) 9–24.

[Milno4] J. Milnor, *On fundamental groups of complete affinely flat manifolds*, Advances in Math. 25 (1977) 178–187.

[MinT] M. Mimura, H. Toda, *Topology of Lie groups. I, II*, Translations of Mathematical Monographs, 91. American Mathematical Society, 1991. iv+451 pp.

[Min] I. Mineyev, *Flows and joins of metric spaces*, Geom. Topol. 9 (2005) 403–482.

[MinMS] I. Mineyev, N. Monod, Y. Shalom, *Ideal bicombings for hyperbolic groups and applications*, Topology 43 (2004) 1319–1344.

[MineY] I. Mineyev, G. Yu, *The Baum-Connes conjecture for hyperbolic groups*, Invent. Math. 149 (2002) 97–122.

[Mins] Y. Minsky, *End invariants and the classification of hyperbolic 3-manifolds*, in *Current developments in mathematics*, 2002, pp, 181–217, Int. Press, 2003.

[Mir1] M. Mirzakhani, *Simple geodesics and Weil-Petersson volumes of moduli spaces of bordered Riemann surfaces*, Invent. Math. 167 (2007) 179–222.

[Mir2] M. Mirzakhani, *Weil-Petersson volumes and intersection theory on the moduli space of curves*, J. Amer. Math. Soc. 20 (2007) 1–23.

[MoeW] C. Moeglin, J. Waldspurger, *Spectral decomposition and Eisenstein series*, in *Cambridge Tracts in Mathematics*, vol. 113, Cambridge University Press, 1995. xxviii+338 pp.

[Mok1] N. Mok, *Aspects of Kähler geometry on arithmetic varieties*, in *Several complex variables and complex geometry*, Proc. Sympos. Pure Math., 52, Part 2, pp. 335–396, Amer. Math. Soc., 1991.

[Mok2] N. Mok, *Metric rigidity theorems on Hermitian locally symmetric manifolds*, World Scientific Publishing Co., 1989. xiv+278 pp.

[MokSY] N. Mok, Y.T. Siu, S. Yeung, *Geometric superrigidity*, Invent. Math. 113 (1993) 57–83.

[MokY] N. Mok, S. Yeung, *Geometric realizations of uniformization of conjugates of Hermitian locally symmetric manifolds*, in *Complex analysis and geometry*, pp. 253–270, Univ. Ser. Math., Plenum, 1993.

[MokZ] N. Mok, J. Zhong, *Compactifying complete Kähler-Einstein manifolds of finite topological type and bounded curvature*, Ann. of Math. 129 (1989) 427–470.

[Mon1] N. Monod, *Superrigidity for irreducible lattices and geometric splitting*, J. Amer. Math. Soc. 19 (2006) 781–814.

[Mon2] N. Monod, *Arithmeticity vs. nonlinearity for irreducible lattices*, Geom. Dedicata 112 (2005) 225–237.

[Mon3] N. Monod, *Continuous bounded cohomology of locally compact groups*, Lecture Notes in Mathematics, vol. 1758, Springer-Verlag, 2001. x+214 pp.

[Mon4] N. Monod, *An invitation to bounded cohomology*, Proc. of International Congress of Mathematicians, Vol. II, 1183–1211, Eur. Math. Soc., Zürich, 2006.

[MonS1] N. Monod, Y. Shalom, *Orbit equivalence rigidity and bounded cohomology*, Ann. of Math. 164 (2006) 825–878.

[MonS2] N. Monod, Y. Shalom, *Cocycle superrigidity and bounded cohomology for negatively curved spaces*, J. Differential Geom. 67 (2004) 395–455.

[More] S. Morel, *Complexes pondérés des compactifications de Baily-Borel. Le cas des varits modulaires de Siegel*, to appear in Journal of Amer. Math. Soc. 21 (2008) 23–61.

[Mori] S. Mori, *Projective manifolds with ample tangent bundles*, Ann. of Math. 110 (1979) 593–606.

[Moro] D. Morrison, *The Kuga-Satake variety of an abelian surface*, J. Algebra 92 (1985) 454–476.

[Morr1] D. Morris, *Introduction to arithmetic groups*, preprint of a book, 2003.

[Morr2] D. Morris, *Ratner's theorems on unipotent flows*, in *Chicago Lectures in Mathematics*, University of Chicago Press, 2005. xii+203 pp.

[MorrK] J. Morrow, K. Kodaira, *Complex manifolds*, Reprint of the 1971 edition with errata. AMS Chelsea Publishing, 2006. x+194 pp.

[MoS] H. Moscovici, R. Stanton, *R-torsion and zeta functions for locally symmetric manifolds*, Invent. Math. 105 (1991) 185–216.

[Most1] G. Mostow, *Strong rigidity of locally symmetric spaces*, Princeton University Press, 1973.

[Most2] G. Mostow, *On a remarkable class of polyhedra in complex hyperbolic space*, Pacific J. Math. 86 (1980) 171–276.

[Most3] G. Mostow, *Factor spaces of solvable groups*, Ann. of Math. 60 (1954) 1–27.

[Most4] G. Mostow, *Some new decomposition theorems for semi-simple groups*, Mem. Amer. Math. Soc. 14 (1955) 31–54.

[MostT] G. Mostow, T. Tamagawa, *On the compactness of arithmetically defined homogeneous spaces*, Ann. of Math. 76 (1962) 446–463.

[Moz] S. Mozes, *Products of trees, lattices and simple groups*, in *Proceedings of the International Congress of Mathematicians*, Vol. II (Berlin, 1998), Doc. Math. 1998, Extra Vol. II, 571–582.

[Mul1] W. Müller, *Weyl's law for the cuspidal spectrum of* $\mathrm{SL}_n$, C. R. Math. Acad. Sci. Paris 338 (2004) 347–352.

[Mul2] W. Müller, *The trace class conjecture in the theory of automorphic forms. II*, Geom. Funct. Anal. 8 (1998) 315–355.

[Mul3] W. Müller, *The trace class conjecture in the theory of automorphic forms*, Ann. of Math. 130 (1989) 473–529.

[Mul4] W. Müller, *Spectral theory and geometry, First European Congress of Mathematics, Vol. I (Paris, 1992)*, pp. 153–185, Progr. Math., 119, Birkhäuser, 1994.

[Mul5] W. Müller, *Spectral geometry and scattering theory for certain complete surfaces of finite volume*, Invent. Math. 109 (1992) 265–305.

[Mul6] W. Müller, *Manifolds with cusps of rank one. Spectral theory and* $L^2$*-index theorem*, Lecture Notes in Mathematics, 1244. Springer-Verlag, 1987. xii+158 pp.

[Mum1] D. Mumford, *A new approach to compactifying locally symmetric varieties*, in *Discrete subgroups of Lie groups and applications to moduli* (Internat. Colloq., Bombay, 1973), pp. 211–224, Oxford Univ. Press, Bombay, 1975.

[Mum2] D. Mumford, *Tata lectures on theta I*, in *Progress in Mathematics*, vol. 28, Birkhäuser, 1983.

[Mum3] D. Mumford, *Hirzebruch's proportionality theorem in the noncompact case*, Invent. Math. 42 (1977) 239–272.

[Mum4] D. Mumford, *An algebraic surface with K ample,* $(K2) = 9$, $p_g = q = 0$, in *Contributions to Algebraic Geometry*, Johns Hopkins Univ. Press, 1979, pp. 233–244.

[Mun] H. Munkholm, *Simplices of maximal volume in hyperbolic space, Gromov's norm, and Gromov's proof of Mostow's rigidity theorem (following Thurston)*, in *Topology Symposium, Siegen 1979*, pp. 109–124, Lecture Notes in Math., 788, Springer, 1980.

[Mur] K. Murty, *Bounded and finite generation of arithmetic groups*, pp. 249–261, CMS Conf. Proc., 15, Amer. Math. Soc., 1995.

[NabW] A. Nabutovsky, S. Weinberger, *Variational problems for Riemannian functionals and arithmetic groups*, Inst. Hautes Études Sci. Publ. Math. No. 92 (2000) 5–62.

[Nad] A. Nadel, *On complex manifolds which can be compactified by adding finitely many points*, Invent. Math. 101 (1990) 173–189.

[NadT] A. Nadel, H. Tsuji, *Compactification of complete Kähler manifolds of negative Ricci curvature*, J. Differential Geom. 28 (1988) 503–512.

[Nag] H. Nagoshi, *Selberg zeta functions over function fields*, J. Number Theory 90 (2001) 207–238.

[Nai1] A. Nair, *Weighted cohomology of arithmetic groups*, Ann. of Math. 150 (1999) 1–31.

[Nai2] A. Nair, *On the Fibres of a Toroidal Resolution*, in Pure and Applied Math Quarterly 3 (2007) 323–346.

[Nam1] Y. Namikawa, *Toroidal compactification of Siegel spaces*, Lecture Notes in Mathematics, Vol. 812, Springer, 1980.

[Nam2] Y. Namikawa, *A new compactification of the Siegel space and degeneration of Abelian varieties. I*, Math. Ann. 221 (1976) 97–141.

[Nam3] Y. Namikawa, *A new compactification of the Siegel space and degeneration of Abelian varieties. II*, Math. Ann. 221 (1976) 201–241.

[Nam4] Y. Namikawa, *Toroidal degeneration of Abelian varieties, Complex analysis and algebraic geometry*, pp. 227–237, Iwanami Shoten, 1977.

[Nas] S. Nasrin, *Criterion of proper actions for 2-step nilpotent Lie groups*, Tokyo J. Math. 24 (2001) 535–543.

[Nay] S. Nayatani, *Patterson-Sullivan measure and conformally flat metrics*, Math. Z. 225 (1997) 115–131.

[Ne] K. Neeb, *Towards a Lie theory of locally convex groups*, Jpn. J. Math. 1 (2006) 291–468.

[New] F. Newberger, *On the Patterson-Sullivan measure for geometrically finite groups acting on complex or quaternionic hyperbolic space*, Geom. Dedicata 97 (2003) 215–249.

[Nib] G. Niblo, *A geometric proof of Stallings' theorem on groups with more than one end*, Geom. Dedicata 105 (2004) 61–76.

[Nic] P. Nicholls, *The ergodic theory of discrete groups*, London Mathematical Society Lecture Note Series, 143. Cambridge University Press, 1989. xii+221 pp.

[Nik] V. Nikulin, *Discrete reflection groups in Lobachevsky spaces and algebraic surfaces*, in *Proc. of the International Congress of Mathematicians*, pp. 654–671, Amer. Math. Soc., 1987.

[NiS] V. Nikulin, I. Shafarevich, *Geometries and groups*, in *Universitext*, Springer Series in Soviet Mathematics, Springer-Verlag, 1987, viii+251 pp.

[NoR] M. Nori, M. Raghunathan, *On conjugation of locally symmetric arithmetic varieties*, in *Proceedings of the Indo-French Conference on Geometry*, pp. 111–122, Hindustan Book Agency, Delhi, 1993.

[NoV] G. Noskov, E. Vinberg, *Strong Tits alternative for subgroups of Coxeter groups*, J. Lie Theory 12 (2002) 259–264.

[Now] P. Nowak, *On exactness and isoperimetric profiles of discrete groups*, J. Funct. Anal. 243 (2007) 323–344.

[Od1] T. Oda, *Hodge structures attached to geometric automorphic forms*, in *Automorphic forms and number theory*, pp. 223–276, Adv. Stud. Pure Math., 7, North-Holland, 1985.

[Od2] T. Oda, *Distinguished cycles and Shimura varieties*, in *Automorphic forms of several variables*, pp. 298–332, Progr. Math., 46, Birkhäuser, 1984.

[Od3] T. Oda, *Periods of Hilbert modular surfaces*, Progress in Mathematics, 19. Birkhäuser, 1982. xvi+123 pp.

[Oh1] H. Oh, *The Ruziewicz problem and distributing points on homogeneous spaces of a compact Lie group*, in *Probability in mathematics*, Israel J. Math. 149 (2005) 301–316.

[Oh2] H. Oh, *Lattice action on finite volume homogeneous spaces*, J. Korean Math. Soc. 42 (2005) 635–653.

[Oh3] H. Oh, *Tempered subgroups and representations with minimal decay of matrix coefficients*, Bull. Soc. Math. France 126 (1998) 355–380.

[OhW1] H. Oh, D. Witte, *New examples of compact Clifford–Klein forms of homogeneous spaces of* $\mathrm{SO}(2,n)$, Int. Math. Res. Notices 5 (2000) 235–251.

[OhW2] H. Oh, D. Witte, *Compact Clifford–Klein forms of homogeneous spaces of* $\mathrm{SO}(2,n)$, Geom. Dedicata 89 (2002) 25–57.

[Ohs] K. Ohshika, *Discrete groups*, American Mathematical Society, 2002. x+193 pp.

[OkBSC] A. Okabe, B. Boots, K. Sugihara, S. Chiu, *Spatial tessellations: concepts and applications of Voronoi diagrams*, Wiley Series in Probability and Statistics. John Wiley & Sons, 2000. xvi+671 pp.

[Ol1] M. Olbrich, *$L^2$-invariants of locally symmetric spaces*, Doc. Math. 7 (2002) 219–237.

[Ol2] M. Olbrich, *Hodge theory on hyperbolic manifolds of infinite volume*, in *Lie theory and its applications in physics, III*, pp. 75–83, World Sci. Publishing, 2000.

[Ol3] M. Olbrich, *Cohomology of convex cocompact groups and invariant distributions on limit sets*, arXiv:math/0207301.

[OlsS] A. Olshanskii, M. Sapir, *Length and area functions on groups and quasi-isometric Higman embeddings*, Internat. J. Algebra Comput. 11 (2001) 137–170.

[Om] H. Omori, *Infinite dimensional Lie transformation groups*, Lecture Notes in Mathematics, Vol. 427. Springer-Verlag, 1974. xi+149 pp.

[OsbW1] M.S. Osborne, G. Warner, *The theory of Eisenstein systems*, Academic Press, 1981.

[OsbW2] M.S. Osborne, G. Warner, *The Selberg trace formula II: partition, reduction, truncation*, Pacific Jour. Math. 106 (1983) 307–496.

[Osh] T. Oshima, *A realization of semisimple symmetric spaces and construction of boundary value maps*, in *Representations of Lie groups*, 1986, pp. 603–650, Adv. Stud. Pure Math., 14, Academic Press, 1988.

[Ot] J.P. Otal, *Le spectre marqué des longueurs des surfaces á courbure négative*, Ann. of Math. 131 (1990) 151–162.

[Pan1] P. Pansu, *Sous-groupes discrets des groupes de Lie: rigidité, arithméticité*, Séminaire Bourbaki, Vol. 1993/94, Astérisque No. 227 (1995), Exp. No. 778, 3, pp. 69–105.

[Pan2] P. Pansu, *Quasiconformal mappings and manifolds of negative curvature*, in *Curvature and topology of Riemannian manifolds*, pp. 212–229, Lecture Notes in Math., Vol. 1201, Springer, 1986.

[Pan3] P. Pansu, *Métriques de Carnot-Carathéodory et quasiisométries des espaces symétriques de rang un*, Ann. of Math. 129 (1989) 1–60.

[Pan4] P. Pansu, *Formules de Matsushima, de Garland et propriété (T) pour des groupes agissant sur des espaces symtriques ou des immeubles*, Bull. Soc. Math. France 126 (1998) 107–139.

[Pap1] P. Papasoglu, *Group splittings and asymptotic topology*, J. Reine Angew. Math. 602 (2007) 1–16.

[Pap2] P. Papasoglu, *Quasi-isometry invariance of group splittings*, Ann. of Math. 161 (2005) 759–830.

[PapW] P. Papasoglu, K. Whyte, *Quasi-isometries between groups with infinitely many ends*, Comment. Math. Helv. 77 (2002) 133–144.

[Par] J. Parker, *On the volumes of cusped, complex hyperbolic manifolds and orbifolds*, Duke Math. J. 94 (1998) 433–464.

[ParP1] W. Parry, M. Pollicott, *Zeta functions and the periodic orbit structure of hyperbolic dynamics*, Astérisque No. 187–188 (1990), 268 pp.

[ParP2] W. Parry, M. Pollicott, *An analogue of the prime number theorem for closed orbits of Axiom A flows*, Ann. of Math. 118 (1983) 573–591.

[Pat1] S. Patterson, *Lectures on measures on limit sets of Kleinian groups*, in *Analytical and geometric aspects of hyperbolic space*, pp. 281–323, London Math. Soc. Lecture Note Ser., 111, Cambridge Univ. Press, 1987.

[Pat2] S. Patterson, *The limit set of a Fuchsian group*, Acta Math. 136 (1976) 241–273.

[Pat3] S. Patterson, *Metric Diophantine approximation of quadratic forms*, in *Number theory and dynamical systems*, pp. 37–48, London Math. Soc. Lecture Note Ser., 134, Cambridge Univ. Press, 1989.

[Pat4] S. Patterson, *The Selberg zeta-function of a Kleinian group*, in *Number theory, trace formulas and discrete groups*, pp. 409–441, Academic Press, 1989.

[PatP] S. Patterson, P. Perry, *The divisor of Selberg's zeta function for Kleinian groups. Appendix A by Charles Epstein*, Duke Math. J. 106 (2001) 321–390.

[Pau1] F. Paulin, *Outer automorphisms of hyperbolic groups and small actions on R-trees*, in *Arboreal group theory*, pp. 331–343, Math. Sci. Res. Inst. Publ., 19, Springer, 1991.

[Pau2] F. Paulin, *Sur les automorphismes extérieurs des groupes hyperboliques*, Ann. Sci. École Norm. Sup. 30 (1997) 147–167.

[Pau3] F. Paulin, *Groupe modulaire, fractions continues et approximation diophantienne en caractéristique p*, Geom. Dedicata 95 (2002) 65–85.

[Pav] M. Pavey, *Class Numbers of Orders in Quartic Fields*, arXiv:math/0602294.

[Pe1] P. Perry, *A Poisson summation formula and lower bounds for resonances in hyperbolic manifolds*, Int. Math. Res. Not. 34 (2003) 1837–1851.

[Pe2] P. Perry, *Spectral theory, dynamics, and Selberg's zeta function for Kleinian groups*, in *Dynamical, spectral, and arithmetic zeta functions*, pp.145–165, Contemp. Math., vol. 290, Amer. Math. Soc., 2001.

[Pe3] P. Perry, *Asymptotics of the length spectrum for hyperbolic manifolds of infinite volume*, Geom. Funct. Anal. 11 (2001) 132–141.

[Pe4] P. Perry, *The Selberg zeta function and a local trace formula for Kleinian groups*, J. Reine Angew. Math. 410 (1990) 116–152.

[Pes1] H. Pesce, *Une formule de Poisson pour les variétés de Heisenberg*, Duke Math. J. 73 (1994) 79–95.

[Pes2] H. Pesce, *Nilvariétés isospectrales*, in *Actes de la Table Ronde de Géométrie Différentielle*, pp.513–530, Sémin. Congr., 1, Soc. Math. France, 1996.

[Pesc] G. Peschke, *The theory of ends*, Nieuw Arch. Wisk. (4) 8 (1990) 1–12.

[PetS] V. Petkov, L. Stoyanov, *Geometry of Reflecting Rays and Inverse Spectral Problems*, Wiley, 1992.

[PhS1] R. Phillips, P. Sarnak, *On cusp forms for co-finite subgroups of* $\mathrm{PSL}(2, R)$, Invent. Math. 80 (1985) 339–364.

[PhS2] R. Phillips, P. Sarnak, *Perturbation theory for the Laplacian on automorphic functions*, J. Amer. Math. Soc. 5 (1992) 1–32.

[PhS3] R. Phillips, P. Sarnak, *Automorphic spectrum and Fermi's golden rule*, J. Anal. Math. 59 (1992) 179–187.

[PhS4] R. Phillips, P. Sarnak, *Geodesics in homology classes*, Duke Math. J. 55 (1987) 287–297.

[PiaS1] I. Piateskii-Shapiro, *Geometry of classical domains and theory of automorphic functions*, Gordon and Breach, 1969.

[PiaS2] I. Piateskii-Shapiro, *Domains of upper half-plane type in the theory of several complex variables*, 1963 Proc. Internat. Congr. Mathematicians (Stockholm, 1962), pp. 389–396 Inst. Mittag-Leffler, Djursholm.

[PiaS3] I. Piateskii-Shapiro, *Selected works of Ilya Piatetski-Shapiro*, Edited and with commentaries by James Cogdell, Simon Gindikin, Peter Sarnak, Pierre Deligne, Stephen Gelbart, Roger Howe and Stephen Rallis, Amer. Math. Soc., 2000. xxvi+821 pp.

[PicR] P. Pickel, J. Roitberg, *Automorphism groups of nilpotent groups and spaces*, J. Pure Appl. Algebra 150 (2000) 307–319.

[Pin1] R. Pink, *Arithmetical compactification of mixed Shimura varieties*, Bonner Mathematische Schriften, vol. 209, 1990. xviii+340 pp.

[Pin2] R. Pink, *On the calculation of local terms in the Lefschetz-Verdier trace formula and its application to a conjecture of Deligne*, Ann. of Math. 135 (1992) 483–525.

[PlR1] V. Platonov, A. Rapinchuk, *Algebraic groups and Number theory*, Academic Press, 1994.

[PlR2] V. Platonov, A. Rapinchuk, *Abstract properties of S-arithmetic groups and the congruence problem*, Russian Acad. Sci. Izv. Math. 40 (1993) 455–476.

[PolS1] M. Pollicott, R. Sharp, *Correlations for pairs of closed geodesics*, Invent. Math. 163 (2006) 1–24.

[PolS2] M. Pollicott, R. Sharp, *A local limit theorem for closed geodesics and homology*, Trans. Amer. Math. Soc. 356 (2004) 4897–4908.

[PolS3] M. Pollicott, R. Sharp, *Poincaré series and comparison theorems for variable negative curvature*, in *Topology, ergodic theory, real algebraic geometry*, pp. 229–240, Amer. Math. Soc. Transl. Ser. 2, 202, 2001.

[PolS4] M. Pollicott, R. Sharp, *Poincaré series and zeta functions for surface group actions on R-trees*, Math. Z. 226 (1997) 335–347.

[Pop1] S. Popa, *Deformation and rigidity for group actions and von Neumann algebras. International Congress of Mathematicians*, Vol. I, pp. 445–477, Eur. Math. Soc., Zürich, 2007.

[Pop2] S. Popa, *Strong rigidity of* $\mathrm{II}_1$ *factors arising from malleable actions of w-rigid groups. I*, Invent. Math. 165 (2006) 369–408.

[Pop3] S. Popa, *Strong rigidity of* $\mathrm{II}_1$ *factors arising from malleable actions of w-rigid groups. II*, Invent. Math. 165 (2006) 409–451.

[Pop4] S. Popa, *Cocycle and Orbit Equivalence Superrigidity for Malleable Actions of w-Rigid Groups*, arXiv:math/0512646.

[Pr1] G. Prasad, *Volumes of S-arithmetic quotients of semi-simple groups*, with an appendix by Moshe Jarden and the author, Inst. Hautes Études Sci. Publ. Math. 69 (1989) 91–117.

[Pr2] G. Prasad, *Lattices in semisimple groups over local fields*, in *Studies in algebra and number theory*, pp. 285–356, Adv. in Math. Suppl. Stud., 6, Academic Press, 1979.

[Pr3] G. Prasad, *Semi-simple groups and arithmetic subgroups*, in *Proceedings of the International Congress of Mathematicians*, pp. 821–832, Math. Soc. Japan, 1991.

[Pr4] G. Prasad, *Strong rigidity of Q-rank* 1 *lattices*, Invent. Math. 21 (1973) 255–286.

[Pr5] G. Prasad, *On some work of Raghunathan*, in *Algebraic groups and arithmetic*, pp. 25–40, Tata Inst. Fund. Res., Mumbai, 2004.

[Pr6] G. Prasad, *Borel's contributions to arithmetic groups and their cohomology*, Gaz. Math. No. 102 (2004) 15–24.

[PrR1] G. Prasad, M. Raghunathan, *Topological central extensions of semisimple groups over local fields*, Ann. of Math. 119 (1984) 143–201.

[PrR2] G. Prasad, M. Raghunathan, *Topological central extensions of semisimple groups over local fields. II*, Ann. of Math. 119 (1984) 203–268.

[PrR3] G. Prasad, M.S. Raghunathan, *On the congruence subgroup problem: determination of the "metaplectic kernel"*, Inv. Math. 71 (1983) 21–42.

[PrRa] P. Prasad, A. Rapinchuk, *Zariski-dense subgroups and transcendental number theory*, Math. Res. Lett. 12 (2005) 239–249.

[PYe1] G. Prasad, S. Yeung, *Fake projective planes*, Invent. Math. 168 (2007) 321–370.

[PYe2] G. Prasad, S. Yeung, *Cocompact arithmetic subgroups of $PU(n-1,1)$ with Euler-Poincaré characteristic $n$ and a fake $P^4$*, preprint, 2006.

[Py] L. Pyber, *Bounded generation and subgroup growth*, Bull. London Math. Soc. 34 (2002) 55–60.

[QiY] N. Qian, C. Yue, *Local rigidity of Anosov higher-rank lattice actions*, Ergodic Theory Dynam. Systems 18 (1998) 687–702.

[Qu] D. Quillen, *Finite generation of the groups $K_i$ of rings of algebraic integers*, in *Algebraic $K$-theory, I: Higher $K$-theories*, 179–198, Lecture Notes in Math., Vol. 341, Springer, 1973.

[Quin1] J. Quint, *Mesures de Patterson-Sullivan en rang supérieur*, Geom. Funct. Anal. 12 (2002) 776–809.

[Quin2] J. Quint, *Groupes convexes cocompacts en rang supérieur*, Geom. Dedicata 113 (2005) 1–19.

[Quin3] J. Quint, *Groupes de Schottky et comptage*, Ann. Inst. Fourier (Grenoble) 55 (2005) 373–429.

[Ra1] M.S. Raghunathan, *On the congruence subgroup problem*, Publ. Math. IHES 46 (1976) 107–162.

[Ra2] M.S. Raghunathan, *On the congruence subgroup problem. II*, Invent. Math. 85 (1986) 73–117.

[Ra3] M.S. Raghunathan, *The congruence subgroup problem*, Proc. Indian Acad. Sci. Math. Sci. 114 (2004) 299–308.

[Ra4] M.S. Raghunathan, *The congruence subgroup problem*, Proceedings of the Hyderabad Conference on Algebraic Groups, pp. 465–494, 1991.

[Ra5] M.S. Raghunathan, *Discrete subgroups of Lie groups*, Springer-Verlag, 1972.

[Ra6] M.S. Raghunathan, *A note on quotients of real algebraic groups by arithmetic subgroups*, Invent. Math. 4 (1967/1968) 318–335.

[Ra7] M.S. Raghunathan, *Cohomology of arithmetic subgroups of algebraic groups. I, II.*, Ann. of Math. 86 (1967) 409–424; ibid. 87 (1967) 279–304.

[Ra8] M.S. Raghunathan, *A note on generators for arithmetic subgroups of algebraic groups*, Pacific J. Math. 152 (1992) 365–373.

[Raj] C. Rajan, *On the non-vanishing of the first Betti number of hyperbolic three manifolds*, Math. Ann. 330 (2004) 323–329.

[RajV] C. Rajan, T. Venkataramana, *On the first cohomology of arithmetic groups*, Manuscripta Math. 105 (2001) 537–552.

[RamRS] J. Ramagge, G. Robertson, T. Steger, *A Haagerup inequality for $\tilde{A}_1 \times \tilde{A}_1$ and $\tilde{A}_2$ buildings*, Geom. Funct. Anal. 8 (1998) 702–731.

[Ramk] D. Ramakrishnan, *Regulators, algebraic cycles, and values of $L$-functions*, in *Algebraic $K$-theory and algebraic number theory*, pp. 183–310, Contemp. Math., 83, Amer. Math. Soc., 1989.

[RamV] D. Ramakrishnan, R. Valenza, *Fourier analysis on number fields*, in *Graduate Texts in Mathematics*, 186, Springer-Verlag, 1999. xxii+350 pp.

[Ran1] A. Ranicki, *On the Novikov conjecture. Novikov conjectures, index theorems and rigidity*, Vol. 1, 1993, pp. 272–337, London Math. Soc. Lecture Note Ser., 226, Cambridge Univ. Press, 1995.

[Ran2] A. Ranicki, *Algebraic L-theory and topological manifolds*, Cambridge Tracts in Mathematics, 102, Cambridge University Press, 1992. viii+358 pp.

[Rap1] A. Rapinchuk, *Congruence subgroup problem for algebraic groups: old and new*, Journées Arithmétiques, 1991 (Geneva.) Astérisque No. 209 (1992) 73–84.

[Rap2] A. Rapinchuk, *On SS-rigid groups and A. Weil's criterion for local rigidity. I*, Manuscripta Math. 97 (1998) 529–543.

[Rap3] A. Rapinchuk, *On the finite-dimensional unitary representations of Kazhdan groups*, Proc. Amer. Math. Soc. 127 (1999) 1557–1562.

[Rap4] A. Rapinchuk, *The congruence subgroup problem for arithmetic groups of bounded generation*, Soviet Math. Dokl. 42 (1991) 664–668.

[Rapp] M. Rapoport, *On the shape of the contribution of a fixed point on the boundary: The case of* $\mathbb{Q}$*-rank* 1 *(with an appendix by L. Saper and M. Stern)*, in *The zeta functions of Picard modular surfaces*, ed. by R. Langlands and D. Ramakrishnan, Les Publications CRM, Montréal, 1992, pp. 479–488.

[RappZ] M. Rapoport, T. Zink, *Period spaces for p-divisible groups*, Annals of Mathematics Studies, vol. 141, Princeton University Press, 1996, xxii+324 pp.

[Rat] J. Ratcliffe, *Foundations of hyperbolic manifolds*, Second edition, Graduate Texts in Mathematics, 149. Springer, 2006. xii+779 pp.

[Re1] B. Remy, *Kac-Moody groups: split and relative theories. Lattices*, in *Groups: topological, combinatorial and arithmetic aspects*, pp. 487–541, London Math. Soc. Lecture Note Ser., 311, Cambridge Univ. Press, Cambridge, 2004.

[Re2] B. Remy, *Groupes de Kac-Moody déployés et presque déployés*, Astérisque No. 277 (2002), viii+348 pp.

[ReR] B. Remy, M. Ronan, *Topological groups of Kac-Moody type, right-angled twinnings and their lattices.* Comment. Math. Helv. 81 (2006) 191–219.

[RiZ] E. Rips, Z. Sela, *Cyclic splittings of finitely presented groups and the canonical JSJ decomposition*, Ann. of Math. 146 (1997) 53–109.

[Rob] G. Robertson, *Torsion in K-theory for boundary actions on affine buildings of type* $\widetilde{A}_n$, $K$-Theory 22 (2001) 251–269.

[RobS1] G. Robertson, T. Steger, *Affine buildings, tiling systems and higher rank Cuntz-Krieger algebras*, J. Reine Angew. Math. 513 (1999) 115–144.

[RobS2] G. Robertson, T. Steger, *Asymptotic K-theory for groups acting on* $\tilde{A}_2$ *buildings*, Canad. J. Math. 53 (2001) 809–833.

[Roe] J. Roe, *Hyperbolic groups have finite asymptotic dimension*, Proc. Amer. Math. Soc. 133 (2005) 2489–2490.

[RogW] J. Rognes, C. Weibel, *Two-primary algebraic K-theory of rings of integers in number fields*, Appendix A by M. Kolster, J. Amer. Math. Soc. 13 (2000) 1–54.

[Roh1] J. Rohlfs, *Projective limits of locally symmetric spaces and cohomology*, J. Reine Angew. Math. 479 (1996) 149–182.

[Roh2] J. Rohlfs, *Lefschetz numbers for arithmetic groups*, in *Cohomology of arithmetic groups and automorphic forms*, pp. 303–313, Lecture Notes in Math., Vol. 1447, Springer, 1990.

[Roh3] J. Rohlfs, *Die maximalen arithmetisch definierten Untergruppen zerfallender einfacher Gruppen*, Math. Ann. 244 (1979) 219–231.

[RohSc1] J. Rohlfs, J. Schwermer, *An arithmetic formula for a topological invariant of Siegel modular varieties*, Topology 37 (1998) 149–159.

[RohSc2] J. Rohlfs, J. Schwermer, *Intersection numbers of special cycles*, J. Amer. Math. Soc. 6 (1993) 755–778.

[RohSp1] J. Rohlfs, B. Speh, *Pseudo-Eisenstein forms and cohomology of arithmetic groups. II*, in *Algebraic groups and arithmetic*, pp. 63–89, Tata Inst. Fund. Res., 2004.

[RohSp2] J. Rohlfs, B. Speh, *Pseudo-Eisenstein forms and cohomology of arithmetic groups. I*, Manuscripta Math. 106 (2001) 505–518.

[RohSp3] J. Rohlfs, B. Speh, *Boundary contributions to Lefschetz numbers for arithmetic groups. I*, in *Cohomology of arithmetic groups and automorphic forms*, pp. 315–332, Lecture Notes in Math., Vol. 1447, Springer, 1990.

[RohSp4] J. Rohlfs, B. Speh, *On limit multiplicities of representations with cohomology in the cuspidal spectrum*, Duke Math. J. 55 (1987) 199–211.

[Ron1] M. Ronan, *Lectures on buildings*, in *Perspectives in Mathematics*, vol. 7, Academic Press, 1989. xiv+201 pp.

[Ron2] M. Ronan, *Buildings: main ideas and applications. I. Main ideas*, Bull. London Math. Soc. 24 (1992) 1–51.

[Ron3] M. Ronan, *Buildings: main ideas and applications. II. Arithmetic groups, buildings and symmetric spaces*, Bull. London Math. Soc. 24 (1992) 97–126.

[Ron4] M. Ronan, *Symmetry and the monster. One of the greatest quests of mathematics*, Oxford University Press, 2006. vi+251 pp.

[Rong] X. Rong, *Collapsed Riemannian manifolds with bounded sectional curvature*, Proceedings of the International Congress of Mathematicians, Vol. II (Beijing, 2002), pp. 323–338, Higher Ed. Press, 2002.

[Ros1] J. Rosenberg, *Algebraic $K$-theory and its applications*, in *Graduate Texts in Mathematics*, 147, Springer-Verlag, 1994. x+392 pp.

[Ros2] J. Rosenberg, *$K$-theory and geometric topology*, in *Handbook of $K$-theory*, pp. 577–610, Springer, 2005.

[Rost] D. Rosenthal, *Continuous control and the algebraic L-theory assembly map*, Forum Math. 18 (2006) 193–209.

[Roy] H. Royden, *Automorphisms and isometries of Teichmüller space*, in *Advances in the Theory of Riemann Surfaces*, pp. 369–383, Ann. of Math. Studies, No. 66. Princeton Univ. Press, 1971.

[RuS] Z. Rudnick, P. Sarnak, *The behaviour of eigenstates of arithmetic hyperbolic manifolds*, Comm. Math. Phys. 161 (1994) 195–213.

[Sa] F. Salein, *Variétés anti–de Sitter de dimension 3 possédant un champ de Killing non trivial*, ENS de Lyons, Ph.D. thesis, Dec. 1999.

[SalD] V. Salvador, G. Daniel, *Topics in the Theory of Algebraic Function Fields*, in *Theory & Applications*, Birkhäuser, 2006, XVI, 652 p.

[San] G. Sankaran, *Fundamental group of locally symmetric varieties*, Manuscripta Math. 90 (1996) 39–48.

[Sap1] L. Saper, *Tilings and finite energy retractions of locally symmetric spaces*, Comment. Math. Helv. 72 (1997) 167–202.

[Sap2] L. Saper, *Geometric rationality of equal-rank Satake compactifications*, Math. Res. Lett. 11 (2004) 653–671.

[Sap3] L. Saper, *On the cohomology of locally symmetric spaces and of their compactifications*, in *Current developments in mathematics, 2002*, pp. 219–289, Int. Press, 2003.

[Sap4] L. Saper, *$\mathcal{L}$-modules and the conjecture of Rapoport and Goresky-MacPherson*, in *Formes Automorphes (I)*, Astérisque 298 (2005), pp. 319–334.

[Sap5] L. Saper, *$\mathcal{L}$-modules and micro-support*, to appear in Ann of Math. arXiv:math/0011225IV3

[SapS] L. Saper, M. Stern, *$L_2$-cohomology of arithmetic varieties*, Ann. of Math. 132 (1990) 1–69.

[SapZ] L. Saper, S. Zucker, *An introduction to $L^2$-cohomology*, in *Several complex variables and complex geometry*, Part 2, pp. 519–534, Proc. Sympos. Pure Math., 52, Part 2, Amer. Math. Soc., 1991.

[Sar1] P. Sarnak, *Some applications of modular forms*, in *Cambridge Tracts in Mathematics*, 99. Cambridge University Press, 1990, x+111 pp.

[Sar2] P. Sarnak, *The arithmetic and geometry of some hyperbolic three-manifolds*, Acta Math. 151 (1983) 253–295.

[Sar3] P. Sarnak, *Spectra of hyperbolic surfaces*, Bull. Amer. Math. Soc. 40 (2003) 441–478.

[Sar4] P. Sarnak, *Selberg's eigenvalue conjecture*, Notices Amer. Math. Soc. 42 (1995) 1272–1277.

[Sar5] P. Sarnak, *Arithmetic quantum chaos*, Israel Math. Conf. Proc. 8(1995) 183–236.

[Sar6] P. Sarnak, *On cusp forms*, in *The Selberg trace formula and related topics*, pp. 93–407, Contemp. Math., 53, Amer. Math. Soc., 1986.

[SarW] P. Sarnak, M. Wakayama, *Equidistribution of holonomy about closed geodesics*, Duke Math. J. 100 (1999) 1–57.

[Sat1] I. Satake, *On representations and compactifications of symmetric spaces*, Ann. of Math. 71 (1960) 77–110.

[Sat2] I. Satake, *On compactifications of the quotient spaces for arithmetically defined discontinuous groups*, Ann. of Math. 72 (1960) 555–580.

[Sat3] I. Satake, *On the compactification of Siegel spaces*, J. India Math. Soc. 20 (1956) 259–281.

[Sat4] I. Satake, *On the arithmetic of tube domains (blowing-up of the point at infinity)*, Bull. Amer. Math. Soc. 79 (1973) 1076–1094.

[Sat5] I. Satake, *Algebraic structures of symmetric domains*, Princeton University Press, 1980.

[Sat8] I. Satake, *Compactifications, past and present*, Sugaku 51 (1999) 129–141.

[Sat9] I. Satake, *On a generalization of the notion of manifold*, Proc. Nat. Acad. Sci. U.S.A. 42 (1956) 359–363.

[SauS] M. du Sautoy, D. Segal, *Zeta functions of groups. New horizons in pro-p groups*, pp. 249–286, Progr. Math., 184, Birkhäuser, 2000.

[Sav] G. Savin, *Limit multiplicities of cusp forms*, Invent. Math. 95 (1989) 149–159.

[ScO] W. Scharlau, H. Opolka, *From Fermat to Minkowski. Lectures on the theory of numbers and its historical development*, in *Undergraduate Texts in Mathematics*, Springer-Verlag, 1985, xi+184 pp.

[Sche1] J. Scherk, *The Borel-Serre Compactification for the Classifying Space of Hodge Structures*, preprint.

[Sche2] J. Scherk, *On the monodromy theorem for isolated hypersurface singularities*, Invent. Math. 58 (1980) 289–301.

[Schl] H. Schlichtkrull, *Hyperfunctions and harmonic analysis on symmetric spaces*, Progress in Mathematics, 49. Birkhäuser, 1984. xiv+185 pp.

[Schm1] W. Schmid, *Variation of Hodge structure: the singularities of the period mapping*, Invent. Math. 22 (1973) 211–319.

[Schm2] W. Schmid, *Automorphic distributions for* $\mathrm{SL}(2,\mathbb{R})$, in *Conférence Moshé Flato 1999*, Vol. I (Dijon), pp. 345–387, Math. Phys. Stud., 21, Kluwer Acad. Publ., 2000.

[Schmu1] P. Schmutz, *Geometry of Riemann surfaces based on closed geodesics*, Bull. Amer. Math. Soc. 35 (1998) 193–214.

[Schmu2] P. Schmutz, *Arithmetic groups and the length spectrum of Riemann surfaces*, Duke Math. J. 84 (1996) 199–215.

[Schmu3] P. Schmutz, *Systoles of arithmetic surfaces and the Markoff spectrum*, Math. Ann. 305 (1996) 191–203.

[Schmu4] P. Schmutz, *A cell decomposition of Teichmller space based on geodesic length functions*, Geom. Funct. Anal. 11 (2001) 142–174.

[Schmu5] P. Schmutz, *Systoles and topological Morse functions for Riemann surfaces*, J. Differential Geom. 52 (1999) 407–452.

[Schu] G. Schumacher, *The theory of Teichmüller spaces. A view towards moduli spaces of Kähler manifolds*, in *Several complex variables, VI*, pp. 251–310, Encyclopaedia Math. Sci., 69, Springer, 1990.

[Schw1] R. Schwartz, *Quasi-isometric rigidity and Diophantine approximation*, Acta Math. 177 (1996) 75–112.

[Schw2] R. Schwartz, *The quasi-isometry classification of rank one lattices*, Inst. Hautes Études Sci. Publ. Math. No. 82 (1995) 133–168.

[Schwe1] J. Schwermer, *On Euler products and residual Eisenstein cohomology classes for Siegel modular varieties*, Forum Math. 7 (1995) 1–28.

[Schwe2] J. Schwermer, *Eisenstein series and cohomology of arithmetic groups: the generic case*, Invent. Math. 116 (1994) 481–511.

[Schwe3] J. Schwermer, *On arithmetic quotients of the Siegel upper half space of degree two*, Compositio Math. 58 (1986) 233–258.

[Schwe4] J. Schwermer, *Holomorphy of Eisenstein series at special points and cohomology of arithmetic subgroups of* $\mathrm{SL}_n(Q)$, J. Reine Angew. Math. 364 (1986) 193–220.

[Schwe5] J. Schwermer, *Kohomologie arithmetisch definierter Gruppen und Eisensteinreihen*, Lecture Notes in Mathematics, Vol. 988, Springer-Verlag, 1983. iv+170 pp.

[Schwe6] J. Schwermer, *Cohomology of arithmetic groups, automorphic forms and L-functions*, in *Cohomology of arithmetic groups and automorphic forms*, Lecture Notes in Math., Vol. 1447, Springer, 1990, pp. 1–29.

[Schwe7] J. Schwermer, *Special cycles and automorphic forms on arithmetically defined hyperbolic 3-manifolds*, Asian J. Math. 8 (2004) 837–859.

[Scot] P. Scott, *The geometries of* 3*-manifolds*, Bull. London Math. Soc. 15 (1983) 401–487.

[ScoS] P. Scott, G. Swarup, *Regular neighbourhoods and canonical decompositions for groups*, Astérisque No. 289 (2003), vi+233 pp.

[ScoW] P. Scott, T. Wall, *Topological methods in group theory*, in *Homological group theory*, pp. 137–203, London Math. Soc. Lecture Note Ser., 36, Cambridge Univ. Press, 1979.

[Scz] R. Sczech, *Eisenstein group cocycles for* $\mathrm{GL}_n$ *and values of L-functions*, Invent. Math. 113 (1993) 581–616.

[Seg1] D. Segal, *On the outer automorphism group of a polycyclic group*, in *Proc. of the Second International Group Theory Conference*, Rend. Circ. Mat. Palermo (2) Suppl. No. 23 (1990) 265–278.

[Seg2] D. Segal, *Polycyclic groups*, Cambridge Tracts in Mathematics, 82. Cambridge University Press, 1983. xiv+289 pp.

[Sel1] Z. Sela, *Structure and rigidity in (Gromov) hyperbolic groups and discrete groups in rank* 1 *Lie groups. II*, Geom. Funct. Anal. 7 (1997) 561–593.

[Sel2] Z. Sela, *Endomorphisms of hyperbolic groups. I. The Hopf property*, Topology 38 (1999) 301–321.

[Sele1] A. Selberg, *Harmonic analysis and discontinuous groups in weakly symmetric Riemannian spaces with applications to Dirichlet series*, J. Indian Math. Soc. (N.S.) 20 (1956) 47–87.

[Sele2] A. Selberg, *On discontinuous groups in higher-dimensional symmetric spaces*, in *Contributions to function theory (internat. Colloq. Function Theory, Bombay, 1960)*, pp. 147–164, Tata Institute of Fundamental Research, Bombay.

[Sen] M. Senechal, *Introduction to lattice geometry*, in *From number theory to physics*, pp. 476–495, Springer, 1992.

[Sele3] A. Selberg, *Collected papers. Vol. I, II*, Springer-Verlag, 1989 & 1991.

[Seri1] C. Series, *The modular surface and continued fractions*, J. London Math. Soc. 31 (1985) 69–80.

[Seri2] C. Series, *The geometry of Markoff numbers*, Math. Intelligencer 7, no. 3 (1985) 20–29.

[Seri3] C. Series, *Geometrical Markov coding of geodesics on surfaces of constant negative curvature*, Ergodic Theory Dynam. Systems 6 (1986) 601–625.

[Serr1] J.P. Serre, *Arithmetic groups*, in *Homological group theory*, pp. 105–136, Cambridge Univ. Press, 1979.

[Serr2] J.P. Serre, *Cohomologie des groupes discrets*, in *Prospects in mathematics*, pp. 77–169, Ann. of Math. Studies, No. 70, 1971.

[Serr3] J-P. Serre, *Groupes discrets-compactifications*, in *Colloque sur les Fonctions Sphiriques et la Thiorie des Groupes*, Exp. No. 6, 4 pp., Inst. Élie Cartan, Univ. Nancy, 1971.

[Serr4] J.P. Serre, *Trees*, Springer-Verlag, 1980.

[Serr5] J.P. Serre, *Le problème des groupes de congruence pour SL2*, Ann. of Math. 92 (1970) 489–527.

[Shaf] I.R. Shafarevich, *Basic notions of algebra*, Springer-Verlag, 1997. iv+258 pp.

[Shah1] F. Shahidi, *Functoriality and small eigenvalues of Laplacian on Riemann surfaces*, in *Surveys in differential geometry*, Vol. IX, pp. 385–400, Int. Press, 2004.

[Shah2] F. Shahidi, *Automorphic L-functions and functoriality*, Proceedings of the International Congress of Mathematicians, Vol. II (Beijing, 2002), pp. 655–666, Higher Ed. Press, 2002.

[Shah3] F. Shahidi, *Langlands functoriality conjecture and number theory*, to appear in *Representation Theory and Automorphic Forms*, Progr. Math., Birkhäuser.

[Shal1] Y. Shalom, *Rigidity of commensurators and irreducible lattices*, Invent. Math. 141 (2000) 1–54.

[Shal2] Y. Shalom, *Rigidity, unitary representations of semisimple groups, and fundamental groups of manifolds with rank one transformation group*, Ann. of Math. 152 (2000) 113–182.

[Shal3] Y. Shalom, *Explicit Kazhdan constants for representations of semisimple and arithmetic groups*, Ann. Inst. Fourier 50 (2000) 833–863.

[Shal4] Y. Shalom, *Bounded generation and Kazhdan's property (T)*, Inst. Hautes Études Sci. Publ. Math. No. 90 (1999) 145–168.

[Shal5] Y. Shalom, *Measurable group theory*, in *European Congress of Mathematics*, pp. 391–423, Eur. Math. Soc., 2005.

[Shal6] Y. Shalom, *Harmonic analysis, cohomology, and the large-scale geometry of amenable groups*, Acta Math. 192 (2004) 119–185.

[Shal7] Y. Shalom, *The growth of linear groups*, J. Algebra 199 (1998) 169–174.

[Shal8] Y. Shalom, *The algebraization of Kazhdan's property (T)*, Proc of International Congress of Mathematicians. Vol. II, pp. 1283–1310, Eur. Math. Soc., Zürich, 2006.

[Shal9] Y. Shalom, *Measurable group theory*, in *European Congress of Mathematics*, pp. 391–423, Eur. Math. Soc., Zürich, 2005.

[Shan] K. Shankar, *Isometry groups of homogeneous spaces with positive sectional curvature*, Differential Geom. Appl. 14 (2001) 57–78.

[ShanSW] K. Shankar, R. Spatzier, B. Wilking, *Spherical rank rigidity and Blaschke manifolds*, Duke Math. J. 128 (2005) 65–81.

[SharV] R. Sharma, T. Venkataramana, *Generations for arithmetic groups*, Geom. Dedicata 114 (2005) 103–146.

[Sharp1] R. Sharp, *Uniform estimates for closed geodesics and homology on finite area hyperbolic surfaces*, Math. Proc. Cambridge Philos. Soc. 137 (2004) 245–254.

[Sharp2] R. Sharp, *Closed geodesics and periods of automorphic forms*, Adv. Math. 160 (2001) 205–216.

[She] N. Shepherd-Barron, *Perfect forms and the moduli space of abelian varieties*, Invent. Math. 163 (2006) 25–45.

[ShT] Y. Shi, G. Tian, *Rigidity of asymptotically hyperbolic manifolds*, Comm. Math. Phys. 259 (2005) 545–559.

[Shi1] G. Shimura, *Introduction to the arithmetic theory of automorphic functions*, Princeton University Press, 1971.

[Shi2] G. Shimura, *Collected papers. Vol. I, II, III, IV*, Springer-Verlag, 2002 & 2003.

[Shi3] G. Shimura, *Algebraic varieties without deformation and the Chow variety*, J. Math. Soc. Japan 20 (1968) 336–341.

[Si1] C.L. Siegel, *Symplectic Geometry*, Academic Press, 1964.

[Si2] C.L. Siegel, *Topics in complex function theory: Vol. I. Elliptic functions and uniformization theory, Vol. II. Automorphic functions and abelian integrals, Vol. III. Abelian functions and modular functions of several variables*, Wiley Classics Library, 1988, xii+186 pp, xii+193, pp, x+244 pp.

[Si3] C.L. Siegel, *Lectures on the geometry of numbers*, Springer, 1989.

[Si4] C.L. Siegel, *Lectures on quadratic forms*, Notes by K. G. Ramanathan. Tata Institute of Fundamental Research Lectures on Mathematics, No. 7, 1967, 192 pp.

[Si5] C.L. Siegel, *Gesammelte Abhandlungen. Bände I, II, III*, Springer-Verlag, 1966.

[SilbV] L. Silberman, A. Venkatesh, *On Quantum unique ergodicity for locally symmetric spaces I*, arXiv:math.RT/0407413.

[Silv] J. Silverman, *The arithmetic of elliptic curves*, GTM 106, Springer-Verlag, 1986.

[Siu1] Y. Siu, *Strong rigidity of compact quotients of exceptional bounded symmetric domains*, Duke Math. J. 48 (1981) 857–871.

[Siu2] Y. Siu, *The complex-analyticity of harmonic maps and the strong rigidity of compact Kähler manifolds*, Ann. of Math. 112 (1980) 73–111.

[SiuY1] Y. Siu, S.T. Yau, *Compactification of negatively curved complete Kähler manifolds of finite volume*, in *Seminar on Differential Geometry*, pp. 363–380, Ann. of Math. Stud., vol. 102, Princeton Univ. Press, 1982.

[SiuY2] Y. Siu, S.T. Yau, *Compact Kähler manifolds of positive bisectional curvature*, Invent. Math. 59 (1980) 189–204.

[Slo] N. Sloane, *The sphere packing problem*, in Proceedings of the International Congress of Mathematicians, Vol. III (Berlin, 1998). Doc. Math. 1998, Extra Vol. III, pp. 387–396.

[Sof] G. Sofier, *Free subgroups of linear groups*, Pure and Applied Math Quarterly 3 (2007), no. 4.

[Sou1] C. Soule, *A bound for the torsion in the $K$-theory of algebraic integers*, Doc. Math. 2003, Extra Vol., pp. 761–788 (electronic.)

[Sou2] C. Soule, *On the 3-torsion in $K_4(Z)$*, Topology 39 (2000) 259–265.

[Sou3] C. Soule, *$K$-theory and values of zeta functions*, in *Algebraic $K$-theory and its applications*, pp. 255–283, World Sci. Publishing, 1999.

[Sou4] C. Soule, *Perfect forms and the Vandiver conjecture*, J. Reine Angew. Math. 517 (1999) 209–221.

[Sou5] C. Soule, *The cohomology of* $\mathrm{SL}_3(Z)$, Topology 17 (1978) 1–22.

[Sou6] C. Soule, *$K$-theory of the integers*, in *Higher algebraic $K$-theory: an overview*, Lecture Notes in Mathematics, vol. 1491, Springer-Verlag, 1992.

[Spa1] R. Spatzier, *An invitation to rigidity theory*, in *Modern dynamical systems and applications*, pp. 211–231, Cambridge Univ. Press, Cambridge, 2004.

[Spa2] R. Spatzier, *Harmonic analysis in rigidity theory*, in *Ergodic theory and its connections with harmonic analysis*, pp. 153–205, London Math. Soc. Lecture Note Ser., vol. 205, Cambridge Univ. Press, 1995.

[Spe] B. Speh, *Representation theory and the cohomology of arithmetic groups*, International Congress of Mathematicians. Vol. II, pp.1327–1335, Eur. Math. Soc., Zürich, 2006.

[SpV] B. Speh, T.N. Venkataramana, *Construction of some generalised modular symbols*, Pure Appl. Math. Q. 1 (2005) 737–754.

[St1] J. Stallings, *On torsion-free groups with infinitely many ends*, Ann. of Math. 88 (1968) 312–334.

[St2] J. Stallings, *Groups of cohomological dimension one*, in *Applications of Categorical Algebra*, Proc. Sympos. Pure Math., Vol. XVIII, pp. 124–128, Amer. Math. Soc., 1970.

[St3] J. Stallings, *Group theory and three-dimensional manifolds*, Yale Mathematical Monographs, 4. Yale University Press, 1971. v+65 pp.

[Sta1] C. Stark, *Topological rigidity theorems*, in *Handbook of geometric topology*, pp. 1045–1084, North-Holland, 2002.

[Sta2] C. Stark, *Surgery theory and infinite fundamental groups. Surveys on surgery theory*, Vol. 1, pp.275–305, Ann. of Math. Stud., 145, Princeton Univ. Press, 2000.

[Sta3] C. Stark, *Arithmetic groups of gauge transformations*, Forum Math. 5 (1993) 49–71.

[Star] H. Stark, *Galois theory, algebraic number theory, and zeta functions*, in *From number theory to physics*, pp. 313–393, Springer, 1992.

[StarT] H. Stark, A. Terras, *Zeta functions of finite graphs and coverings*, Adv. Math. 121 (1996) 124–165.

[Stark] A. Starkov, *Fuchsian groups from the dynamical viewpoint*, J. Dynam. Control Systems 1 (1995) 427–445.

[Ste] M. Stern, *$L^2$-index theorems on locally symmetric spaces*, Invent. Math. 96 (1989) 231–282.

[Stu1] U. Stuhler, *p-adic homogeneous spaces and moduli problems*, Math. Z. 192 (1986) 491–540.

[Stu2] U. Stuhler, *On the cohomology of* $\mathrm{SL}n$ *over rings of algebraic functions*, in *Algebraic K-theory, Part II*, pp. 316–359, Lecture Notes in Math., 967, Springer, 1982.

[Stu3] U. Stuhler, *Homological properties of certain arithmetic groups in the function field case*, Invent. Math. 57 (1980) 263–281.

[Sul1] D. Sullivan, *Disjoint spheres, approximation by imaginary quadratic numbers, and the logarithm law for geodesics*, Acta Math. 149 (1982) 215–237.

[Sul2] D. Sullivan, *Entropy, Hausdorff measures old and new, and limit sets of geometrically finite Kleinian groups*, Acta Math. 153 (1984) 259–277.

[Sun] T. Sunada, *Closed geodesics in a locally symmetric space*, Tohoku Math. J. 30 (1978) 59–68.

[SunTW] C. Sung, L. Tam, J. Wang, *Spaces of harmonic functions*, J. London Math. Soc. (2) 61 (2000) 789–806.

[Sur] B. Sury, *Bounded generation does not imply finite presentation*, Comm. Algebra 25 (1997) 1673–1683.

[Ta] K. Takeuchi, *A characterization of arithmetic Fuchsian groups*, J. Math. Soc. Japan 27 (1975) 600–612.

[Tav] O. Tavgen, *Bounded generation of normal and twisted Chevalley groups over the rings of S -integers*, pp. 409–421, Contemp. Math., 131, Part 1, Amer. Math. Soc., 1992.

[Te1] A. Terras, *Harmonic analysis on symmetric spaces and applications. I*, Springer-Verlag, 1985. xiv+341 pp.

[Te2] A. Terras, *Harmonic analysis on symmetric spaces and applications. II*, Springer-Verlag, 1988. xii+385 pp.

[Th1] W. Thurston, *Three-dimensional geometry and topology*, Vol. 1. Princeton Mathematical Series, 35. Princeton University Press, 1997. x+311 pp.

[Th2] W. Thurston, *On the geometry and dynamics of diffeomorphisms of surfaces*, Bull. Amer. Math. Soc. (N.S.) 19 (1988) 417–431.

[Tie] A. Tiemeyer, *A local-global principle for finiteness properties of S-arithmetic groups over number fields*, Transform. Groups 2 (1997) 215–223.

[Tit1] J. Tits, *Buildings of spherical type and BN-pairs*, Lect. Notes in Math. 386, Springer-Verlag, 1974.

[Tit2] J. Tits, *On Buildings and Their Applications*, Proc. ICM, Vancouver, 1974, pp. 209–220.

[Tit3] J. Tits, *Free subgroups in linear groups*, J. Algebra 20 (1972) 250–270.

[Tit4] J. Tits, *Sur le groupe des automorphisms d'un arbe*, in *Essays on topology and related topics*, pp. 188–211, ed. by A. Haefliger and R. Narasimhan, Springer, 1970.

[Tit6] J. Tits, *Uniqueness and presentation of Kac-Moody groups over fields,* Journal of Algebra 105 (1987) 542–573.

[Tit7] J. Tits, *Sur le groupe des automorphismes d'un arbre*, in *Essays on topology and related topics*, pp. 188–211. Springer, 1970.

[To] W. To, *Hermitian metrics of seminegative curvature on quotients of bounded symmetric domains*, Invent. Math. 95 (1989) 559–578.

[Tom1] G. Tomanov, *The virtual solvability of the fundamental group of a generalized Lorentz space form*, J. Differential Geom. 32 (1990) 539–547.

[Tom2] G. Tomanov, *Actions of maximal tori on homogeneous spaces*, in *Rigidity in dynamics and geometry*, pp. 407–424, Springer, 2002.

[Tom3] G. Tomanov, *Orbits on homogeneous spaces of arithmetic origin and approximations*, in *Analysis on homogeneous spaces and representation theory of Lie groups*, pp. 265–297, Adv. Stud. Pure Math., 26, Math. Soc. Japan, 2000.

[TomW] G. Tomanov, B. Weiss, *Closed orbits for actions of maximal tori on homogeneous spaces*, Duke Math. J. 119 (2003) 367–392.

[ToW] T. Tong, S. Wang, *Geometric realization of discrete series for semisimple symmetric spaces*, Invent. Math. 96 (1989) 425–458.

[Ts] R. Tsushima, *On the spaces of Siegel cusp forms of degree two*, Amer. J. Math. 104 (1982) 843–885.

[Us1] S. Usui, *Complex structures on partial compactifications of arithmetic quotients of classifying spaces of Hodge structures*, Tohoku Math. J. 47 (1995) 405–429.

[Us2] S. Usui, *Images of extended period maps*, J. Algebraic Geom. 15 (2006) 603–621.

[V] S. Vaes, *Rigidity results for Bernoulli actions and their von Neumann algebras*, Séminaire Bourbaki, 58éme année, 2005–2006, *n°* 961.

[Va1] A. Valette, *Group pairs with property (T), from arithmetic lattices*, Geom. Dedicata 112 (2005) 183–196.

[Va2] A. Valette, *Nouvelles approches de la propriété (T) de Kazhdan*, Astérisque No. 294 (2004) 97–124.

[vand] G. van der Geer, *Hilbert modular surfaces*, Springer-Verlag, 1988.

[VandMS] G. van der Geer, B. Moonen, R. Schoof, *Number Fields and Function Fields – Two Parallel Worlds*, in *Progress in Mathematics*, Vol. 239, Birkhäuser, 2005, XIV, 318 p.

[vanE] W. van Est, *A generalization of the Cartan-Leray spectral sequence. I, II*, Indag. Math. 20 (1958) 399–413.

[vanG] B. van Geemen, *Kuga-Satake varieties and the Hodge conjecture*, in *The arithmetic and geometry of algebraic cycles*, pp. 51–82, NATO Sci. Ser. C Math. Phys. Sci., 548, Kluwer Acad. Publ., 2000.

[Vas] A. Vasiu, *Integral canonical models of Shimura varieties of preabelian type*, Asian J. Math. 3 (1999) 401–518.

[Ve1] T. Venkataramana, *Lefschetz properties of subvarieties of Shimura varieties*, in *Currents trends in number theory*, pp. 265–270, Hindustan Book Agency, New Delhi, 2002.

[Ve2] T. Venkataramana, *On cycles on compact locally symmetric varieties*, Monatsh. Math. 135 (2002) 221–244.

[Ve3] T. Venkataramana, *Restriction maps between cohomologies of locally symmetric varieties*, in *Cohomology of arithmetic groups, L-functions and automorphic forms*, pp. 237–251, Tata Inst. Fund. Res. Stud. Math., vol.15, 2001.

[Ve4] T. Venkataramana, *Cohomology of compact locally symmetric spaces*, Compositio Math. 125 (2001) 221–253.

[Ve5] T. Venkataramana, *On systems of generators of arithmetic subgroups of higher rank groups*, Pacific J. Math. 166 (1994) 193–212.

[Ven] A. Venkov, *Spectral theory of automorphic functions and its applications*, in *Mathematics and its Applications (Soviet Series)*, vol. 51. Kluwer Academic Publishers Group, 1990. xiv+176 pp.

[Vig] M. Vignéras, *Arithmétique des algèbres de quaternions*, Lecture Notes in Mathematics, 800. Springer, 1980. vii+169 pp.

[Vin] E. Vinberg, *Discrete reflection groups in Lobachevsky spaces*, in *Proc. of the International Congress of Mathematicians*, pp. 593–601, Warsaw, 1984.

[ViS] E. Vinberg, O. Shvartsman, *Discrete groups of motions of spaces of constant curvature*, in *Geometry II*, pp. 139–248, Encyclopaedia Math. Sci., 29, Springer, 1993.

[Vla] S. Vladut, *Kronecker's Jugendtraum and modular functions*, Studies in the Development of Modern Mathematics, 2, Gordon and Breach Science Publishers, 1991. x+411 pp.

[VogaZ] D. Vogan, G. Zuckerman, *Unitary representations with nonzero cohomology*, Compositio Math. 53 (1984) 51–90.

[Vogt1] K. Vogtmann, *Automorphisms of free groups and outer space*, Geom. Dedicata 94 (2002) 1–31.

[Vogt2] K. Vogtmann, *The cohomology of automorphism groups of free groups*, Proc. of International Congress of Mathematicians. Vol. II, pp. 1101–1117, Eur. Math. Soc., Zürich, 2006.

[Voi1] C. Voisin, *Hodge theory and complex algebraic geometry. I.*, Cambridge Studies in Advanced Mathematics, 76, Cambridge University Press, 2002. x+322 pp.

[Voi2] C. Voisin, *Hodge theory and complex algebraic geometry. II.*, Cambridge Studies in Advanced Mathematics, 77, Cambridge University Press, 2003. x+351 pp.

[Voi3] C. Voisin, *A generalization of the Kuga-Satake construction*, Pure Appl. Math. Q. 1 (2005) 415–439.

[Vos] V. Voskresenskii, *Algebraic groups and their birational invariants*, Translations of Mathematical Monographs, 179. American Mathematical Society, 1998. xiv+218 pp.

[Wag1] J. Wagoner, *Diffeomorphisms, $K_2$, and analytic torsion*, in *Algebraic and geometric topology*, Proc. Sympos. Pure Math., vol. XXXII, Part 1, pp. 23–33, 1978.

[Wag2] J. Wagoner, *$K_2$ and diffeomorphisms of two and three dimensional manifolds*, in *Geometric topology*, pp. 557–577, Academic Press, 1979.

[Wal1] N. Wallach, *Automorphic forms*, Notes by Roberto Miatello, in *New developments in Lie theory and their applications*, Progr. Math., vol. 105, pp.1–25, Birkhduser, 1992.

[Wal2] N. Wallach, *Square integrable automorphic forms and cohomology of arithmetic quotients of* $\mathrm{SU}(p, q)$, Math. Ann. 266 (1984) 261–278.

[Wal3] N. Wallach, *The spectrum of compact quotients of semisimple Lie groups*, in *Proc. of the International Congress of Mathematicians (Helsinki, 1978)*, pp. 715–719, 1980.

[Wal4] N. Wallach, *An asymptotic formula of Gelfand and Gangolli for the spectrum of* $\Gamma\backslash G$, J. Differential Geometry 11 (1976) 91–101.

[Wal5] N. Wallach, *On the Selberg trace formula in the case of compact quotient*, Bull. Amer. Math. Soc. 82 (1976) 171–195.

[Wal6] N. Wallach, *Harmonic analysis on homogeneous spaces*, Marcel Dekker, 1973. xv+361 pp.

[Wal7] N. Wallach, *Homogeneous positively pinched Riemannian manifolds*, Bull. Amer. Math. Soc. 76 (1970) 783–786.

[Wal8] N. Wallach, *Compact homogeneous Riemannian manifolds with strictly positive curvature*, Ann. of Math. 96 (1972) 277–295.

[Wan1] S. Wang, *On isolated points in the dual spaces of locally compact groups*, Math. Ann. 218 (1975) 19–34.

[Wan2] S. Wang, *On subgroups with property P and maximal discrete subgroups*, Amer. J. Math. 97 (1975) 404–414.

[Wan3] S. Wang, *Topics on totally discontinuous groups: Short courses presented at Washington University*, in *Symmetric spaces*, ed. by W. Boothby and G. Weiss, Pure and Applied Mathematics, Vol. 8., pp. 459–487, Dekker, 1972.

[WanZ1] M. Wang, W. Ziller, *Symmetric spaces and strongly isotropy irreducible spaces*, Math. Ann. 296 (1993) 285–326.

[WanZ2] M. Wang, W. Ziller, *Existence and nonexistence of homogeneous Einstein metrics*, Invent. Math. 84 (1986) 177–194.

[War] F. Warner, *Foundations of differentiable manifolds and Lie groups*, Graduate Texts in Mathematics, 94. Springer-Verlag, 1983. ix+272 pp.

[Wat] Y. Watatani, *property (T) of Kazhdan implies property FA of Serre*, Math. Japon. 27 (1982) 97–103.

[We] C. Weibel, *Algebraic $K$-theory of rings of integers in local and global fields*, in *Handbook of $K$-theory*, pp. 139–190, Springer, 2005.

[Wei] A. Weil, *Basic number theory*, Reprint of the second (1973) edition, Classics in Mathematics. Springer, 1995. xviii+315 pp.

[Wein1] S. Weinberger, *Computers, rigidity, and moduli. The large-scale fractal geometry of Riemannian moduli space*, Princeton University Press, 2005. xii+174 pp.

[Wein2] S. Weinberger, *Aspects of the Novikov conjecture*, in *Geometric and topological invariants of elliptic operators*, pp. 281–297, Contemp. Math., 105, Amer. Math. Soc., 1990.

[Wein3] S. Weinberger, *The topological classification of stratified spaces*, Chicago Lectures in Mathematics. University of Chicago Press, 1994. xiv+283 pp.

[Weis] B. Weiss, *Divergent trajectories and $\mathbb{Q}$-rank*, Israel J. Math. 152 (2006) 221–227.

[Wey] H. Weyl, *Symmetry*, Reprint of the 1952 original, Princeton University Press, Princeton, 1989. viii+168 pp.

[Wh] K. Whyte, *The large scale geometry of the higher Baumslag-Solitar groups*, Geom. Funct. Anal. 11 (2001) 1327–1343.

[Wie] A. Wienhard, *The action of the mapping class group on maximal representations*, Geom. Dedicata 120 (2006) 179–191.

[Wi1] B. Wilking, *Positively curved manifolds with symmetry*, Ann. of Math. 163 (2006) 607–668.

[Wi2] B. Wilking, *Rigidity of group actions on solvable Lie groups*, Math. Ann. 317 (2000) 195–237.

[Woe] W. Woess, *Random walks on infinite graphs and groups*, Cambridge Tracts in Mathematics, 138. Cambridge University Press, 2000. xii+334 pp.

[Wol1] J. Wolf, *Spaces of constant curvature*, Third edition, Publish or Perish, 1974, xv+408 pp.

[Wol2] J. Wolf, *Fine structure of Hermitian symmetric spaces*, in *Symmetric spaces: short courses presented at Washington University*, ed. by W. Boothby and G. Weiss, Marcel Dekker, 1972, pp. 271–357.

[Wolp1] S. Wolpert, *Disappearance of cusp forms in special families*, Ann. of Math. 139 (1994) 239–291.

[Wolp2] S. Wolpert, *Spectral limits for hyperbolic surfaces. I, II*, Invent. Math. 108 (1992) 67–89, 91–129.

[Wolp3] S. Wolpert, *Geometry of the Weil-Petersson completion of Teichmüller space*, in *Surveys in differential geometry*, Vol. VIII, pp. 357–393, Int. Press, Somerville, 2003.

[Wolp4] S. Wolpert, *The length spectra as moduli for compact Riemann surfaces*, Ann. of Math. 109 (1979) 323–351.

[Wolp5] S. Wolpert, *The eigenvalue spectrum as moduli for compact Riemann surfaces*, Bull. Amer. Math. Soc. 83 (1977) 1306–1308.

[Wolp6] S. Wolpert, *Weil-Petersson perspectives*, in *Problems on mapping class groups and related topics*, pp. 269–282, Proc. Sympos. Pure Math., 74, Amer. Math. Soc., 2006.

[Xi1] X. Xie, *Quasi-isometric rigidity of Fuchsian buildings*, Topology 45 (2006) 101–169.

[Xi2] X. Xie, *The Tits boundary of a* CAT(0) *2-complex*, Trans. Amer. Math. Soc. 357 (2005) 1627–1661.

[Xi3] X. Xie, *Groups acting on* CAT(0) *square complexes*, Geom. Dedicata 109 (2004) 59–88.

[Ya] T. Yamazaki, *On a generalization of the Fourier transformation*, J. Fac. Sci. Univ. Tokyo Sect. IA Math. 25 (1978) 237–252.

[Yas1] D. Yasaki, *An explicit spine for the Picard modular group over the Gaussian integers*, to appear in Journal of Number Theory.

[Yas2] D. Yasaki, *On the existence of spines for Q-rank 1 groups*, Selecta Math. (N.S.) 12 (2006) 541–564.

[Yau1] S.T. Yau, *Calabi's conjecture and some new results in algebraic geometry*, Proc. Nat. Acad. Sci. U.S.A. 74 (1977) 1798–1799.

[Yau2] S.T. Yau, *A splitting theorem and an algebraic geometric characterization of locally Hermitian symmetric spaces*, Comm. Anal. Geom. 1 (1993) 473–486.

[Yau3] S.T. Yau, *Harmonic functions on complete Riemannian manifolds*, Comm. Pure Appl. Math. 28 (1975) 201–228.

[YaZ1] S.T. Yau, F. Zheng, *Negatively $\frac{1}{4}$-pinched Riemannian metric on a compact Khler manifold*, Invent. Math. 103 (1991) 527–535.

[YaZ2] S.T. Yau, F. Zheng, *On a borderline class of non-positively curved compact Khler manifolds*, Math. Z. 212 (1993) 587–599.

[Ye1] S. Yeung, *Integrality and Arithmeticity of Co-compact Lattice Corresponding to Certain Complex Two-ball Quotients of Picard Number One*, Asian Jour. Math. 8 (2004) 107–130.

[Ye2] S. Yeung, *Virtual first Betti number and integrality of compact complex two-ball quotients*, Int. Math. Res. Not. 38 (2004) 1967–1988.

[Ye3] S. Yeung, *Compactification of Kähler manifolds with negative Ricci curvature*, Invent. Math. 106 (1991) 13–25.

[Ye4] S. Yeung, *Representations of semisimple lattices in mapping class groups*, Int. Math. Res. Not. 31 (2003) 1677–1686.

[Ye5] S. Yeung, *Quasi-isometry of metrics on Teichmller spaces*, Int. Math. Res. Not. 4 (2005) 239–255.

[Yo1] M. Yoshida, *Fuchsian differential equations: with special emphasis on the Gauss-Schwarz theory*, Friedr. Vieweg & Sohn, 1987.

[Yo2] M. Yoshida, *Hypergeometric functions, my love: modular interpretations of configuration spaces*, Friedr. Vieweg & Sohn, 1997.

[Yos] T. Yoshino, *A solution to Lipsman's conjecture for* $\mathbb{R}^4$, Int. Math. Res. Notices 80 (2004) 4293–4306.

[Yu1] G. Yu, *The Novikov conjecture for groups with finite asymptotic dimension*, Ann. of Math. 147 (1998) 325–355.

[Yu2] G. Yu, *The coarse Baum-Connes conjecture for spaces which admit a uniform embedding into Hilbert space*, Invent. Math. 139 (2000) 201–240.

[Yu3] G. Yu, *Hyperbolic groups admit proper affine isometric actions on* $l^p$*-spaces*, Geom. Funct. Anal. 15 (2005) 1144–1151.

[Yue1] C. Yue, *Mostow rigidity of rank* 1 *discrete groups with ergodic Bowen-Margulis measure*, Invent. Math. 125 (1996) 75–102.

[Yue2] C. Yue, *Dimension and rigidity of quasi-Fuchsian representations*, Ann. of Math. 143 (1996) 331–355.

[Yue3] C. Yue, *Rigidity and dynamics around manifolds of negative curvature*, Math. Res. Lett. 1 (1994) 123–147.

[Ze1] S. Zelditch, *Selberg trace formulae and equidistribution theorems for closed geodesics and Laplace eigenfunctions: finite area surfaces*, Mem. Amer. Math. Soc. 96 (1992), no. 465, vi+102 pp.

[Ze2] S. Zelditch, *Trace formula for compact* $\Gamma\backslash\mathrm{PSL}_2(R)$ *and the equidistribution theory of closed geodesics*, Duke Math. J. 59 (1989) 27–81.

[Ze3] S. Zelditch, *Uniform distribution of eigenfunctions on compact hyperbolic surfaces*, Duke Math. J. 55 (1987) 919–941.

[Ze4] S. Zelditch, *The inverse spectral problem, With an appendix by Johannes Sjöstrand and Maciej Zworski*, in *Surveys in differential geometry*, Vol. IX, 401–467, Int. Press, 2004.

[Zh] X. Zhou, *Some results related to group actions in several complex variables*, in *Proceedings of the International Congress of Mathematicians*, Vol. II (Beijing, 2002), pp. 743–753, Higher Ed. Press, 2002.

[Zi1] R. Zimmer, *Ergodic theory and semisimple groups*, Birkhäuser, 1984.

[Zi2] R. Zimmer, *Actions of semisimple groups and discrete subgroups*, in *Proceedings of the International Congress of Mathematicians*, pp. 1247–1258, Amer. Math. Soc., 1987.

[Zi3] R. Zimmer, *Discrete groups and non-Riemannian homogeneous spaces*, J. Amer. Math. Soc. 7 (1994) 159–168.

[Zi4] R. Zimmer, *Strong rigidity for ergodic actions of semisimple Lie groups*, Ann. of Math. 112 (1980) 511–529.

[Zu1] S. Zucker, $L_2$ *cohomology of warped products and arithmetic groups*, Invent. Math. 70 (1982) 169–218.

[Zu2] S. Zucker, *Satake compactifications*, Comment. Math. Helv. 58 (1983) 312–343.

[Zu3] S. Zucker, *On the reductive Borel-Serre compactification: $L^p$-cohomology of arithmetic groups*, Amer. J. Math. 123 (2001) 951–984.

[Zu4] S. Zucker, *$L^2$-cohomology of Shimura varieties*, in *Automorphic forms, Shimura varieties, and L-functions*, Vol. II, Perspect. Math., 11, ed. by L. Clozel and J. Milne Academic Press, 1990, pp. 377–391.

[Zuk] A. Zuk, *Property (T) and Kazhdan constants for discrete groups*, Geom. Funct. Anal. 13 (2003) 643–670.

[Zwo1] M. Zworski, *Dimension of the limit set and the density of resonances for convex co-compact hyperbolic surfaces*, Invent. Math. 136 (1999) 353–409.

[Zwo2] M. Zworski, *Poisson formula for resonances in even dimensions*, Asian J. Math. 2 (1998) 609–617.

# Index